3rd International Conference on Research Methodology for Built Environment and Engineering 2017

IOP Conference Series: Earth and Environmental Science
Volume 117

Shah Alam, Malaysia
8 - 9 November 2017

Editors:

Siti Aekbal Salleh
Norazian Mohamad Yusuwan

ISBN: 978-1-5108-5949-4
ISSN: 1755-1307

Printed from e-media with permission by:

Curran Associates, Inc.
57 Morehouse Lane
Red Hook, NY 12571

Printed by Curran Associates, Inc. (2018)

For permission requests, please contact the Institute of Physics at the address below.

Institute of Physics
Dirac House, Temple Back
Bristol BS1 6BE UK

Phone: 44 1 17 929 7481
Fax: 44 1 17 920 0979

techtracking@iop.org

Additional copies of this publication are available from:

Curran Associates, Inc.
57 Morehouse Lane
Red Hook, NY 12571 USA
Phone: 845-758-0400
Fax: 845-758-2633
Email: curran@proceedings.com
Web: www.proceedings.com

Table of contents

Volume 117

3rd International Conference on Research Methodology for Built Environment and Engineering 2017

8–9 November 2017, Shah Alam, Selangor, Malaysia

Accepted papers received: 1 February 2018
Published online: 12 March 2018

Preface

Papers

IOP Conf. Series: Earth and Environmental Science **117** (2018) 011001 doi:10.1088/1755-1315/117/1/011001

Preface

The 3rd International Conference on Research Methodology in Built Environment and Engineering (ICRMBEE 2017), held on 8th to 9th November 2017 at Best Western Hotel I-City, Shah Alam, Malaysia. This year mark the 3rd instalments of ICRMBEE series that carries the same focus on the method adopted in the area of built environment and engineering. With the theme, '*Regenerative Sustainability in Built Environment for Human Livability*, ICRMBEE 2017 hope to provides an opportunity for interaction and exchange of ideas on the most appropriate ways of building bridges across universities, research institutions, government, industry and experts from around the world in signifying built environment and engineering research in the societal livelihood. We are proud to announce that all accepted papers will be published in the IOP Conference Series: Earth and Environmental Science indexed in SCOPUS.

Research and innovation has helped the globalization, modernization and the development of human capital in many ways. This year ICRMBEE is trying to connect human liveability into the equations, not just the quantitative and qualitative means of research methodology. We hope to see many research methods are carved and contour with regards to benefit our nations livability that add up to a community's quality of life, including the built and natural environments, economic prosperity, social stability and equity, educational opportunity, and cultural, entertainment and recreation possibilities. The organizing committee of ICRMBEE 2017 aims to make the conference grows into a renowned avenue and platform for global intellectual discourse.

Sr Dr Siti Aekbal Salleh
Conference Chair ICRMBEE 2017
Faculty of Architecture Planning and Surrveying
Universiti Teknologi MARA
40450 Shah Alam, Selangor
Malaysia

Organizing Committees

Guest Editors:

ACKNOWLEDGEMENT

The organizing committee would like to acknowledge the following for their contribution and effort to this conference:

IOP Conf. Series: Earth and Environmental Science **117** (2018) 011002 doi:10.1088/1755-1315/117/1/011002

Peer review statement

All papers published in this volume of *IOP Conference Series: Earth and Environmental Science* have been peer reviewed through processes administered by the proceedings Editors. Reviews were conducted by expert referees to the professional and scientific standards expected of a proceedings journal published by IOP Publishing.

IOP Conf. Series: Earth and Environmental Science **117** (2018) 012001 doi:10.1088/1755-1315/117/1/012001

Moderate Imaging Resolution Spectroradiometer (MODIS) Aerosol Optical Depth Retrieval for Aerosol Radiative Forcing

A Asmat[1], K A Jalal[1], and N Ahmad[2]

[1]Faculty of Applied Sciences, Universiti Teknologi MARA Shah Alam, Selangor
[2]National Space Agency, Putrajaya

*arnisasmat@gmail.com

Abstract. The present study uses the Aerosol Optical Depth (AOD) retrieved from Moderate Imaging Resolution Spectroradiometer (MODIS) data for the period from January 2011 until December 2015 over an urban area in Kuching, Sarawak. The results show the minimum AOD value retrieved from MODIS is -0.06 and the maximum value is 6.0. High aerosol loading with high AOD value observed during dry seasons and low AOD monitored during wet seasons. Multi plane regression technique used to retrieve AOD from MODIS (AOD_{MODIS}) and different statistics parameter is proposed by using relative absolute error for accuracy assessment in spatial and temporal averaging approach. The AOD_{MODIS} then compared with AOD derived from Aerosol Robotic Network (AERONET) Sunphotometer ($AOD_{AERONET}$) and the results shows high correlation coefficient (R^2) for AOD_{MODIS} and $AOD_{AERONET}$ with 0.93. AOD_{MODIS} used as an input parameters into Santa Barbara Discrete Ordinate Radiative Transfer (SBDART) model to estimate urban radiative forcing at Kuching. The observed hourly averaged for urban radiative forcing is -0.12 Wm^{-2} for top of atmosphere (TOA), -2.13 Wm^{-2} at the surface and 2.00 Wm^{-2} in the atmosphere. There is a moderate relationship observed between urban radiative forcing calculated using SBDART and AERONET which are 0.75 at the surface, 0.65 at TOA and 0.56 in atmosphere. Overall, variation in AOD tends to cause large bias in the estimated urban radiative forcing.

1. Introduction

The urban pollution sources are quite variable in space and time. The impacts of urban aerosol towards environment and the regional and global air quality. Atmospheric aerosols play an important role in the global energy balance by scattering and absorbing the solar and terrestrial radiation (direct effect), as well as by acting as cloud condensation (indirect effect) [1]. It can be derived from natural as well as anthropogenic sources especially the urban pollution which is quite variable in space and time. These aerosols are known to affect the air quality, human health and Earth radiation budget [2]. The direct effect due to aerosols as effect of both natural and anthropogenic on the radiation budget primarily due to scattering and absorption of radiation by aerosols and is measured in terms of watts per square meter (Wm^{-2}) or also known as aerosol radiative forcing (ARF) [3]. The values of ARF at top of atmosphere (TOA) and surface are important parameter in the quantification of the impact of aerosols towards climate change. The largest uncertainties in estimating the changes in radiation budget due to high spatial variability of aerosols and lack of an adequate information regarding aerosol

IOP Conf. Series: Earth and Environmental Science **117** (2018) 012001 doi:10.1088/1755-1315/117/1/012001

radiative properties [1]. Therefore, an understanding of the changes urban aerosols over time are very important to predict the effect of warming and cooling towards the regional area.

Different ground and satellite based remote sensing techniques helps to provide a systematic retrieval of aerosol optical properties on the global and regional scale [2]. Satellite remote sensing is the only technology help to observe the spatial and temporal variations of aerosol properties and have been used broadly in estimating aerosol ARF. Based on the remote sensing observations, Moderate Resolution Imaging Spectroradiometer (MODIS) data significantly can be used to estimate ARF value under clear sky conditions [4]. While ground based remote sensing such as Aerosol Robotic Network (AERONET) is useful in estimating continuous optical properties for aerosol in real time monitoring and also can be used for validation [5]. One of the important parameter in assessing the ARF value is Aerosol Optical Depth (AOD). It defined as extinction coefficient of aerosol loading measured in vertical column of the atmosphere [6]. As stated by [7], the changes in surface ARF value can primarily governed by the magnitude of AOD values which consistent with the aerosol loading.

The objective of this study is to investigate the influence of urban aerosols on the estimation of ARF by using AOD retrieved from MODIS (AOD_{MODIS}) over Kuching area. In this study, AOD_{MODIS} are retrieved by multi plane regression technique. The combination technique with minimum distance technique and the usage of relative absolute error are introduced for accuracy assessment in spatial and temporal averaging approach and compared with AOD derived from AERONET ($AOD_{AERONET}$). Next, the AOD_{MODIS} used as an input parameter into radiative transfer model which is Santa Barbara Discrete Ordinate Radiative Transfer (SBDART) for conducting quantitative estimation of urban radiative forcing.

2. Data measurement and study site

Terra MODIS data Level 2 (Collection 5) provides data used to study aerosol climatology and highly suitable for characterization of aerosol properties spatial resolution of 10km x 10km and can be downloaded at http://ladsweb.nascom.nasa.gov/data/search.html. AERONET is a ground based monitoring aerosol that assess aerosol optical properties and to validate satellite retrievals of aerosol optical properties. The Level 2 AOD at 500nm was used for the analysis and can be downloaded at http://aeronet.gsfc.nasa.go. The uncertainty in retrieval of AOD under cloud free conditions is $<\pm 0.01$ for $\lambda > 440$nm and $<\pm 0.02$ for shorter wavelengths [5]. The data used for this study is from 2011 until 2015 both for MODIS and AERONET, respectively.

The experimental site for this study is at Kuching with tropical rainforest climate which moderately hot and humid. The wettest seasons are during the North-East Monsoon months (November to March) and dry seasons from June until September every year. Figure 1 shows the location of the study site and location of AERONET Sunphotometer.

Figure 1. Study site of Kuching, Sarawak and location of AERONET Sunphotometer

3. Methodology

3.1. Spatial and temporal averaging approach
Firstly, a time window of ±20 min was chosen with respect to MODIS overpass and a temporal average was compared with the time availability of AERONET. Since air masses are constantly in motion, an air mass captured by MODIS across a certain horizontal span over AERONET station will be sampled by the during a certain time period [8]. Minimum distance technique was performed for MODIS data to calculate the closest pixel of latitude and longitude to the AERONET site and the output were given in column and row pixel. The equation for minimum distance technique in the following way:

$$[R, C] = \left(\min\left[((x - x1)^2 + (y - y1)^2)^{\frac{1}{2}}\right]\right) \tag{1}$$

Based on Equation 1, x is a longitude for MODIS and AERONET and y is the latitude for MODIS and AERONET. Smaller window sizes (11 x 11 pixels) are used to reduce undesirable errors due to topographic or aerosol types heterogeneity. The spatial average of AOD_{MODIS} over the AERONET were collocated from pixels lying in ±(1/4°) of latitude (1.491°) and longitude (110.349°) of AERONET station in Kuching, Sarawak.

A table of AOD along with the corresponding latitude and longitude that fall into these pixels was prepared for each of the MODIS data. Based on the MODIS algorithm for the retrieval of AOD, the valid range allowed for AOD was between -0.05 to 5.0. These values are indicative of very clean conditions and allowed in order to avoid an artificial bias in statistics [9]. Based on the tabulated table, the five nearest values lying within the lowest relative absolute error were chosen to compare which produced less error values.

3.2. Multiple Regression Plane Technique
Next, the selected five points values for each of MODIS data available was plotted by using scatter plot with independent variables, latitude and longitude, on the X and Y axes and dependent variable, AOD, on the Z axis in order to perform multiple regression plane technique. These points have been fitted in the form of Equation 2.

$$\mathrm{Z = aX + bY + c} \tag{2}$$

By using Equation 2, the predicted AOD_{MODIS} data for each of the Julian Days can be calculated. The AOD_{MODIS} data was at the wavelength range 550nm while $AOD_{AERONET}$ was at 500nm. For easy comparison with AOD_{MODIS}, $AOD_{AERONET}$ were interpolated by using Equation 3.

$$\mathrm{AOD_{550nm}} = \mathrm{AOD_{500nm}} \left(\frac{550}{500}\right)^{-\alpha} \tag{3}$$

Where α is the angstrom exponent at wavelength 440/870nm obtained from AERONET data [10]. After that, the relationship between AOD_{MODIS} is established with $AOD_{AERONET}$ based on least square regression and it output presented in correlation coefficient (R^2).

3.3. Estimation of aerosol radiative forcing
Finally, AOD_{MODIS} used as an input parameter into SBDART for urban radiative forcing estimation [11]. The net flux at the TOA and surface were computed separately, within the wavelength range from 0.2 to 4.0 μm over 24 h with 1h time intervals. The outcome from SBDART model is the value of instantaneous irradiance which is downwards and upward solar fluxes. Calculation of ARF can be estimated by using Equation 4 with negative values of ARF corresponds to cooling effect while positive values of ARF corresponds to warming effect either at the surface or at TOA [6].

$$\Delta F = (F_a \downarrow - F_a \uparrow) - (F_n \downarrow - F_n \uparrow) \quad (4)$$

Where $F_a \downarrow$ and $F_a \uparrow$ are downward and upward solar fluxes at the surface in the presence of aerosols and $F_n \downarrow$ and $F_n \uparrow$ are the same quantity but without aerosols and ΔF is the ARF value. The difference between the TOA and surface gives the atmospheric forcing (ΔF_{atm}) which represents the energy absorbed in the atmosphere.

4. Results and Discussions

4.1 Variations of Aerosol Optical Depth

The relationship between AOD_{MODIS} and $AOD_{AERONET}$ can be analyzed for easy comparison because of the same wavelength range at 550nm.

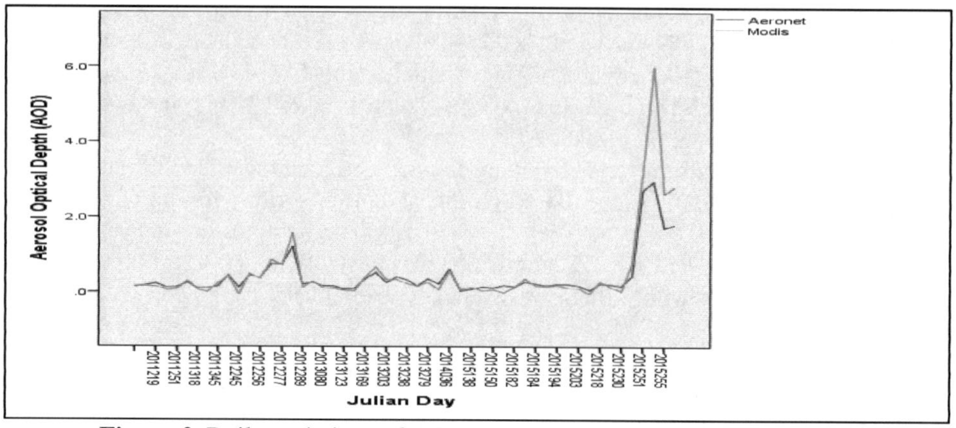

Figure 2. Daily variations of AOD_{MODIS} and $AOD_{AERONET}$ at 550nm

Based on Figure 2, the minimum values for AOD_{MODIS} is -0.06 and the maximum value estimated is 6.0 while for $AOD_{AERONET}$ the range values were from 0.04 to 2.95. High concentration of aerosol loading presence in Kuching area causes at certain Julian Days were obtained high value for AOD. As resulting by the occurrence of dry season which usually occurs during June until September yearly and also comes from haze episodes and also from urbanization. Study by [12] classified that In Kuching area dominant aerosols occurred in Kuching area urban and continental aerosols. It can be shown that at Julian Day 253 in 2015 there are extremely high value of AOD_{MODIS} was retrieved with 6.0 while only 2.9 for $AOD_{AERONET}$. The reason for the overestimation of AOD from MODIS may be due to dry season where the surface is very dry causes higher value for surface reflectance. An error of ±0.006 in measuring the surface reflectance yields an error of ±0.06 in retrieval of AOD [10]. Extremely low AOD value obtained during the wettest seasons from November to March. The inconsistency between aerosol microphysical and optical properties and surface reflectance used in the MODIS are possible reasons for the underestimation of AOD during wet season [13]. The comparison between AOD_{MODIS} and $AOD_{AERONET}$ can be analyzed by using linear regression technique and shown in Figure 3.

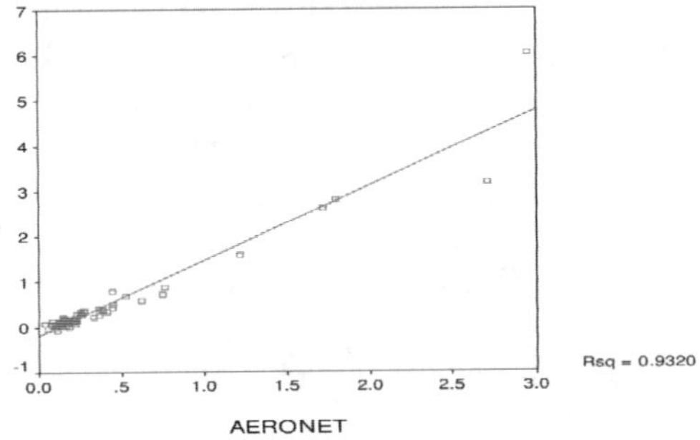

Figure 3. Linear regression between AOD_{MODIS} and $AOD_{AERONET}$ at 550nm

Based on Figure 3, the retrieval algorithm performance can be validated by the statistical parameters of the linear regression which are intercept (A), slope (B) and correlation coefficient (R^2). From the result, it shows that the non-zero intercept at A= -0.1181 corresponds that the retrieval algorithm is biased at low AOD value due association with a sensor calibration error or an improper assumption on ground surface reflectance and larger errors in surface reflectance lead to larger intercept values [14]. [9] studied on the comparison AOD using MODIS and AERONET where the underestimation of AOD probably due to incorrect characterization of the local aerosols and the predominantly low AOD values being observed. A slope that is different from unity indicates that there are inconsistency between aerosol microphysical and optical properties used in the retrieval algorithm [14] and represents systematic biases in the MODIS retrievals [13]. Slope higher than unity for AOD_{MODIS} at 1.6469 indicates an overestimation of AOD_{MODIS} with respect to AERONET retrieval. There is a strong relationship between AOD_{MODIS} and $AOD_{AERONET}$ with the value for R^2 is 0.932 and the mean absolute percentage error (MAPE) was found to be 12% and root mean square error (RMSE) is 0.47.

4.2 Comparison between AOD estimated by SBDART and AERONET

The reliability of the ARF calculated by the AOD_{MODIS} using SBDART can be compared with ARF derived from AERONET and can be shown in Figure 4.

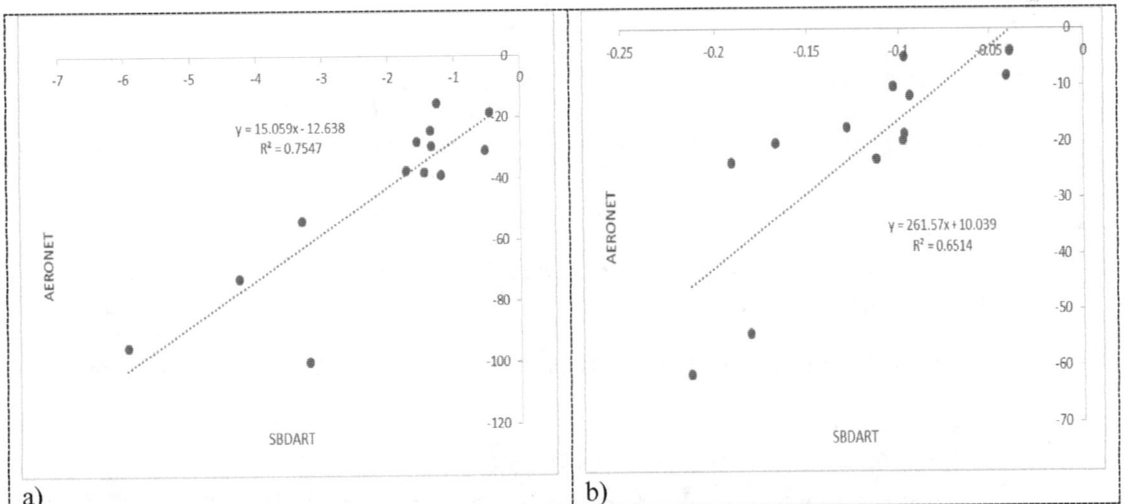

Figure 4. Scatterplot of the ARF value between SBDART and AERONET a) at the surface and b) at the top of atmosphere

Figure 4 is the hourly averaged ARF from SBDART model and being compared with the ARF derived from AERONET. The R^2 value of ARF estimates calculated by SBDART and AERONET are 0.75 at the surface, 0.65 at TOA and 0.56 in atmosphere. The observed hourly averaged for urban radiative forcing is -0.12 Wm^{-2} for top of atmosphere (TOA), -2.13 Wm^{-2} at the surface and 2.00 Wm^{-2} in the atmosphere. In general, the ARF value at TOA and surface are negative, whereas ARF in atmosphere is positive because energy have been absorbed into the aerosol layer [4]. These results show that ARF derived from SBDART by using AOD_{MODIS} is in moderate agreement with that calculated by AERONET. It can be suggested that the AOD_{MODIS} could also be used to estimate ARF using SBDART model with some improvements in obtaining accurate ARF values.

5. Conclusion

The aerosol concentration and urban radiative forcing over a period of five years had been analyzed at urban area of Kuching. The spectral variations of AOD at Kuching are based on the seasonal distinctive features with high AOD observed during dry seasons (June to September) and low concentration of AOD during wet seasons from November to March. The urban radiative forcing within the atmosphere is higher during dry seasons when AOD values also higher indicating higher absorption in the atmosphere. The comparison of the urban radiative forcing estimated from SBDART model with that obtained from AERONET is also an important aspect of the present study. The comparison reveals that urban radiative forcing estimated from SBDART are highly correlated at the surface but moderate relationship obtained for urban radiative forcing at TOA and in the atmosphere. In this comparison not all the ARF value been taken during the observation period. This is because there are only few months of data with the flux by derived by the AERONET and lack of AOD data. Though the results are preliminary and comes from limited observational results, the outcome from the study might significantly influence the urban radiative forcing over Kuching thus contributing the warming and cooling effect at Malaysia area.

Acknowledgement

The authors are grateful to the MODIS teams at NASA for the provision of satellite data, and thank the PI's their effort in establishing and maintaining Kuching sites for the provision of AERONET data. This study was supported by Universiti Teknologi MARA under 600-IRMI/DANA 5/3BESTARI (0012/2016)

References

[1] Intergovernmental Panel on Climate Change (IPCC) 2007
[2] Kaufman Y J, Tanre D and Boucher O 2002 *Nature* **419 (6903)** 215-223
[3] Srivastava A K, Singh S, Tiwari S and Bisht D S 2012 *Environ. Sci. Pollut. Res.* **19(4)** 1144-1158
[4] Xu H, Guo J, Ceamanos X, Roujean J L, Min M and Carrer D 2016 *Atmospheric Environment* **141** 186-196
[5] Holben B N, Eck T F, Slutsker I, Tanre D, Buis J P, Setzer A, et al 1998 *Remote Sensing of Environment* **66(1)** 1-16
[6] Alam K, Trautmann T, Blaschke T and Majid H 2012 *Atmospheric Environment* **50** 234-245
[7] Dhar P, De B K, Banik T, Gogoi M M, Babu S S and Guha A 2017 *Science of the Total Environment* **580** 499-508
[8] Ichoku C, Chu D A, Mattoo S, Kaufman Y J, Remer L A, Tanre D, Slutsker I and Holben B N 2002 *Geophysical Research Letters* **29(12)** 1-4
[9] Lanzaco B L, Olcese L E, Palancar G G and Toselli B M 2016 *Aerosol and Air Quality Research* **16** 1509-1522
[10] Tripathi S N, Dey S Chandel A, Srivastava S, Singh R P and Holben B N 2005 *Ann. Geophys* **23** 1093-1101
[11] Ricchiazzi P, Yang S, Gautier C and Sowle D 1998 *Bulletin of the American Meteorological*

IOP Conf. Series: Earth and Environmental Science **117** (2018) 012001 doi:10.1088/1755-1315/117/1/012001

Society **79(10)** 2101-2114
[12] Jalal K A, Asmat A and Ahmad N 2012 *Advanced Materials Research* **518-523** 5734-5737
[13] More S, Pradeep Kumar P, Gupta P, Devara P C S and Aher G R 2013 *Aerosol and Air Quality Research* **13(1)** 107-121
[14] Zhao T X P, Stowe L L, Smirnov A, Crosby D, Sapper J and McClain C R 2002 *American Meteorological Society* **59** 294-312

IOP Conf. Series: Earth and Environmental Science **117** (2018) 012002 doi:10.1088/1755-1315/117/1/012002

Effective mechanisms for environmental awareness enhancement of the Thai public company

P Tilokwan[1] and S Limjirakan[2]

[1]The Environment, Development and Sustainability Program, Graduate School, Chulalongkorn University, Phayathai Road, Pathumwan, Bangkok, 10330, Thailand
[2] The Environmental Research Institute, Chulalongkorn University, Phayathai Road, Pathumwan, Bangkok, 10330, Thailand

*sangchan.l@chula.ac.th

Abstract. The purpose of this paper is to explore effective mechanisms towards environmental awareness enhancement of the Thai public company. The environmental awareness has been strongly mentioned in the international agenda such as Agenda 21 to achieve sustainable development and be implemented at the global communities in all sectors. Thailand's environmental awareness has been set up as the national policy and continuously promoted in the business sectors. The selected study area is one of Thai industrial public companies mainly utilizing natural resources for its business. Data collection was conducted by using questionnaires with a stratified sampling method comprising of 28 managerial and 134 operational levels. Descriptive statistics were be used for data analysis presented in terms of percentage. The study found that the effective mechanisms towards environmental awareness enhancement include employees' participation in environmental friendly field trip, environmental knowledge provided by the company, supply chain involvement, law enforcement, and international environmental standard applied by the company. This study would recommend that public participation should be involved in order to make the mechanisms effectively.

1. Introduction

Environmental problems have been globally concerned by all countries. The issues concerned are, for example, food, biodiversity, land, climate change, energy, technology and waste management according to The United Nations Environment Programme (UNEP) [1]. Such concerning issues have been originated from human activities. The Fifth Assessment Report of the Intergovernmental Panel on Climate Change [2] summarized that 95 percent of human activities have created the global warming to the world. Those activities are expressed from the degree of people's awareness towards the environmental problems. The awareness related to the environment has been strongly pointed out in many parts in Agenda 21 [3] as one of tools to cope with the environmental issues which the environmental awareness is needed to be raised, and to be continuously implemented under the Sustainable Development Goal (SDG). Although, all stakeholders such as public and private companies, governments, and non-governmental organizations in the global communities have given more focuses on such issues, their environmental awareness have not been fully developed and be fully implemented. Corporate Social Responsibility or "CSR" in short is defined as the integration processes between business itself and its operations impact to the society and environment as a whole [4]. The CSR concept can be one mechanism to stimulate the environmental awareness inside people's mindset. In Thailand, the environmental awareness has been set up as the country policy and promoted under the Thailand Business Council for Sustainable Development (TBCSD). The TBCSD has focused on the sustainable movement, raising the environmental awareness and capacity building on environmental management [5]. Nevertheless, the current environmental awareness in Thailand has been realized in some companies since it is not compulsory policy that company must comply with. To enhance environmental awareness, an effective mechanism technique towards increasing environmental awareness is needed to be raised the level of people environmental concerned, particularly at the company's level. Additionally, this mechanism would encourage people to shape their behavior as environmental friendly concerned.

2. Literature Review

2.1. Relationship between environmental problems and human being

According to a primary assumption that most of the environmental issues originated from human being, this assumption was referred in "70 Ways the United Nations, Makes a Difference" [6]. This article summarized what United Nations has been dealt with global problems for the past 70 years and what the future actions that United Nations is going to take. However, UN achievements in the aspect of environmental issues were raised and the linkage between environmental problems and human being was also mentioned; for example, the Intergovernmental Panel on Climate Change (IPCC) reported in 2007 that human activities were a primary cause of climate change.

2.2. Environmental awareness framework

Dourish and Bellotti [7] described a definition of awareness as activities proceeding by others that needed to be conceived, and would be a part of individual activity. In other words, awareness was a pre-stage of thinking before taking actions that would be perceived from others' actions. Hence, the environmental awareness could be described as the issues of environmental concern by people from different perceive such as individual direct experiences related to the impact from the environmental problems and indirect experiences like environmental issues raised by outsiders including community and society.

2.3. Mechanisms towards increasing the environmental awareness

Effective mechanisms related to environmental awareness means a system, technique, process, tool that could increasingly change people's attitude and behavior to be more environmental friendly concerned. Brennan, et al. [8] studied on the environmental sustainability concept which was widely recognized by stakeholders in the business. They found that the motivation of the environmental issues from top management level to operational level was an important mechanism to change employees' environmental attitude and behavior in the future. Moreover, the moral of doing business in terms of environmental issues concerned was another effective mechanism that could help increasing the level of environmental awareness. They also revealed that most businesses did not consider the environmental issues as a part of their strategies of becoming a successful company. Agnes [9] also stated other effective mechanisms regarding environmental awareness enhancement, including a well-written environmental policy, internal and external environmental audits, environmental training programs for employees, and environmental performance goals or indicators. Remmen and Lorentzen [10] indicated the mechanism on the environmental awareness was the employee participation in the environmental activities. The employee participation could become an initiating point of changes in terms of environmental attitude and behaviour. It was highlighted that the best way to get the most employee participation was the cooperation at a management level on solving the environmental problems. Only information provided by the environment teams was not sufficient to create the environmental activities, the discussion and brainstorming sessions were also required to implement such activities.

Gadenne, et al. [11] reported that the environmental awareness enhancement mechanisms that employers could receive as the indirect force were suppliers, customers, and laws. These matters could indirectly force them to change their operations to be more environmental friendly process. The supporting fund from either external environmental company or the government were also considered as mechanisms to motivate business to become more actions oriented in aspect of environmental friendly concerned. Abdul-Wahab [12] suggested that the public environmental awareness motivation was also needed to be increased in order to motivate all stakeholders taking more concern on the environmental issues of the society. One result from this study clearly showed that the major cause of environmental problems was "Lack of environmental awareness amongst some citizens".

Eaganl and Streckewald [13] reported that another success factor to increase the environmental awareness was from end-customer's requisition regarding the environmental products. In this regard, what the producer would definitely do was applying more environmental friendly concept into the internal production process. Additionally, it was also noticeable that another key success factor was promptly support by the top management.

3. Research Methodology

A quantitative research approach was used focusing on the operational and managerial level's point of view in environmental awareness concerned of one of the Thai public companies which mainly utilizes natural resources for its business. Deductive method was applied to examine whether mechanisms are effective and well-functioned for real implementation at the company. Questionnaires were designed to collect primary data. The sample size of interview was used a stratified sampling method under the purposive research approach in order to avoid the bias that would be occurred. There were a total of 162 respondents participated in the questionnaire process. Among these respondents, 28 are in the managerial level (17.28%) and 134 are in the operational level (82.72%). Descriptive statistics was applied for data analysis. The study result is presented in terms of percentage by calculating the 5-point rating scale scoring by the respondents.

4. Results and Discussions

The study found that 89.81% of respondents viewed that they perceived and understood company's mission and vision related to environmental policy, while 84.71% of them indicated that such management has conducted mission and vision based on the company's effective actions. Meanwhile, respondents (88.54%) understood the purpose of environmental friendly activities originated by management. In aspect of environmental friendly activities, the study revealed that 92.99% of respondents perceived that the company has initiated the environmental friendly activities continuously. In the meanwhile, about 82% of them seen that the company usually monitor, evaluate the result of environmental friendly activities, and communicate to employees afterwards. From the lower percentage of agreements on company's monitoring and evaluating process (82%). This would be implied that there would be some communication gaps exist between managerial and operational level that the management needs to take concern with. The study also found that 94.27% of respondents noted that they voluntarily participated in the environmental friendly activity. In addition, 90.45% of them agreed to adapt what they perceived from participating in those activities at workplace to be more environmental friendly behaviour in their personal lives.

The study also revealed that 85.99% of respondents agreed that if they were educated more on environmental knowledge, their level of environmental awareness would increase. On the other hand, their environmental awareness would enhance if they participated in the environmental friendly field trip. The number of respondents agreed with this statement was 89.81%. This would be implied that respondents would like to join more practical activities rather than the formal activity such as classroom training. The study also showed that 87.90% of respondents agreed on another effective tool to increase their level of environmental awareness as if the company collaborated with external organization by setting up seminar, meeting, or environmental concerned field trip. The finding revealed that 81.53% of respondents agreed if the company complied with the international standard related to environmental concerned such as ISO14001:2004. Lastly, employees' environmental awareness would enhance, if the company provided reward/recognition and an incentive in form of money for the outstanding employees whom explicitly perform their environmental concerned behavior. The study found that 84.08% of respondents agreed on reward/recognition provided by the company would increase their level of environmental awareness. Additionally, the number of agreed respondents decreased to 73.25% for the financial incentive provided by the company, whilst 7.64% of them disagreed with this statement. This would be implied that the financial incentive provided by the company might not be an appropriate approach to arouse employees to increase their environmental awareness in the sustainable way.

5. Conclusion and Recommendations

5.1. Conclusion

Effective mechanisms towards environmental awareness enhancement can be included motivation and cooperation from top management level of the company, effective environmental friendly policy applied, environmental training program and knowledge provided, employees' participation in environmental activities, supply chain involvement, law enforcement, and public participation.

In addition, this study also affirms that those finding mechanisms are really effective as reflecting by percentage of agreement. Such mechanisms include company's mission and vision on environmental friendly company, the management intention to initiate and develop environmental friendly activities continuously such as providing more environmental knowledge, collaborating with external institutes for organizing environmental friendly trip, presenting reward or recognition for the outstanding environmental friendly employees, and complying with the international standard related to environmental friendly consideration.

5.2. Recommendations

- The mechanisms can be applied to any organization including government and private sectors, as appropriate.
- The mechanisms towards environmental awareness enhancement should be further studied to other types of case study such as governmental sector, service sector, public organization, and non-profit organization which indirectly uses the natural resources as their primary sources to increasing the environmental awareness.
- Stakeholder's involvement and public participation would be another effective mechanism for environmental awareness enhancement.

Acknowledgements

This study was fully completed through the excellent support from the corresponding author. This study also received a partial funding support by the Environment, Development and Sustainability (EDS) Program and the Graduate School, Chulalongkorn University.

References

[1] United Nations Environment Programme (UNEP) 2012 *21 Issues for 21st Centuries, Results of the UNEP Foresight Process on Emerging Environmental Issues* [Online]. Available from: http://www.unep.org/publications/ebooks/foresightreport/Portals/24175/pdfs/Foresight_Report-21_Issues_for_the_21st_Century.pdf [2012, February]

[2] Intergovernmental Panel on Climate Change (IPCC) 2014 *Climate Change 2014: Synthesis Report. Contribution of Working Groups I, II and III to the Fifth Assessment Report of the Intergovernmental Panel on Climate Change* [Core Writing Team, R.K. Pachauri and L.A. Meyer (eds.)]. IPCC, Geneva, Switzerland, 151 pp

[3] United Nations (UN) 1992 *Agenda 21* [Online]. Available from: http://sustainabledevelopment.un.org/content/documents/Agenda21.pdf
[4] United Nations Institute for Training and Research (UNITAR) 2012 *Introduction to Corporate Social Responsibility* [Online]. Available from: http://www.unitar.org/event/introduction-corporate-social-responsibility
[5] Thailand Business Council for Sustainable Development (TBCSD) 2014 *The 1/2014 Joint Meeting of Council and Associates of Thailand Business Council for Sustainable Development (TBCSD)* [Online]. Available from: http://www.tei.or.th/tbcsd/reports/140325-presentation-2014.pdf
[6] United Nations (UN) 2015 *70 Ways the UN Makes a Difference* [Online]. Available from: http://www.un.org/un70/en/content/70ways#environment
[7] Dourish P and Bellotti V 1992 *Conference on Computer Supported Cooperative Work. Awareness and coordination in shared workspaces* (Toronto: In J. Turner & R. Kraut (Eds.), Proc. of ACM CSCW'92) p. 107
[8] Brennan L, Binney W, McCrohan J and Lancaster N 2011 *Australasian Marketing Journal. Implementation of environmental sustainability in business: Suggestions for improvement* **19** 52
[9] Agnes N. 2008 *Journal of Cleaner Production. Consistency and awareness gaps in the environmental behavior of Hungarian companies* **16** 322
[10] Remmen A and Lorentzen B 2000 *Journal of Cleaner Production. Employee participation and cleaner technology: learning processes in environmental teams* **8** 365
[11] Gadenne D, Kennedy J, and McKeiver C 2009 *Journal of Business Ethics. An Empirical Study of Environmental Awareness and Practices in SMEs* **84** 45
[12] Abdul-Wahab S 2010 *Climatic Change. Level of environmental awareness towards depletion of the ozone layer among distributors and consumers in the solvent sector: a case study from Oman* **103** 503
[13] Eaganl P and Streckewald K 1997 *J. Ckaner Prod. Striving to improve business success through increased environmental awareness and design for the environment education. Case study: AMP incorporated* **5** 219

Key factors of low carbon development strategy for sustainable transport

K Thaveewatanaseth[1] and S Limjirakan[2]

[1] The Environment, Development and Sustainability Program, Graduate School, Chulalongkorn University, Phayathai Road, Phatumwan, Bangkok, 10330, Thailand
[2] The Environmental Research Institute, Chulalongkorn University, Phayathai Road, Phatumwan, Bangkok, 10330, Thailand

*sangchan.l@chula.ac.th

Abstract. Cities become more vulnerable to climate change impacts causing by urbanization, economic growth, increasing of energy consumption and carbon dioxide (CO_2) emissions. People who live in the cities have already been affected from the impacts in terms of socio-economic and environmental aspects. Sustainable transport plays the key role in CO_2 mitigation and contributes positive impacts on sustainable development for the cities. Several studies in megacities both in developed and developing countries support that mass transit system is an important transportation mode in CO_2 mitigation and sustainable transport development. This paper aims to study key factors of low carbon development strategy for sustainable transport. The Bangkok Mass Rapid Transit System (MRT) located in Bangkok was the study area. Data collection was using semi-structured in-depth interview protocol with thirty respondents consisting of six groups i.e. governmental agencies, the MRT operators, consulting companies, international organizations, non-profit organizations, and experts. The research findings highlighted the major factors and supplemental elements composing of institution and technical capacity, institutional framework, policy setting and process, and plan of implementation that would support more effective strategic process for low carbon development strategy (LCDS) for sustainable transport. The study would highly recommend on readiness of institution and technical capacities, stakeholder mapping, high-level decision-makers participation, and a clear direction of the governmental policies that are strongly needed in achieving the sustainable transport.

1. Introduction

Cities are the first priority areas to concentrate in economic development, social activities, extensive infrastructures and governmental operations. They have a great potential for improving people's lives, economic, infrastructure, and accessibility in healthcare and education. However, urbanization process can cause several negative impacts on quality of life, natural resource, and environment. Consequently, this causes the changing of climate. Socio-economic development in many cities in Asia has been growing very rapidly such as economic growth, urbanization, production and consumption, urban employment, and motorization. As results of these developments, this region has the fastest producing greenhouse gas (GHG) emissions in the world that would generate GHG emissions up to 40-54 percent of global CO_2 emissions as of business-as-usual (BAU) scenario in 2050 [1]. Such GHG emissions, leading temperatures increasing, urban climate variable, and extreme events have already affected people who living in the cities in terms of socio-economic and environmental aspects leading unsustainable city. The Intergovernmental Panel on Climate Change (IPCC) reported that countries in Asia region had high vulnerable to the impacts of global climate change [2]. Sustainable transport has been raised up into the mainstreaming of low carbon development and sustainable city frameworks since first recognized at the United Nations Conference on Environment and Development (UNCED), in the outcome document namely the Agenda 21 [3]. Nowadays, sustainable transport is one of the major approaches for the mega cities to achieving Sustainable Development Goals (SDGs) as the 2030 Agenda for Sustainable Development [4]. Bangkok is one of the cities in Asia where become more vulnerable areas to climate change impacts and unsustainable city patterns [5]. As results from inefficient public transport system, lack of land-use planning as well as lack of mobility management, Bangkok has more environmental, social and economic problems. Mass rapid transit system is presently considered as the key transportation mode to support sustainable transport and low carbon city with regard to CO_2 mitigation, air quality improvement, energy saving and quality of life improvement. In Thailand, the mass rapid transit master plans have been developed as the key role in developing Bangkok toward low carbon and sustainable city, but there still lack of sustainable transport concept applied in such plans. The research aims to study key factors of low carbon development strategy for sustainable transport.

2. Literature review

2.1. City, urban development, and climate change

The United Nations Department of Economic and Social Affairs (UN DESA) reported that South-Eastern Asia countries have gradually increased in the proportions of urban population since 1950 and it is projected to increase to 65 per cent of the total region population by 2050 [6]. In addition, Institute of Southeast Asian Studies (ISEAS) noted that small cities and lack of urban management capacity lead to difficulty in improving the local economy and develop infrastructure [7]. Regarding to cities and environment, the effects of urbanization create unprecedented negative impacts upon quality of life, social stability, economic, and environment. Furthermore, global cities are critical areas in mitigate climate change [8]. The United Nations Human Settlements Programme (UN HABITAT) also noted that there is the interrelationship between climate change and urban development [9]. While economic development under BAU scenario leads to higher GHG emissions leading climate change impacts. In order to facing of climate change, urban areas must improve planning and building capacities for adaptation and mitigation. There has been become more and more cities in developed and developing countries doing climate change mitigation policy and implementation [8, 10].

2.2. Low carbon development concepts

Low carbon development is a development paradigm that contributes to addressing the challenges in economic development and environmental sustainability by keeping GHG emissions low without intervention. Low carbon development is at the intersection of development and mitigation by harmonizing the two imperatives of economic growth and environmental sustainability [11]. According to changing in a development model from treating environmental protection as an economic burden to a driver for global and national economic development, green low carbon growth concept has changed as essential for long-term economic and environmental sustainability [12]. Additionally, Tilburg et al. [11] reported that the development of low carbon development planning needs six elements including (1) readiness refers to available in data of the current situation, experts, and awareness of stakeholders, (2) building blocks refer to assessing the current situation in data and capacity; and identification of policy aims, actions, and interventions, (3) technical and political refer to technical assessments and national political process and decision-makers, (4) continuous process refers to planning, analysis and design, implementation, and evaluation, (5) purpose and (6) outcomes.

2.3. Sustainable transport

Sustainable transport has been concerned in many international agreements. Sustainable transport was firstly considered in the Agenda 21 [3]. In addition, the Future We Want, the outcome document from the United Nations General Assembly (UNGA) in 2012 or Rio+20, mentioned sustainable transport as the central to sustainable development [13]. Nowadays, sustainable transport is one of the major approaches for the mega cities to achieving Sustainable Development Goals (SDGs) on climate actions and sustainable cities [4, 14]. Several studies in megacities both in developed and developing countries support that public transportation in mass transit system plays the key role in CO_2 mitigation, efficient energy use and quality of life improvement through sustainable transport [15-17]. Moreover, sustainable transport could contribute to enhancing the three pillars of sustainable development including environment, economic, and social aspects as well as climate change mitigation and adaptation to the city [18].

3. Research methodology

Bangkok is located in the central part of Thailand which has high density in the inner city areas locating by official places and commercial business areas. In terms of motorization, Bangkok had number of registered cars about 7.5 million at an increasing rate of 27 percent between 2008 and 2012 while the road capacity available was only for 1.6 million cars [19]. The transportation system study selected is the Bangkok Mass Rapid Transit System. Regarding the Mass Rapid Transit Master Plan in Bangkok Metropolitan Region or M-Map proposed by the Office of Transport and Traffic Policy and Planning (OTP) is the current master plan to develop an urban rail transit network system serving in Bangkok [20]. It was designated eight primary routes and four feeder lines and plan to be constructed within a development period of twenty years (2010–2029).

The nature of this research was designed as exploratory and descriptive researches. A qualitative research method was used in the study. Purposive sampling method was used in order to select the key respondents who play the major roles and responsibilities in the policies and plans. Source triangulation was used to increase the conformability of the result findings. In this study, various sources of data and information were integrated in order to build a coherent for themes and to confirm the validity. The instruments of data collection consisted of two approaches: (1) desk study - literature reviews based on low carbon development strategy and sustainable transport including secondary relevant document assessments such as policies and plans in order to understand the current situations according to the existing problems, their consequent effects, and management. (2) the in-depth interview applied semi-structured interview and the interview protocol was open-ended and descriptive.

Total numbers of thirty respondents are from various organizations with different levels of operation and multi-fields of expertise consisting of six groups including 40% are governmental agencies, 17% are the MRT operators, 13% are consulting companies, another 13% are international organizations, 10% are non-profit organizations, and 7% are experts.

4. Results and discussions

4.1. Factors and elements of low carbon development strategy (LCDS)

The research findings indicate factors and elements for the LCDS as follows:

- Readiness - The study found that there are inadequate in data availability and sharing as well as data organized for support sustainable transport and low carbon development plan while data gaps and overlapping are also mentioned. Additionally, readiness in experts, human resources, and technology in such development plan are inadequate. For process and policy, shifting political priorities has influence on policy processes and unpredictably changing in politics could result in different policy priority setting and discontinuity in transport policy implementation. In addition, the study further found that lack of stakeholders' awareness, engagement, and ownership to the development plan is still exists leading to less their buy-in for the strategy and its implementation. Furthermore, high-level decision-makers participation is required in an important role.

- Building blocks – For capacity assessment, the capacity to analyze the data is needed due to insufficiency of stakeholders' knowledge, skills, and technology. Such capacity is still limited caused by budgeting supports and the government support policies. In terms of stakeholder mapping, unclear roles and responsibilities of stakeholders are still exist in the cross-sectoral strategies implementation. This could result in lacking of their awareness, engagement, and ownership to the development plan leading to less stakeholders' coordination and communication. The study further found that there is inter-ministerial committee which set up with key sector ministries, departments and agencies for the sustainable transport; however, high-level decision-makers and inter-ministerial participation are inadequate. In long-term vision, targets and actions of the policy, Thailand policies have long-term vision in transportation development; nevertheless, concrete targets and actions are lacked to achieving the long-term goals. Additionally, financial sources and private sector investment supports are needed in developing the sustainable transport. Moreover, the research findings indicate that government supporting policies and effective relevant regulations enforcement are the supplemental elements for institutional and technical capacity of sustainable transport development.
- Technical and political process - The LCDS should ensure to align with technical and policy process for confirming that the strategy could be implemented properly. In terms of technical assessments, the findings indicated lack of data availability, reliability, and analysis due to less data sharing, data disorganized, data gaps and overlapping. In addition, the numbers of experts, human resources and technology are inadequate for technical capacity. For national political process and decision-makers, the study found that shifting political priorities has influence on policy processes.
- Continuous process - The research findings from secondary data assessments show that the plans and strategies have policy development process include planning, analysis and design, implementation, and evaluation. However, the study found that continuous and effective implementation plans of stakeholder groups are inefficient.
- Purpose – Although, the LCDS has different purposes depending on the stakeholders and apply the strategy to their different purposes. However, the study found that there are different transport policies and plans in Thailand context which proposed by different stakeholders but such policies and plans have common purposes in reducing traffic congestion and improving transportation mobility.
- Outcomes - Based on the secondary data assessments, the study found that Thailand policies and plans have been adjusted regarding changes of condition and circumstance. For example, Thailand's Transport Development Strategy: 20 years plan proposed by the OTP has developed the strategic plans in order to achieving the 20-year National Strategic Plan goals by recognizing the importance of the current transport infrastructure development. This strategic plan has three operational periods: critical issue, medium, and long-term periods to ensure the optimization of implementations and outcomes [21].

The study findings are consistent with Tilburg et al. [11]. Additionally, the study further found more important factors with relevant elements that would make more effective low carbon development planning.

4.2. The LCDS factors to achieve sustainable transport

Table 1, six factors of the LCDS elements include an inter-ministerial participation and stakeholders coordination and communication as (F1), integration and consistency of policies and plans as (F2), budget and financial support as (F3), clarification of role, responsibility and awareness of stakeholders as (F4), readiness of knowledge, skills, and technology of stakeholders as (F5) and readiness of data availability, reliability, and analytical data as (F6).

Table 1. The LCDS factors to achieve sustainable transport.

LCDS factors	Governmental agencies	MRT operators	Consulting companies	International organizations	Experts	Non-profit organizations	Average
F1	88%	76%	80%	80%	100%	93%	86%
F2	90%	76%	80%	80%	90%	80%	83%
F3	92%	88%	95%	65%	70%	67%	79%
F4	83%	32%	70%	70%	100%	93%	75%
F5	75%	56%	65%	70%	90%	80%	73%
F6	77%	32%	70%	70%	80%	87%	69%

To achieve sustainable transport, all respondent groups strongly agreed in two major factors including inter-ministerial participation and stakeholders' coordination and communication (86%) and integration and consistency of policies and plans (83%). However, it is obviously noticed that the MRT operators did not emphasized on the factors including readiness of knowledge, skills, and technology of stakeholders (56%); clarification of role, responsibility and awareness of stakeholders (32%); and readiness of data availability, reliability, and analytical data (32%).

5. Conclusion and recommendations

According to the research findings to achieve sustainable transport, key LCDS factors compose of institution and technical capacity, institutional framework, policy setting and process, and plan of implementation as shown in figure 1.

In terms of institution and technical capacity, readiness of data availability, quality, and reliability is required to provide input data of the current situation and related information to the policy maker and decision as well as relevant stakeholders to analyzing and assessing for setting their policies and actions. Readiness of experts, human resources, and technology is also needed to increasing capacity in generating knowledge, skills, and technology for develop and implement filling information gaps of the countries in developing the low carbon strategic plan. In addition, sources of finance are essential for high investment in transport infrastructure and budget allocation to key ministries sectors, departments, and agencies. Finally, awareness and engagement of stakeholders are important elements to make their involvement and buy-in to the process. Therefore, institution and technical capacity is a prerequisite in the LCDS process.

Institutional framework is the second factor of the LCDS process which calls for high-level attention and participation to develop a stronger awareness and buy-in among decision-makers and a participatory process of high-level political support from the start could provide strong high-level leadership within the government. The framework needs inter-ministerial committee setting up effectively coordinates with key sector ministries, departments and agencies. Shifting political priorities has influence on policy processes resulting in different policy priority setting and discontinuity in transport policy implementation. Therefore, consistency of political and policy are important for institutional framework. Additionally, regulations and government intervention could support governmental agencies in good management and control on the traffic and transportation as well as the city planning. Not only sources of finance is mentioned as a requirement to ensure the institutional and technical capacity, but also budget allocation and financial support are required in institutional framework for the LCDS. In terms of stakeholder mapping, clear roles and responsibilities of them are necessary in setting the institutional framework for the cross-sectoral development strategic plans. Such unclear roles and responsibilities could result in lacking of their awareness, engagement, and ownership to the development plans.

For policy setting and process, based on institutional framework part, this could lead to support policy setting and process for the LCDS. With high-level and inter-ministerial participation, these could lead long-term concrete plan and strategy as well as policy integration and inclusive plan.

Consistency of political and policy leads to policy and strategic orientation. Additionally, stakeholder mapping and clear roles and responsibilities are major elements to create stakeholders involvement and buy-in in the policy setting and process.

In plan of implementation, effective regulations enforcement and administration supports are needed to promote efficiency and concrete action and implementation. Iterative process and continuity in implementing plan would provide more effective in such strategic process. In addition, stakeholder collaboration and communication are demanded for supporting in the cross-sectoral strategic implementation.

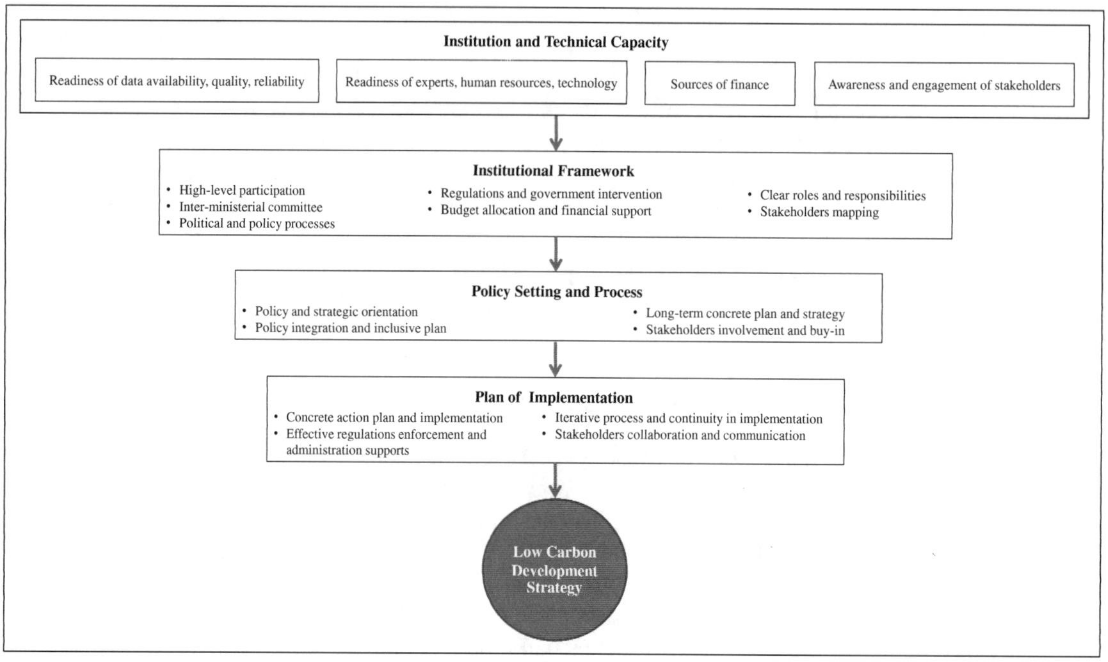

Figure 1. Key factors of low carbon development strategy for sustainable transport.

Based on the findings on key factors to support effective strategic process for the LCDS on sustainable transport, the further recommendations are as follows: (1) the governmental agencies and relevant stakeholders such as the MRT operators should have readiness of institution and technical capacities as a prerequisite component of achieving the LCDS process, (2) the governmental agencies should have stakeholder mapping and clarification of their roles and responsibilities to enable the strategy to be successfully implemented plan, (3) high-level decision-makers would have participation and consistency of political and policy process to ensure concrete long-term vision and supportive intervention, and (4) those decision-makers would have a clear direction of the governmental policies that would be strongly support in achieving the LCDS on sustainable transport.

Acknowledgments

I would like to extend my sincere thanks to Dr. Sangchan Limjirakan, for her insightful comments and valuable suggestion for this research paper. I also thank the Environment, Development and Sustainability (EDS) Program and the Graduate School, Chulalongkorn University for a partial funding support for this research.

References

[1] Kainuma M, Shukla P R and Jiang K 2012 *Energy Economics. Framing and Modeling of a Low Carbon Society: An Overview* **34** S316

[2] Intergovernmental Panel on Climate Change (IPCC) 2007 *Climate Change, Impact, Adaptation and Vulnerability* (Cambridge: Cambridge University Press)

[3] United Nations Conference on Environment and Development (UNCED) 1992 *Agenda 21* United Nations Division for Sustainable Development New York

[4] United Nations Secretary-General's High-level Advisory Group on Sustainable Transport (UNSGs) 2014 *Global Sustainable Transport Outlook Report: Mobilizing Sustainable Transport for Development* United Nations

[5] Banister D 2011 *J. of Transport Geography. Cities, Mobility and Climate Change* **19** 1538

[6] United Nations Department of Economic and Social Affairs (UN DESA) 2012 *World Urbanization Prospects: The 2011 Revision* UN DESA, Population Division, New York

[7] Institute of Southeast Asian Studies (ISEAS) 2009 *Urbanization in Southeast Asian Countries* The ASEAN Studies Centre, Singapore

[8] Hoornweg D, Freire M, Lee M J, Bhada-Tata P and Yuen B 2011 Cities and Climate Change: Responding to an Urgent Agenda *Urban Development Series* The World Bank

[9] United Nations Human Settlements Programme (UN HABITAT) 2010 Part 5: The Urban Environment and Climate Change *The State of Asian Cities 2010/11* pp.162-205

[10] Hoornweg D 2010 Cities and Climate Change: An Urgent Agenda *Urban Development Series* The World Bank

[11] Tilburg X V, Würtenberger L, Coninck H D and Bakker S 2011 *Paving the Way for Low-Carbon Development Strategies* Energy research Centre of the Netherlands (ECN)

[12] The World Bank 2011 *Thailand: Clean Energy for Green Low-Carbon Growth* The World Bank Group and the Office of the National Economic and Social Development Board (NESDB) Thailand

[13] United Nations General Assembly (UNGA) 2012 *The Future We Want* United Nations Conference on Sustainable Development (Rio+20) Rio de Janeiro Brazil 20-22 June 2012 A/RES/66/288 United Nations

[14] United Nations General Assembly (UNGA) 2015 *Transforming our world: the 2030 Agenda for Sustainable Development* A/RES/70/1 United Nations

[15] Kennedy C, Miller E, Shalaby A, Maclean H and Coleman J 2005 *Transport Reviews. The Four Pillars of Sustainable Urban Transportation* **25** 393

[16] Zhao P 2010 *Habitat International. Sustainable Urban Expansion and Transportation in a Growing Megacity: Consequences of Urban Sprawl for Mobility on the Urban Fringe of Beijing* **34** 236

[17] Nakamura K and Hayashi Y 2012 *Transport Policy. Strategies and Instruments for Low-carbon Urban Transport: An International Review on Trends and Effects* **29** 264

[18] Ang G and Marchal V 2013 Mobilising Private Investment in Sustainable Transport: The Case of Land-Based Passenger Transport Infrastructure *OECD Environment Working Papers* No. 56 OECD Publishing

[19] National Information Centre (NIC) 2013 Registered Vehicles in Bangkok Ministry of Information and Communication Technology [Online]. Available from: http://www.nic.go.th /search1/showDetail.asp?url1=http://ws.mot.nic.go.th/wservice/public/4778&tGraph=1.

[20] Office of Transport and Traffic Policy and Planning (OTP) 2010 *Mass Rapid Transit Master Plan in Bangkok Metropolitan Region (2010-2029) Executive Summary* Ministry of Transport Thailand

[21] Office of Transport and Traffic Policy and Planning (OTP) 2017 *Thailand's Transport Development Strategy: 20 years plan (2017-2036)* Ministry of Transport Thailand

IOP Conf. Series: Earth and Environmental Science **117** (2018) 012004 doi:10.1088/1755-1315/117/1/012004

Confirmatory factor analysis of Post-Occupancy Evaluation Model (POEM) for sustainable neighborhood development

R Yaman[1, 3], S Thadaniti[1], J Abdullah[3], N Ahmad[2], N M Ishak[3]
[1]The Environment, Development and Sustainability Program, Graduate School, Chulalongkorn University, Phayathai Road, Pathumwan, Bangkok, 10330, Thailand
[2]Kulliyah of Architecture and Environmental Design, International Islamic University Malaysia, Gombak, Selangor, Malaysia
[3] Faculty of Architecture, Planning and Surveying, Universiti Teknologi MARA, 40450 Shah Alam, Selangor, Malaysia

* rostamyaman1@yahoo.co.uk

Abstract. The sustainable urban development growth in the ASEAN region has accelerated tremendously. More demand on the sustainable development has led to bigger market driven certified green neighborhood and buildings. However, there is a lack of post-occupancy evaluation study conducted in assessing the end-users perspective on the certified sustainable neighborhood development. This paper aims to investigate the end-users point of view on sustainable dimension pillar (SDP) adaptation based on environment dimension, social dimension and economic dimension using Post-Occupancy Evaluation Model (POEM) framework. The research methodology employed stakeholders-Inclusion Approach survey questionnaires in order to obtained the sustainable dimensional adaptation score. The results show that there is sustainable dimension gap in POEM evaluation, hence, suggested the pre-occupancy criteria did not fulfill the SDP adaptation and pre-occupancy criteria and variables if differ from post-occupancy criteria and variables.

1. Introduction

The Malaysian construction and development sector has been over the years emerged and thrived in the direction of a more progressive sustainable urban agenda. The subject of sustainable neighbourhood or township benchmarking approach in the field of urban development in Malaysia is relatively new. Although GBI Township Assessment Criteria has been developed and implemented but there is no post-occupancy evaluation being conducted in assessing the performance, effectiveness and sustainability level on the certified development. There has been no study done to measure the greenness' of the tools and the post certified township development. Even though there is study conducted on environment dimension, however fewer so are conducted on what really signifies socially or/and economically, and/or its application in the sustainable urban development context. Do sustainable neighborhood or township evaluation criteria and frameworks in common signify sustainable holistically? Therefore, proficiency and comprehension on urban sustainability benchmarking criteria could strengthen neighborhood sustainable indicators, effectiveness and sustainability level among the Malaysian sustainable urban development actors are vastly still low. Studies has indicated one of the main obstacles that hinder sustainable township/neighborhood development in ASEAN nations is the deficiency of knowledge in sustainable concerns subjects in relation to the involved building profession [1].

Past study done on sustainable building rating system potential in Malaysia also shows that Malaysia construction and development sector key stakeholders have insufficient understanding on sustainability development evaluation, benchmarking & indexing approach [2]. Due to this many green certified urban development project in Malaysia claim sustainability merely for label advertisement, marketing tools and higher premiums instead of fully addressing the sustainable pillars. Certified GBI township projects normally a high-end urban development projects, study has indicated that sustainable certification do improve leasing and selling rate of developed properties, but this outcome is further substantial for end-users who are more innately concerned with sustainability, or pushing their 'green' appearance [3]. Even though the noble foundation for sustainable township/neighborhood is to promote and applied sustainable development, but there is no study was undertaken to address and gauge this issues. Thus, this embarks the problem of the research: whether the existing GBI Township Assessment Criteria and certified urban development project fulfilled the post-occupancy performances, effectiveness and sustainability level according to dimensions Pillars of Sustainability.

2. Profile of the Respondents

The case study of this research is three GBI certified sustainable neighbourhood of Ken Rimba Development, Bandar Rimbayu Development and Sunway Resort City in Greater KL. In this Confirmatory Factor Analysis study, preliminary data analysis is conducted prior to it. Preliminary data analysis involves the total number of end-users/household respondents and the conduct test of reliability measurement in ensuring the consistency of the applicable data.

The tabulation of Phase 3 survey respondents is as shown in Table 1. The detail composition of the respondents based on case studies are Ken Rimba 122 respondents, where 59 respondents are homeowners (15.60%), 52 respondents are tenants (13.75%) and 11 respondents are workers (2.91%). For Bandar Rimbayu case study, 71 respondents are homeowners (18.78%), 38 respondents are tenants (10.05%) and 16 respondents are workers (4.23%). The third or last case study is Sunway Resort City where 70 respondents are homeowners (18.51%), 39 respondents are tenants (10.31%) and 22 respondents are workers (5.82%). The description for such composition is that homeowners is the main type of users reside in the certified sustainable neighbourhood development. This is followed by tenants as the second type of users who occupied the chosen case studies and the balance is the workers. Since sustainable neighborhood development in the Klang Valley/Greater KL is a high-end property development, it is justifiable that most of the end-user respondents is the homeowners where they plan for permanent or long-term stay instead of tenants or workers.

Table 1. Number of Respondents by User Type / Neighborhood

User/Neighborhood	No. Of Respondents	Percentage (%)
Homeowners	59	15.60%
Tenants	52	13.75%
Workers	11	2.91%
Ken Rimba	122	32.27%
Homeowners	71	18.78%
Tenants	38	10.05%
Workers	16	4.23%
Bandar Rimbayu	125	33.06%
Homeowners	70	18.51%
Tenants	39	10.31%
Workers	22	5.82%
Sunway Resort City	131	34.65%
Total	378	100.00%

3. Theoretical Framework

The Framework of Research in this POEM for SND study using Stakeholders-Inclusion Approach – End-User/Household Survey comprises of one latent exogenous construct which is Post-Occupancy Evaluation Model for Sustainable Neighborhood Development (POEM for SND), and three dependent variable of Environment Dimension Pillar (EnP), Social Dimension Pillar (EnP), and Economic Dimension Pillar (EnP). The theoretical relationship between the constructs is modelled and shown in Figure 1.

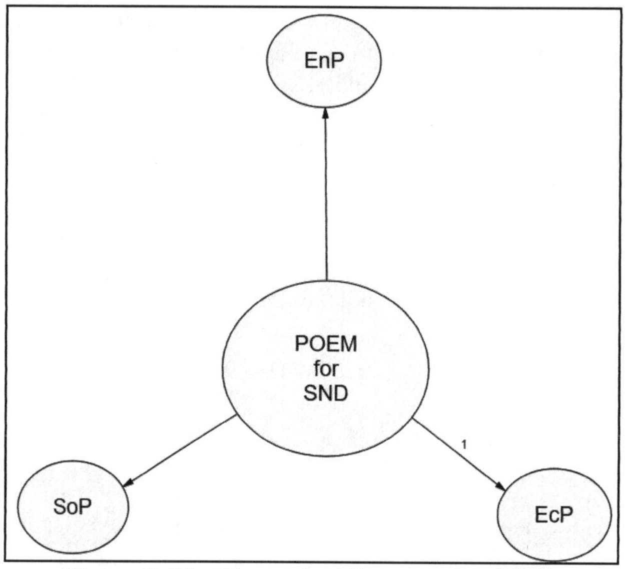

Figure 1. The Theoretical Framework of CFA

As presented in Figure 1, the latent construct is Post-Occupancy Evaluation Model for Sustainable Neighborhood Development (POEM for SND). The three (3) observed variables in the study model are Environment Dimension Pillar (EnP), Social Dimension Pillar (EnP), and Economic Dimension Pillar (EcP). The first exogenous construct is Environment Dimension Pillar (EnP) which consists of six (6) endogenous variables. The second listed exogenous construct is Social Dimension Pillar (EnP), which contains of eight (8) endogenous variables. The last or third listed exogenous construct is Economic Dimension Pillar (EcP) which consists of seven (7) endogenous variables.

All endogenous variables is assessed by means of an interval scale which vary between from 1 (Very Low Sufficiency/Practice) to 5 (Very High Sufficiency/Practice) with the given end-user occupancy sustainable experience statement in POEM Assessment Criteria. The measurement of model for every exogenous construct in this study is adapted and simplified from previous POEM for SND Study (phase 1 & phase2). For this Stakeholders-Inclusion Approach – End-User/Household Survey model of POEM for SND, the study could assess the CFA measurement using one model for all constructs in order to achieve the relevant thresholds for validity and reliability (Awang, 2014; 2015; Awang et al., 2015; Kashif et al., 2015; 2016).

Hence, in this Stakeholders-Inclusion Approach – End-User/Household Survey model of POEM for SND study decided to conduct a single CFA procedure for all exogenous construct. According to Awang (2012; 2013; 2014; 2015) [4], Awang et al. (2015) [5] and Kashif et al. (2015; 2016) [6], the study prerequisites are to verify that all constructs in the theoretical model are discriminant of each other or it's are not highly correlated particularly amongst the exogenous constructs prior to model the full structural model and implementation of Structural Equation Modelling. In the case where there are two or more exogenous constructs are highly correlated (correlation above than 0.85). Highly correlation will results of a significant problem called Multicollinearity or a phenomenon in which one predictor variable in a multiple regression model can be linearly predicted from the others with a substantial degree of accuracy

3.1 Reliability Analysis

This POEM for SND study construct consists of three exogenous constructs named as Environment Dimension Pillar (EnP), Social Dimension Pillar (SoP) and Economic Dimension Pillar (EcP). EnP exogenous construct is measured using six (6) endogenous variables of POEM for SND evaluation criteria in the questionnaire which were Sufficient Designated Green Area (Q8.1), Sufficient Street or Park Lighting (Q8.2), Generate or Use Renewable Energy (Q8.3), Reduced or Recycle Water Practice (Q8.4), Bio-diversity Reserved Availability (Q8.5) and Infrastructure Services Efficiency (Q8.7). SoP exogenous construct is measured using eight (8) endogenous variables of POEM for SND evaluation criteria in the questionnaire which were Sufficient Communal Space (Q8.9), Population Density & Development Level (Q8.10), Universal Accessibility Availability (Q8.12), Security and Safety Experiences (Q8.13), Public Health Concerns (Q8.14), Recycling Facilities or Practices (Q8.15), Community Diversification (Q8.16) and Community Engagement & Management (Q8.17). Lastly, EcP exogenous construct is measured using seven (7) endogenous variables of POEM for SND evaluation criteria in the questionnaire which were Public Transport Reliability (Q8.19), Sufficient Pedestrian & Cycling Networks (Q8.20), Low Impact & Regional Materials (Q8.21), Promotion of Sustainable Construction (Q8.22), Construction Waste & Sedimentation (Q8.23), Sufficient Commercial Amenities (Q8.24) and Innovative Development (Q8.25). Figure 2 below show the conceptual path model of POEM for SND CFA analysis.

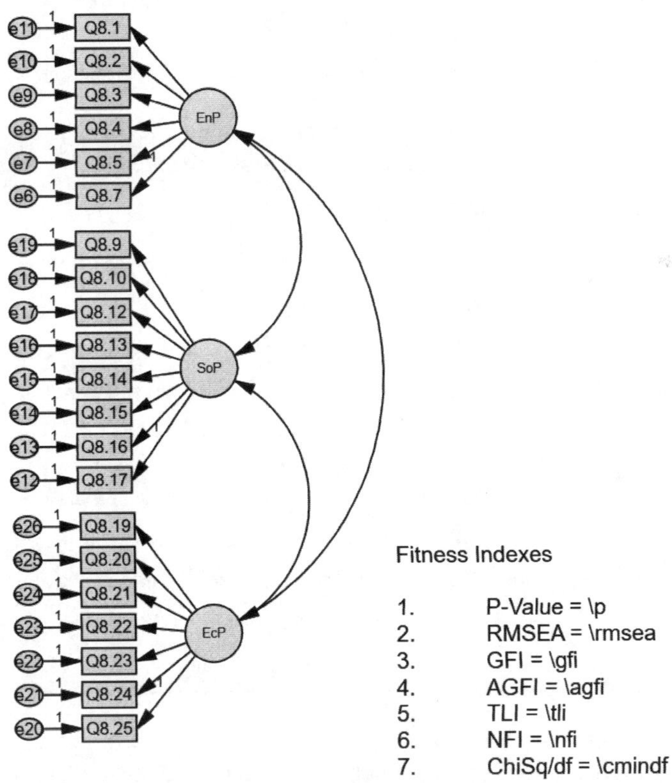

Figure 2. The exogenous constructs and measuring endogenous variables of POEM - SND

This study employed the internal consistency approach in order to assess the reliability using Cronbach's Alpha correlation-coefficient for evaluating internal consistency. It is by standard agreement that a Cronbach's Alpha is 0.70 or higher, however, according to Hair (2010) in the exploratory research study Cronbach's Alpha value can be lower up to 0.60.

Considering the importance POEM for SND evaluation criteria and its Sustainable Dimension Pillars (SDP) adaptation in relationship with post-occupied certified sustainable neighbourhood, a single latent construct (POEM for SND) has generated with these three sustainable dimension pillars influence factors as its endogenous variables. POEM for SND construct acted as the dependent variable. Each exogenous construct comprises of question items to which the respondents would opted their sustainable related experience based on a 5-scale point that suggest the extent to their conformity or disconformity with each given assertion. Table 2 details down the constructs, survey items and Cronbach's Alpha indexes employed in this CFA of POEM for SND study. The findings suggested that the highest internal consistency is Social Dimension Pillar (SoP), followed by Economic Dimension Pillar (EcP) while the lowest is Environment Dimension Pillar (EnP).

Table 2. Measurement of the Variables of the Hypothesized Model

Construct	Item	Survey questions	Mean	SD	Alph
EnP (Environment Dimension Pillar)	Q8.1	Sufficient Designated Green Area	3.751	.0740	0.533
	Q8.2	Sufficient Street or Park Lighting	3.791	0.736	
	Q8.3	Generate or Use Renewable Energy	2.822	0.914	
	Q8.4	Reduced or Recycle Water Practice	3.029	0.969	
	Q8.5	Bio-diversity Reserved Availability	4.148	1.263	
	Q8.7	Infrastructure Services Efficiency	3.624	0.575	
SoP (Social Dimension Pillar)	Q8.9	Sufficient Communal Space	3.825	0.754	0.725
	Q8.10	Population Density & Development Level	3.447	0.616	
	Q8.12	Universal Accessibility Availability	3.164	0.671	
	Q8.13	Security and Safety Experiences	3.695	0.651	
	Q8.14	Public Health Concerns	3.510	0.695	
	Q8.15	Recycling Facilities or Practices	3.343	0.957	
	Q8.16	Community Diversification	3.484	0.703	
	Q8.17	Community Engagement & Management	3.227	0.880	
EcP (Economic Dimension Pillar)	Q8.19	Public Transport Reliability	3.187	0.678	0.613
	Q8.20	Sufficient Pedestrian & Cycling Networks	2.735	1.417	
	Q8.21	Low Impact & Regional Materials	3.425	0.652	
	Q8.22	Promotion of Sustainable Construction	3.261	0.796	
	Q8.23	Construction Waste & Sedimentation	2.079	1.717	
	Q8.24	Sufficient Commercial Amenities	3.854	0.657	
	Q8.25	Innovative Development	3.957	0.838	

3.2 Full Measurement of CFA

In this study, a full measurement path model is acquired by using the Confirmatory Factor Analysis (CFA) method on the applied endogenous variables scales. The measurement path of the model is determined to investigate the level of the observed endogenous variables measure and the exogenous latent constructs. It indicates how the observable endogenous variables capture the hypothetical exogenous latent constructs intended in this study. Confirmatory Factor Analysis (CFA) is applied to assess all the construct via the overall measurement of path model as shown in Figure 3. Prior to full model measurement, the exogenous constructs (dependent variables) and their endogenous variables (independent variables) were selected corresponding to its underlying study theories. The factors denote exogenous constructs, and the supporting study questions denote their endogenous variables. Table 3 specifies the detail breakdown of CFA measurement of the listed constructs and variables.

In executing the full CFA model measurement, firstly, the appropriateness of fit for every measure was assessed in order to equate how adequate the model describes the data rationally. In this Phase 3 study, Chi-square (χ^2), RMSEA (Root Mean Square Error of Approximation) TLI (Tucker Lewis Index) and CFI (Comparative Fit Index) are applied to decide whether the model is acceptable. Secondly, the estimated path model outcomes are assessed to identify transgressing estimates. This is referring to the results signs and the statistical significance of overall study estimated parameters, which are the regression weights, the standardized regression weights and the probability value (P-value). For this study IBM SPSS Amos 22 was used, the full CFA measurement path model was designed and executed efficaciously which has produced the affirmative outcomes as presented in the Figure 3.

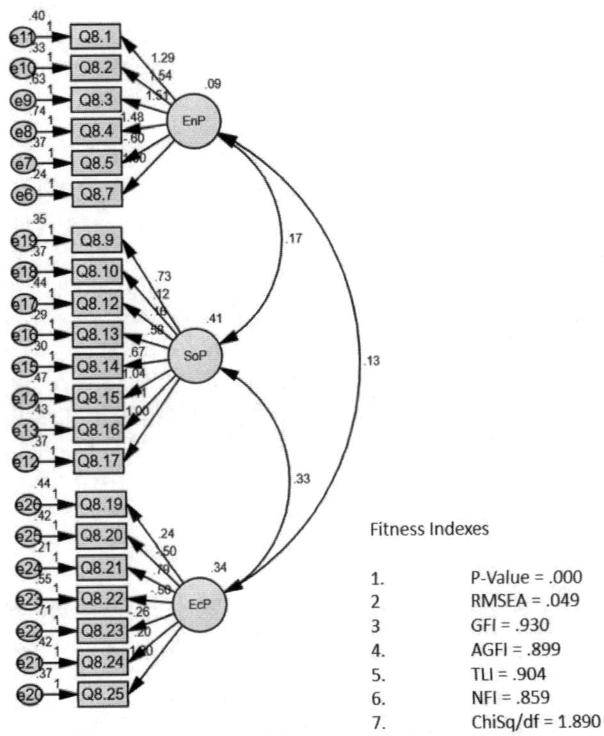

Figure 3. Results of Confirmatory Factor Analysis (CFA) procedure

Table 3. Parameter Estimates of Full Measurement Model

Construct &Variables			*(a)Unstd*	*(b)Std*	*S.E.*	*C.R.*	*P*
Q8.7	<---	EnP	1.000	.545			
Q8.5	<---	EnP	-.714	-.354	.127	-5.639	***
Q8.4	<---	EnP	1.292	.421	.197	6.563	***
Q8.3	<---	EnP	1.310	.453	.190	6.881	***
Q8.2	<---	EnP	1.520	.646	.172	8.823	***
Q8.1	<---	EnP	1.218	.515	.160	7.635	***
Q8.17	<---	SoP	1.000	.721			
Q8.16	<---	SoP	.424	.383	.060	7.014	***
Q8.15	<---	SoP	1.018	.677	.084	12.141	***
Q8.14	<---	SoP	.622	.567	.061	10.233	***

Q8.13	<---	SoP	.577	.563	.057	10.144	***
Q8.12	<---	SoP	.193	.191	.060	3.239	.001
Q8.10	<---	SoP	.112	.116	.053	2.110	.035
Q8.9	<---	SoP	.740	.623	.066	11.283	***
Q8.25	<---	EcP	1.000	.711			
Q8.24	<---	EcP	.246	.222	.064	3.844	***
Q8.23	<---	EcP	-.250	-.174	.082	-3.044	.002
Q8.22	<---	EcP	-.491	-.367	.077	-6.372	***
Q8.21	<---	EcP	.749	.684	.066	11.325	***
Q8.20	<---	EcP	-.464	-.390	.069	-6.745	***
Q8.19	<---	EcP	.218	.191	.065	3.332	***

Note:
(a) Estimated unstandardized regression co-efficient
(b) Standardized regression co-efficient
(c) Standard error of estimated unstandardized co-efficient.
(d) Probability of a t value equal to or greater than actual t value in a two-tailed test for significance of coefficient under the null hypothesis that the true value is zero. The symbol *** indicates that the null hypothesis is rejected at the .001 level of significance.

4. Main Results and Discussions

The overall measurement of parameter estimates are shown in Table 3. The P-values of all specific parameter estimates are highly significant where the value is at 0.001 (***) level amongst the construct except for Q8.10-->SoP and Q8.23-->EcP are statistically significant where the P-values are $\geq$ 0.001 but $\leq$ 0.05 (refer to Table 6.4). The affirmative co-efficient sign with highly significant P-value (***) and statistically significant P-value $\leq$ 0.05 means all hypothesized paths within the CFA model indicates a highly significant relationship, therefore, all hypotheses are supported.
The Confirmatory Factor Analysis (CFA) model generated the overall data (relative Chi-square = 1.890; RMSEA = 0.049; GFI = 0.930; TLI = 0.904) which fulfilled their significant outscores. The values of AGFI (0.899) and NFI (0.859) are acceptable for a satisfactory model although these are slightly less than the suggested condition level of 0.90 (Hair, 2010). Thus, the model justifies that the data were ideally well represented as most of the indices yield a fit values of standardized loading estimates.

4.1 Environment Dimension Pillar (EnP)

EnP has six variables as indicated in Table 3. These variables consisted of information regarding environment dimension pillar post-occupancy evaluation criteria, which concerns on Sufficient Designated Green Area (Q8.1), Sufficient Street or Park Lighting (Q8.2), Generate or Use Renewable Energy (Q8.3), Reduced or Recycle Water Practice (Q8.4), Bio-diversity Reserved Availability (Q8.5) and Infrastructure Services Efficiency (Q8.7). The P-values for all specific parameter estimates of Climate, Energy & Water Construct are statistically high significant value at 0.001 (***) level, signifying that the paths model created from the construct to all six (6) variables are accepted. The standardized regression weights of the construct and variables are range in-between -0.354 to 0.646. The relationship between construct to EnP (Environment Dimension Pillar) indicated less significant value as indicated by Hair (2010), where the regression weights generated are below than 0.7. Furthermore one of the construct indicated negative value at -0.354 (Q8.5) against EnP.

It is distinctly indicated that Environment Dimension Pillar is less significant and moderately sufficient or experienced by the end-users/households of the selected case studies of certified sustainable neighborhood in the Klang Valley/Greater KL. Nevertheless, the correlation concerning construct to Q8.1 (Sufficient Designated Green Area), Q8.2 (Sufficient Street or Park Lighting), Q8.3 (Generate or Use Renewable Energy), Q8.4 (Reduced or Recycle Water Practice), and Q8.7 (Infrastructure Services Efficiency) standardized regression weight (0.421 – 0.646) denoted less significant but closed to the considerable level indicated by Hair (2010). However for Q8.5 (Bio-diversity Reserved Availability), the correlation concerning construct is negative value (-0.354) indicated that this provision is not available or experiences by the end-users/household of the selected certified case studies.

4.2 Social Dimension Pillar (SoP)
SoP construct has eight (8) variables as presented in Table 3. These variables consisted of information regarding social dimension pillar post-occupancy evaluation criteria, which regards to fulfil sustainable community related criteria of Sufficient Communal Space (Q8.9), Population Density & Development Level (Q8.10), Universal Accessibility Availability (Q8.12), Security and Safety Experiences (Q8.13), Public Health Concerns (Q8.14), Recycling Facilities or Practices (Q8.15), Community Diversification (Q8.16) and Community Engagement & Management (Q8.17). The P-values of all specific individual parameter estimates for EEC Construct are highly significant at 0.001 (***) value and statistically significant at $\geq$ 0.001 but $\leq$ 0.05 (refer to Table 4), representing that all eight (8) variables of the construct paths are accepted. The regression weights of the construct between the variables are range from 0.116 to 0.721 which is less than significant level except for Q8.17 (Community Engagement & Management) variables is significant where the generated values higher than 0.7.

4.3 Economic Dimension Pillar (EcP)
EcP has seven variables as shown in the Table 3. These variables consisted of information regarding economic dimension pillar post-occupancy evaluation criteria, which concerns on Public Transport Reliability (Q8.19), Sufficient Pedestrian & Cycling Networks (Q8.20), Low Impact & Regional Materials (Q8.21), Promotion of Sustainable Construction (Q8.22), Construction Waste & Sedimentation (Q8.23), Sufficient Commercial Amenities (Q8.24) and Innovative Development (Q8.25). The P-values of all specific individual parameter estimates for EcP Construct are highly significant at 0.001 (***) value and statistically significant at 0.002 value (Q8.23). The hypothesized model paths shows significant correlation, thus hypotheses of variables are all supported. For all variables of EcP Construct, the relationship co-efficient are yielding from -0.390 to 0.711 in standardized regression weight, which on standard are less significant except for Q8.25 (0.711) where the value is above significant level. All these variables generated the loading estimates below 0.7 value which are less ideal regression weight (Hair, 2010). Three variables yielded negative standardized regression weight value (Q8.20 at -0.390; Q8.22 at -0.367; and Q8.23 at -0.174). The balance three (3) variables yielded positive standardized regression weight value but below suggested acceptance level of above 0.7 value (Q8.19 at 0.191; Q8.21 at 0.684; Q8.24 at 0.222).

5.Conclusions
The CFA analysis and findings above reflect end-users opinion on SDP adaptation in certified sustainable neighborhood/township development in Greater KL. Summary of main findings of this study indicated overall theoretical model for POEM is significance where the P-value is at .0001 for all evaluation criteria. The conclusion of the study suggested there is SDP gap in SND development upon occupancy even though the pre-occupancy did adapted SDP. The full measurement CFA of this study indicated that the highest adapted SDP dimensions is Social Dimension (SoP), followed by Environmental Dimension (EnP) and finally Economic Dimension (EcP). It is also suggested that the pre-occupancy assessment criteria and post-occupancy assessment criteria theory and variables are differs. Hence, the study concluded that the existing GBI Township Assessment Criteria and certified urban development project did not fully fulfilled the post-occupancy performances, effectiveness and sustainability level according to dimensions Pillars of Sustainability.

Acknowledgments
The authors would like to thank EDS of Chulalongkorn University for providing a platform in conducting this study and the UiTM for supporting this research and all informants who provided inputs in this study through workshops, surveys, and interviews.

References

[1] Yahya WK and Hashim NH 2011 *International Conference on IEEE. The role of public awareness and government regulations in stimulating sustainable consumption of Malaysian consumers.* Business, Engineering and Industrial Applications (ICBEIA)

[2] Shari, Z., et al. *2008 International Journal on Sustainable Tropical Design Research & Practice. The Potential of sustainable building rating System in the Malaysian building industry ALAM CIPTA* **3** 3

[3] Sullivan, L. J., et al 2014 *Neighbourhood sustainability frameworks-a literature review*

[4] Awang, Z. 2012 *A handbook on SEM. Structural Equation Modeling.*

[5] Awang, Z 2015 *SEM made simple: A gentle approach to learning Structural Equation Modeling*. MPWS Rich Publication.

[6] Awang Z, Afthanorhan WMAW, & Asri MAM 2015 *Modern Applied Science. Parametric and non parametric approach in structural equation modeling (SEM): The application of bootstrapping* **9** 58.

[7] Delaney DJ, Bernstein M, Farrow M and Stein LAR 2017 *Alcoholism-Clinical And Experimental Research . The Brief Situational Confidence Questionnaire: Validating The Measure With An Incarcerated Adolescent Sample Using Structural Equation Modeling* **41** pp. 87A (USA: WILEY)

Modelling the *Kampungkota*: A quantitative approach in defining Indonesian informal settlements

D B Anindito[1], F K Maula[1], and R Akbar[1]

[1]Urban Planning and Design Research Group, Department of Regional and City Planning, Institut Teknologi Bandung, Ganesha Street 10, Bandung, West Java, Indonesia

*dhimasanindito@sappk.itb.ac.id

Abstract. Bandung City is home to 2.5 million inhabitants, some of which are living in slums and squatter. However, the terms conveying this type of housing is not adequate to describe that of Indonesian called as *kampungkota*. Several studies suggest various variables in constituting *kampungkota* qualitatively. This study delves to define *kampungkota* in a quantitative manner, using the characteristics of slums and squatter. The samples for this study are 151 villages (*kelurahan*) in Bandung City. Ordinary Least Squares, Geographically Weighted Regression, and Spatial Cluster and Outlier Analysis are employed. It is suggested that *kampungkota* may have distinguished variables regarding to its location. As *kampungkota* may be smaller than administrative area of *kelurahan*, it can develop beyond the jurisdiction of *kelurahan*, as indicated by the clustering pattern of *kampungkota*. *Keywords: kampongkota, informal urbanism, housing system, Indonesia.*

1. Introduction to *kampungkota* in Indonesia

Slums and squatter has been issues of housing in Indonesia. Cities are not merely consisted of slum area but also planned residential area, trading and services centers, offices, and other tertiary sectors. However, the particular social values and the dynamics of local neighborhoods have contributed in generating *kampungkota*. This term refers to a specific neighborhood which at some extent has unique social values amidst the heterogeneity of urban dwellers. The definition is not settled though, as there is not suitable word in English; several *kampungkota* are built upon legal property and some are not considered slums (Obermayr, 2017). This condition is exacerbated as legal framework in Indonesia only recognized the management of slums (*permukiman kumuh*).

The continuing struggles in constituting the definition of *kampungkota* motivates further studies to gather evidentiary supports thoroughly in order to better grasp what the term actually is. Hence, this study aims to investigate the characteristics of *kampungkota* as the common issue in Indonesian cities. Quantitative methods are employed in this study, in which analyses are done using GIS. Taking case in Bandung as both the third most inhabited cities with population around 2.5 million citizens and the second highest population living in slum (Tarigan et al, 2015), this study should suggest more suitable definition of *kampungkota*. In bigger picture, this study should encourage the studies in informal urbanism, which is sometimes having less quantitative data. Also, this study should identify basic model of how *kampungkota* will manifest in urban perspective of Indonesia.

2. Previous studies pertained to *kampungkota*

The emergence of slums and squatter has been widely witnessed in mostly developing countries, including in Bangladesh (Paul, 2006), Albania (Pojani, 2013), The Philippines, and Vietnam (Minnery et al., 2013). However, the terms of 'slums' and 'squatter' is too vague to discern, leading to the use of more comprehensive term which is called "marginalized settlements" (Obermayr, 2017). Several evidences found by Obermayr (2017) suggest that the dwellers of squatter could be lower and middle class and most of them have legal ownership of their properties, while a large part of them don't have clear ownership over their property (Özdemirli, 2014). Another findings by Pugh (2000) suggests that squatters typically do not only have low access to water, but also possess high risk of mortality, particularly the children.

The vulnerability of slums' dwellers and squatters' inhabitants encourages various global initiatives to upgrade the livability in slums and squatters throughout the world. Kool et al. (1989) conclude that vast implementation of slums upgrading—particularly done in low to moderate standards—diminishes the needs of displacement. Milone (1993) investigated the implementation of *Kampung* Improvement Program in small and medium-sized cities in Java, Indonesia and found that improving the accessibility of water supply and sanitary facilities correlates with higher livability. However, Minnery et al (2013) suggest that security of tenure, government and regulation setting, public participation, up-scaling and extension program, and the connections between slum upgrading and sustainable livelihoods in implementing effectual slum upgrading programs and integrating them into urban planning.

A comprehensive study by Obermayr (2017) identifies two polarized types of Indonesian squatter; the first is located in inner-city which could be either *kampungkota* with improved infrastructure or *kampungkota* with insufficient infrastructures while the second is located in the periurban area. *Kampungkota* in peri-urban area may be constructed illegally which characteristics are more rural (Obermayr, 2017). These lead to the recognition that some characteristics of both slums and squatters can be used in indicating which neighborhood is actually *kampungkota*. These characteristics may include the land use surrounding kampungkota, the number of poor people living there, and number of slum housing.

3. Methodology

This study employs three quantitative methods, which are Ordinary Least Squares (OLS) linear regression, Geographically Weighted Regression (GWR), and Clusters and Outliers Analysis (COA). Clusters and Outliers Analysis (COA) is a method which identifies the clusters of objects having similarities not only in value but also in geographical proximity (ESRI, 2012). Also, this analysis recognizes the outliers that having no similarity with surrounding or similar objects. Both clustering objects and outliers can be defined by looking the *z*-score, *p*-value, and Moran's Index (Moran's I).

OLS linear regression generates predictions or models a dependent variable in terms of its relationships to a set of explanatory variables. However, if any misspecification resulted by trying to model nonstationary variables using OLS, then GWR may be used to improve predictions and to better understand the nonstationarity (regional variation) inherent in explanatory variables (ESRI, 2012). GWR models spatially varying relationships, whose coefficients are functions of spatial location (Fotheringham et al, 2002; Chi et al, 2013) give a general form of a basic GWR model as:

$$y_i = \beta_{ik}x_{ik} + \varepsilon_{ik} \quad (1)$$

where y_i is the dependent variable at location i; x_{ik} is the k-th independent variable at location i; β_{ik} is the local regression coefficient for the k-th independent variable at location i; and ε_i is the random error at location i.

Firstly, ratio of slum area in *kelurahan* in Bandung City are analyzed using COA. This will reveal the likeliness of *kelurahan* having high ratio of slum area to cluster with another *kelurahan* which has relatively similar condition. Secondly, GWR is employed to explore what factors influencing the *kampungkota*. Several variables used in GWR will be described in Table 1.

3.1. *Variables*

The objects to analyze in this study are the possible clusters of *kelurahan* (This term is equivalent to "village". There is also *kecamatan* which is equal to "subdistricts". *Kelurahan* is hierarchically lower than *kecamatan*) which have significant relationship in term of ratio of slum area in each *kelurahan* and the relationship between the area of slum in each *kelurahan* and its predictor variables. The predictors are selected after the reiterations of GWR using various variables inputted to reach the highest possible value of R-square and the lowest value of Akaike Information of Criterion (AIC). R-square indicates how accurate the model is while AIC looks for the model having a good fit to the actual condition but minimum parameters.

The data used in this study are gained from *Potensi Desa* 2014. There is also the calculation of land use ratio compared to the area of *kelurahan* whose data are attained from the land use map of Bandung in 2015, with scale 1:5000. The area of slum in each *kelurahan* in Bandung City is collected by the Mayor Decree 648/2015 about The Location of Slums Area in Bandung City.

Table 1. Variables used in GWR

Variable	Definition
AS	The area of slum in *kelurahan* (m^2)
RHDRA	Ratio of high density residential area in *kelurahan* (m^2)
RMDRA	Ratio of medium density residential area in *kelurahan* (m^2)
RCA	Ratio of commercial area in *kelurahan* (m^2)
PD	Population density in *kelurahan* (persons/km^2)
NSH	Number of slum housing in *kelurahan* (unit)
SKTM	Number of Identification Letter of Being Poor (*Surat Keterangan Tidak Mampu*) issued in *kelurahan* (household)

3.2. *Model*

Based on the basic GWR model, the relationship between the area of slum and the variables aforementioned is mathematically expressed as:

$$AS_i = \beta_{0i} + \beta_{1i}RHDRA_i + \beta_{2i}RMDRA_i + \beta_{3i}RCA_i + \beta_{4i}PD_i + \beta_{5i}NSH_i + \beta_{6i}SKTM_i + \varepsilon_i \quad (2)$$

4. Results

Regarding to the ratio of slums area in each *kelurahan*, the analysis shows that several *kelurahan* do cluster into two groups, along with two outliers (see Table 2 and Figure 1). From this analysis, there are three *kelurahan* which have the highest *z*-score and having COType of HH (High-High), indicating the similarity of each *kelurahan* in terms of ratio of slums area. It is indicated that the characteristics in these *kelurahan* is relatively similar.

From the OLS and GWR results (see Table 3), it is found that GWR outperforms OLS by the higher R-square and Adjusted R-square of GWR than that of OLS, though AIC values of both models imply that two models are good fit with minimum parameters. However, it is also found that the significant variables in OLS are only the intercept and the ratio of high density residential area.

Table 2. The result of spatial clusters and outliers analysis

Kelurahan	*Kecamatan*	Local Moran's I	*z*-score	*p*-value	COType
Garuda	Andir	0.006514	2.216655	0.026647	HH
Dungus Cariang	Andir	0.010308	3.772482	0.000162	HH
Maleer	Batununggal	0.020147	5.087551	0	HH

Kelurahan	*Kecamatan*	Local Moran's I	*z*-score	*p*-value	COType
Babakan Sari	Kiaracondong	0.027240	10.117238	0	HH
Binong	Batununggal	0.021094	5.400121	0	HH
Babakan Surabaya	Kiaracondong	0.013313	4.890801	0.000001	HH
Kebon Gedang	Batununggal	0.056198	13.513579	0	HH
Kebon Jayanti	Kiaracondong	0.083101	24.312960	0	HH
Kebon Kangkung	Kiaracondong	0.016754	4.900997	0.000001	HH
Sukahaji	Babakan Ciparay	0.015053	5.077553	0	HH
Warung Muncang	Bandung Kulon	0.016778	5.786343	0	HH
Cibuntu	Bandung Kulon	-0.009798	-3.372949	0.000744	LH
Panjunan	Astana Anyar	-0.009685	-2.190693	0.028474	LH

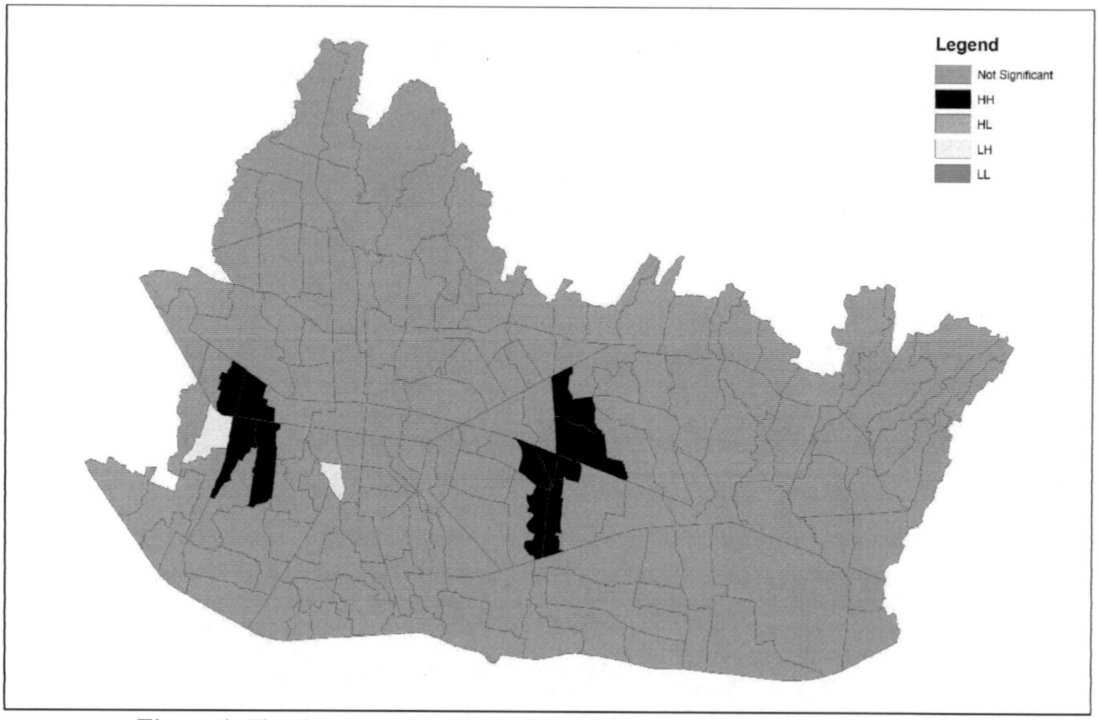

Figure 1. The clusters and outliers of *kelurahan* in terms of slums area ratio.

Table 3. The result of OLS and GWR

Explanatory Variables	OLS		GWR				
	β	Significance (p-value <0.05)	Min.	1st Quartile	Median Value	3rd Quartile	Max.
Constant	0.071924	Yes	-0.021176	0.042298	0.067999	0.084537	0.141261
RHDRA	0.188206	Yes	-0.182355	0.182699	0.251664	0.584486	1.683953
RMDRA	-0.071623	No	-2.256099	-0.188014	-0.069691	-0.035128	0.043207
RCA	-0.019501	No	-0.301951	-0.174507	-0.033771	0.136894	0.698841
PD	0.000001	No	-0.000007	0.000001	0.000002	0.000002	0.000004
NSH	-0.000105	No	-0.000373	-0.000183	-0.000115	-0.000076	-0.000019
SKTM	0.000002	No	-0.000048	-0.000001	0.000005	0.000027	0.000106
Residual sum of squares					1.631524		

Explanatory Variables	OLS		GWR				
	β	Significance (p-value <0.05)	Min.	1st Quartile	Median Value	3rd Quartile	Max.
Akaike Information of Criterion	-166.191220				-173.649031		
R-square	0.134172				0.462420		
Adjusted R-square	0.098096				0.268034		

From the GWR, the residuals of the models vary regarding to its spatial location. Ratio of slums area in 69 *kelurahan* can be modeled by the GWR using the same variables as this study does as the residuals are minimum. On the other hand, Figure 2 further suggests that there are more explanatory variables to be included in the GWR (as the area gets red and darker) or even excluded (as the area gets blue and darker).

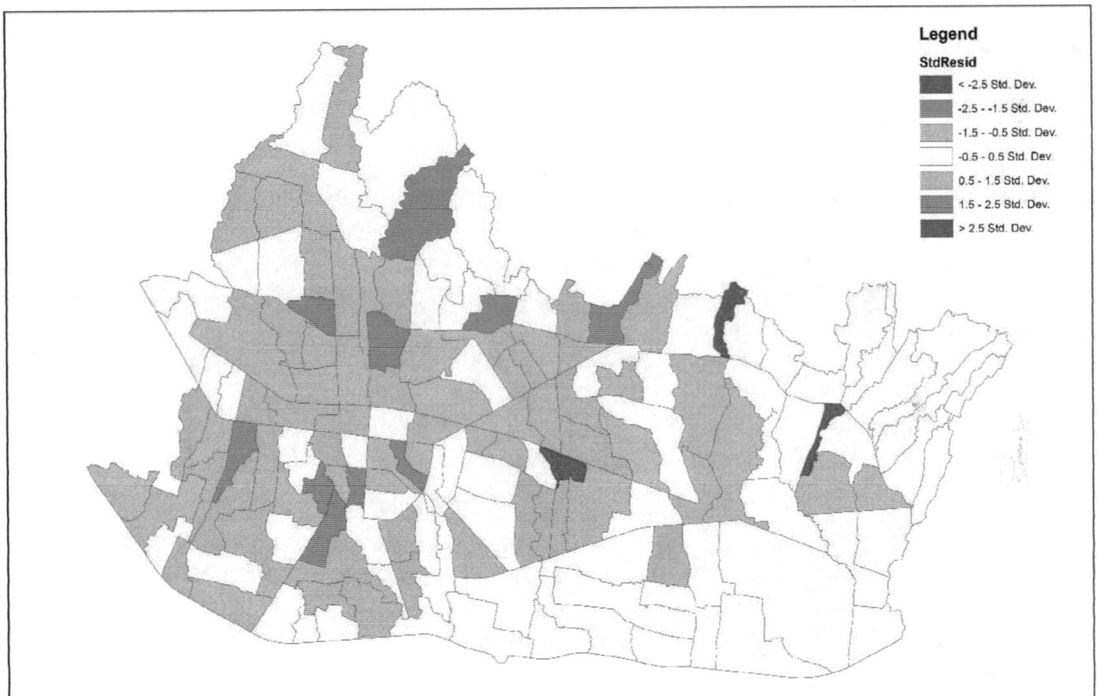

Figure 2. The residuals of GWR.

5. Discussion and Concluding Remarks

From the results, although a *kelurahan* may have a high ratio of slums area, it may be statistically unrelated with its surrounding *kelurahan* in terms of *kampungkota*. There are high yet significant relationships of *kelurahans* clustering by high ratio of slums area. The clustering *kelurahans* indicates that the *kampungkota* may expand beyond the administrative area of *kelurahan*. However, the relatively weak accuracy of GWR models in this study to better define *kampungkota* still needs to be explored. It should also be noted that *kampungkota* is not only slums and squatters, as there is still unclear definition of variables constituting it.

In city scale, there is a palpable difference of *kampungkota* in city centers and periphery area, as the models in both areas are distinctive. It can be noticed that the models are relatively similar in *kelurahans* located on the periphery area, while the models of *kelurahan* located in city centers vary considerably. This slightly corroborates the classification of informal settlements as seen in other cities such as Surakarta (Obermayr, 2017). This may be caused by the lack of exploratory variables that better represent the socio-economic and socio-demographic conditions, land use, historical background, and legal framework. Noting that spatial location does influence the variables, it can be concluded that variables of *kampungkota* are different from one to another.

The construction of model is somehow limited by the data availability. *Kampungkota* in this study consists only of the ratio of several land uses—accommodating the location proximity to jobs of slum dwellers by the ratio of commercial area in each *kelurahan* and the tendency to build intensified buildings in a narrow space as represented by the ratio of high density residential area and population density of each *kelurahan*. There is also several socio-economic elements such as the numbers of slums housing and the number of Identification Letter of Being Poor issued in *kelurahan*. There should be more relevant variables in conveying the condition of *kampungkota*, for example the ratio of legal area in *kelurahan* to be built as housings and the growth rate of the housings in each *kelurahan*. These facts suggest the construction of better *kampungkota* information system to integrate this area into city planning (Minnery et al. 2013).

Acknowledgements
This study is conducted by research grant from Institut Teknologi Bandung entitled "*Program Penelitian, Pengabdian Kepada Masyarakat, dan Inovasi*".

References
[1] Chi S, Grigsby-Toussaint DS, Bradford N, and Choi J 2013 *Applied Geography. Can Geographically Weighted Regression improve our contextual understanding of obesity in the US? Findings from the USDA Food Atlas.* **44** 134
[2] ESRI 2012 *ArcGIS Desktop Release 10.1*. Redlands, CA: Environmental Systems Research Institute.
[3] Fotheringham AS, Brunsdon C and Charlton M 2002 *Geographically weighted regression: the analysis of spatially varying relationships* (Chichester: Wiley)
[4] Kool M, Verboom D, and van der Linden J 1989 *Habitat Intl. Squatter Settlement Improvement and Displacement: A Review of Concepts, Theory and Comparative Evidence* **13** 187
[5] Milone PD 1993 *Review of Urban and Regional Development Studies. Kampung Improvement in the Small and Medium Sized Cities of Central Java.* **5** 74
[6] Minnery J, Argo T, Winarso H, Hau D, Veneracion CC, Forbes D and Childs I 2013 *Habitat International. Slum upgrading and urban governance: Case studies in three South East Asian cities.* **39** 162
[7] Obermayr C 2017 *Sustainable City Management: Informal Settlements in Surakarta, Indonesia.* (Cham: Springer)
[8] Özdemirli YK 2014 *Cities. Alternative strategies for urban redevelopment: A case study in a squatter housing neighborhood of Ankara* **38** 37
[9] Paul BK 2006 *Urban Geography. Fear of eviction: the case of slum and squatter dwellers in Dhaka, Bangladesh.* **27** 567
[10] Pojani D 2013 *Housing Studies. From Squatter Settlement to Suburb: The Transformation of Bathore, Albania.* **28** 805
[11] Pugh C 2000 *Cities. Squatter settlements: Their sustainability, architectural contributions, and socio-economic roles* **17** 325
[12] Tarigan AK.M, Sagala S, Samsura DAA, Fiisabiilillah DF, Simarmata HA, and Nababan M 2015 *Cities. Bandung City, Indonesia* **50** 100

IOP Conf. Series: Earth and Environmental Science **117** (2018) 012006 doi:10.1088/1755-1315/117/1/012006

The potential inclusion of value management subject for postgraduate programmes in Malaysia

M Che Mat[1], S B Abd Karim[2] and N A E Amran[2]

[1]Centre of Studies in Quantity Surveying, Faculty of Architecture, Planning and Surveying, Universiti Teknologi MARA, 40100 Shah Alam, Selangor.
[2] Centre for Building, Construction and Tropical Architecture, Faculty of Built Environment, University of Malaya, 50603 Kuala Lumpur, Malaysia

*masnizan@gmail.com

Abstract: The development of construction industry is increasing tremendously. To complement with this scenario, Value Management (VM) is needed to achieve the optimum function by reducing or eliminating the unnecessary cost that does not contribute to the product, system or service. As VM has been increasingly applied to enhance and improve value in construction projects, the purpose of this study is to implement VM as a subject for master's students at selected public universities in Malaysia. The research is conducted to investigate the potential inclusion of VM as a subject at master degree programmes in Malaysia. Questionnaire survey was designed and delivered to existing master students to explore the current understanding of VM as well as the possibility of introducing VM as a subject. The results showed that the level of awareness on VM is high, yet the understanding of VM is low. This research presents the result of implementing VM as a subject learning for master's level programme at selected public universities in Malaysia.

1. Value management education

Value Management (VM) was first introduced in Malaysia in the late 1980s and early 1990s [1],[2],[3] and since then various efforts have been made to spread the knowledge and to promote the implementation VM in Malaysia. According to Abd-Karim [3], VM is presently rapidly applied in construction industries not only in Malaysia but other countries as well. It has also been regarded as an up-and-coming concept [4]. Additionally, the implementation of VM in Malaysia is now made mandatory for all government projects and programmes of RM50 million, guided by the VM implementation guide [5],[6].

VM as a subject or part of subject was taught at undergraduate level at some of the major public universities such as Housing, Building and Planning, University of Science, Malaysia (USM) in 1991, Department of Quantity Surveying, University of Technology, Malaysia (UTM), 1993/19994, University of Malaya (UM), 1996, The Faculty of Architecture, Planning and Surveying, University of Technology (UiTM), 1997, International Islamic University of Malaysia (IIUM), 2000, and Northern University of Malaysia (UUM). However, the inclusion of VM as a subject at postgraduate level particularly master degree is still at infancy stage in our country. The pedagogy of VM in local universities need to be developed first and implemented. This research provides insights on the VM in educational framework focusing on master degree programmes offered by selected public universities in Malaysia.

1.1 The curriculum

In Malaysia, VM has been introduced as a subject for Master of Project Management at UM. The curriculum was designed based on the requirement set by the various VM-related institutions such as the Institute of Value Management Malaysia (IVMM), SAVE International, as well as the UK's Institute of Value Management. Meanwhile, the Hong Kong Polytechnic University has also introduced VM subject that is currently being offered in the Faculty of Construction and Environment [7]. [8] highlighted that the postgraduate VM course should not only cover the traditional value engineering (VE) concepts and techniques, but also to include managerial knowledge. To implement VM in master's level programme, Leung (2001) suggested that VM course should be divided into two main sections: Traditional VE and Soft VM. The traditional VE aspects cover the basic principles of VE, overall VE procedures and VE techniques while soft VM include VM concept in system theory, major characteristics of VM and role of facilitator. Students are expected to understand this two major area in VE / VM Course. Basic understanding of VE is essential to ensure the understanding of master students towards VM is achieved. After VE concepts are applied, the master students will enable to learn the theory and major characteristics of VM. A comprehensive understanding towards the overall concepts of VM helps the students to implement VM in their nature of the project. Additionally, [8] also proposes ten subjects in the postgraduate VM course which covers VE techniques, understanding of value, decision analysis, systematic project management, soft VM, characteristics of VM, facilitation skills, VM workshop, research methodology and research project.

A proper curriculum contents are needed to ensure VM can be implemented as an educational framework for master students in the universities. In developing curriculum content, [9] pointed out that the learning outcomes should focus on the high value skills, high status knowledge and the real world application. The curriculum contents also follow the goals and objectives of the curriculum, study guide, reference readings and power point slides that are produced or chosen for the learning and teaching activities. According to [10], under the new outcomes based curriculum, the learning outcomes of the VM should cover on the understanding the VM methodology, use VM tools/techniques such as function analysis technique in workshop, organise and manage VM workshops, conduct practical creativity skills and work with a team of stake holder, ensure value for money for projects and, implement VM methodology and techniques in real life. Lecturers will give a great impact to the students. The roles and responsibilities of the lecturers are to know their subject matter but also need the pedagogical contents knowledge on how to deliver the contents to the respective students. According to [11], students learn more and retain the information longer when they are put in active roles than they do passively receiving the information from others. Problem–Based Learning (PBL) should be adopted and combining PBL and group work will facilitate the investigation of more complex problems and promote the development of the student's interpersonal and communication skills.

2. The research

This research paper aims to propose VM as a subject for postgraduate master degree programme at selected public universities in Malaysia. The objectives of this paper are to identify the potential benefits of implementing VM at master's level programme in public universities; to assess the awareness of VM among the master students in selected public universities and; to develop VM educational framework for master students in selected universities. This research focuses mainly on the master degree programmes at seven (7) public universities in Malaysia namely, UM, UiTM, UTM, USM, IIUM, UUM and also Tun Hussein Onn University, Malaysia (UTHM). These universities were selected based on the availability of their construction-related master degree programmes which are currently being offered.

2.1 Research methodology

The research was conducted using quantitative methodology whereby questionnaire survey was used as its tool for data collection and all data were analysed using Statistical Package for Social Science (SPSS). The questionnaire survey was designed, piloted and distributed to ensure that the information related to the implementation of VM for construction related master degree programmes in Malaysia. A total of 41 questions were included in the questionnaire. Through intensive desktop studies, the research found twenty-four (24) construction-related master degree programmes which are currently offered by these universities. These are tabulated in Table 1 below. Additionally, a total estimated population of 700 (N:700) master students were identified through secondary information, mainly from the researchers' personal network and individual contacts including information requested from the programmes head or coordinators. Using [12], the sample size is determined based on 95% confidence level with 5% margin of error, for which it gives a sample size of 248.

The various head of related departments or programmes were contacted for assistance in distributing the questionnaires to their master students. Alternatively, the students were also contacted through personal networks for the purpose of conducting this research. The research managed to get 248 sets of questionnaires were sent to the master students both through email as well as hand delivery.

Table 1. Master's programme offered by the universities

No.	University	Master Programme (by coursework)	Estimated Number of Students (N)
1	UiTM	a) Master of Science in Geographical Information Science b) Master of Property Investment c) Master of Science in Urban Development and Management d) Master of Science Integrated Construction Project Management e) Master of Science in Heritage and Conservation Management f) Master of Science in Facilities Management	150
2	UM	a) Master of Architecture b) Master of Facilities Management c) Master of Real Estate d) Master of Science in Project Management	150
3	UIA	a) Master of Science in Quantity Surveying & International Procurement b) Master of Urban Management c) Master of Quantity Surveying & International Procurement d) Master of Science in Asset and Facilities Management	100
4	UTM	a) Master of Architecture b) Master of Science in Quantity Surveying c) Master of Science in Construction Contract Management d) Master of Science in Transport Planning	124
5	USM	a) Master of Science – Building Technology b) Master of Science – Housing c) Master of Science – Landscape Architect d) Master of Science – Planning e) Master of Science – Project Management	110
6	UTHM	a) Master of Science in Construction Technology Management	36
7	UUM	a) Master of Science (Management)	30

2.1.1 Data collection and analysis

Data collection was conducted for a period of one month. Out of the 248 set of questionnaire distributed, only 90 responses (36%) were received. The reasons of poor respondents may be due to the limitation of time for the research to reach the respondents. Additionally, the respondents may be occupied with their work and studies commitments that lead to the insufficient of time to return the questionnaires. Besides that, the distance of each university that is far from the author's location also contributed to the poor respondents from the students as the author only can reach the students via email. Table 2 shows the total of the response rate of the survey and Table 4 indicates the detail responses according to the universities.

Table 2. Response rate of the survey

Response of the Survey	No. of Respondents (Master's Students)	Percentage (%)
Responded Questionnaire	90	36
Non – response	158	64
Total	**248**	**100**

Table 3. The distribution of sent and received questionnaires according to the universities

No	University	Total N	Sent Questionnaires (S)	Received Questionnaires	Percentages of Sent Questionnaires (%)	Percentages of Received Questionnaires (%)
1	UM	150	67	25	27.02	27.78
2	UiTM	150	50	20	20.16	22.22
3	UTM	124	25	15	10.08	16.67
4	UIA	100	30	15	12.10	16.67
5	UTHM	36	36	10	14.52	11.11
6	USM	110	20	5	8.06	5.56
7	UUM	30	20	0	8.06	0.00
Total		**700**	**248**	**90**	**100**	**100**

2.1.2 Findings

(i) *Understanding of VM*

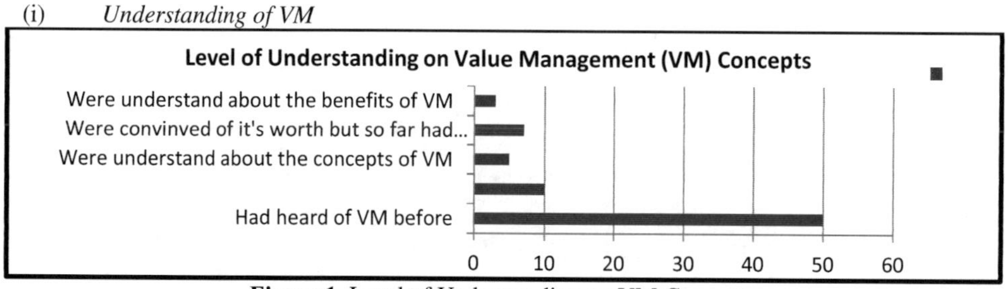

Figure 1. Level of Understanding on VM Concepts

Figure 1 shows the level of understanding on VM concepts among the master's students in selected public universities in Malaysia. Out of 90 respondents, 50 respondents have heard about the concepts of VM before. It is also found that most respondents have heard of VM without understanding what VM is all about. Only 3 respondents indicated that they understand about the benefits of VM and 10 respondents were convinced of its worth and practice in daily life. Meanwhile, 7 respondents were convinced of its worth but had little experience in VM while 5 respondents were understand about the concept of VM. They are 15 respondents were categorised as missing items which means that the respondents never heard of VM before. These results indicate that the understanding and awareness of VM has progressed well after being introduced in the late 1980s and early 1990s [1],[2],[3].

(ii) *VM as a subject to the master's programme*

From the survey, the research found that majority of them (67%) indicates that their current institutions did not offered VM as a subject for their master's programme while 33% stated that VM is offered as a subject. The respondents were also asked on the relevance to include VM as a subject at master level. The results show that 67.8% of them agreed that VM is a relevant subject for their master's programme while 32.2% believed that VM is not a relevant subject.

(iii) *VM educational framework*

Table 4. Feedback of subject learning outcomes by respondents

Learning Outcomes	Strongly Agree	Agree	Neutral	Disagree	Strongly Disagree	Mean
1) Have an understanding of VM methodology	5	38	32	12	3	2.67
2) Students are able to implement value management in daily life	6	38	33	8	5	2.64
3) Students are able to conduct function analysis (FAST) and relate to their nature of project	8	36	25	17	14	2.70
4) Student are able to apply life cycle costing to their nature of project	8	30	37	13	2	2.68
5) Ensure value for money for project by using value management tools	12	31	34	9	4	2.58
6) VM helps to improve team working	11	29	38	9	3	2.60
7) VM helps to provide problem solving skills	11	32	33	11	3	2.59
8) VM helps to enhance communication skills	9	30	38	9	4	2.66
9) VM helps to make the project efficient and effective	15	32	30	10	3	2.49
10) VM helps to enrich the leadership skills	9	29	35	14	3	2.70
11) Student are able to exercise practical creativity skills to work with a team of stakeholders	10	33	35	8	4	2.56

Table 4 above shows the feedback of VM learning outcomes by the respondents. Based on the tabulation above, 15 out of 90 respondents were convinced that VM helps to make project efficient and effective. It leads to the highest number of the respondents who strongly agreed with this statement and only 3 respondents who strongly disagreed. Meanwhile, there is only 5 respondents were strongly agreed that by learning VM, they will understand VM methodology. It contributes to the lowest number of respondents. 14 respondents were strongly disagreed that through VM learning outcomes, students are able to conduct FAST and can relate to their nature of project. This contributed to the highest number of respondents. However, most of the respondents strongly believed that VM will give benefits and added values to them. Besides that, most of the respondents preferred to be "neutral" at almost all the learning outcomes statements. The main cause of this problem is that respondents are not familiar with the concept of VM even though some of the respondents had heard about VM. Additionally, some of them are not aware and not exposed with the actual concept of VM. These findings conform to suggestions by [9] which identified the various elements of VM or VE concepts which will lead to the better understanding and implementation of VM.

Table 5. Feedback of curriculum contents by respondents

Curriculum Contents	Strongly Agree	Agree	Neutral	Disagree	Strongly Disagree	Mean
1) Describe the terminologies of VM	5	28	45	8	4	2.76
2) Describe the components of VM	6	33	40	7	4	2.67
3) Describe the advantages and benefits of value management	10	33	37	6	4	2.57
4) Describe the principles of VM	12	26	39	9	4	2.63
5) Describe the implementation of VM to the nature of project	7	30	43	6	4	2.67

Table 5 above shows the feedback of VM curriculum contents by the respondents. The main purpose of the questionnaires is to get the feedback from the respondents on the curriculum contents of VM as a subject at master's level programme. The research found that, majority of the respondents opted to be "neutral" for almost all the statement provided at the table. Only 12 respondents were strongly agreed that they could describe the principles of VM while 5 respondents strongly agreed that curriculum contents will describe the terminologies of VM. However, 4 respondents from each component of curriculum contents were strongly disagreed on the curriculum contents mentioned on the tabulation above. This indicates that majority of the respondents are still lacking in terms of exposure and understanding about the overall concept and benefits of VM in education and it curriculum contents. Additionally, the respondents were unable to foresee the advantages of VM itself towards their nature of the project. To overcome this problem, workshop, seminar and conferences should be given to the master's student to enhance and improve the knowledge on the level of awareness towards VM and the development of VM in education. Nonetheless, in developing the curriculum, the university must look into the information and suggestion by [9] as well as by [11] on the PBL to ensure the effectiveness of the syllabus.

3. Conclusion

As VM has been rapidly applied to enhance the values in most of the construction projects, it is timely that VM should be being offered as a subject, particularly at the master degree level. To begin with, VM as a subject can be introduced at the major public universities and later followed by other public and private universities. Designing the curriculum requires careful consideration and planning to ensure that the students are equipped with not only the technical knowledge which is considered as hard VM but also the non-technical VM-related knowledge or soft VM. Additionally, under the new outcome-based of curriculum contents, the learning outcomes, pedagogy of teaching VM, assessment methods to assess the effectiveness and efficiency of VM and teaching and learning process of VM must be formulated systematically. Based on the research, most of the respondents agreed that VM is beneficial for them and provide good platform for them to implement VM in their nature of project. They also agreed that VM helps to provide problem solving skills and make the project efficient and effective, and also helps to improve team working and enrich the leadership skills. However, the understanding of VM still requires attention for which VM knowledge should be intensively promoted. It may take some time for the development of VM in educational framework to embed in the university sector especially at the postgraduate level but the first step has to start.

References

[1] Jaapar A and Torrance JV 2007 *Prototype Value Management Guidelines for the Malaysian Construction Industry*. Paper presented at the QSIC 2007- Enhancing & Empowering the Profession, Crown Plaza Mutiara Hotel, Kuala Lumpur, Malaysia.

[2] Jaapar A and Torrance JV 2005 Value Management and its current status in Malaysia. The Malaysian Surveyor 39.2 (February): pp.14-25

[3] Abd-Karim SB 2016 The application of Value Management on Private Projects. QS Link, Board of Quantity Surveyors Malaysia Bulletin, 2/2016: p 33-36.

[4] Ellis R, Wood G and Keel D 2005 Value Management practices of leading UK cost consultants. Construction Management and Economics 23:p.483-493

[5] EPU 2011 Panduan Pelaksanaan, Pengurusan Nilai Dalam Program / Projek Kerajaan .Putrajaya: Jabatan Perdana Menteri.

[6] EPU 2015 Pekeliling Bil. 1/2015 – Penambahbaikan Pelaksanaan Pengurusan Nilai dan Garis Panduan dan Peraturan bagi Perancangan Bangunan dalam Program/Projek Persekutuan. Economic Planning Unit, Prime Minister's Department, Malaysia.

[7] Ramly ZM, Shen GQ, Ann TY, Zhao Y and Jacky KC 2013 Learning by doing: a hands-on value management workshop for postgraduate students. Value World. 36 (Spring 2013). 3-11.

[8] Leung M 2001 Educational framework for MSc in Value Management. Paper presented at a conference on Towards a Value Management and Value Engineering Educational Framework at Masters Level held at Florida, USA.

[9] Parker WC, Nanomiya A and Cogan JJ 2002 Educating World Citizens: Toward Multinational Curriculum Development, in Merryfield M.M. Education for Democracy: Context, Curricular, Assessments. Vol.2, Research in Social Education. Connecticut: Information Age Publishing.

[10] Ann TW, Yu and Shen GQ 2008 Pedagogy of Teaching Value Management in Postgraduate Construction and Property Program. Hong Kong

[11] Kauchak D and Eggen P 2012 Learning and Teaching – Research Based Methods. 6th Ed. Boston: Pearson Edu.

[12] Yamane T 1967 Statistics: an introductory analysis. New York: Harper and Row.

Urban farming activity towards sustainable wellbeing of urban dwellers

N Othman[1], M Mohamad[2], R A Latip[3]and M H Ariffin[1]

[1]Faculty of Architecture, Planning and Surveying, Universiti Teknologi MARA, 40450 Shah Alam, Malaysia
[2]National Landscape Department, 50400 Kuala Lumpur, Malaysia
[3]Postgraduate Studies, Faculty of Architecture, Planning and Surveying, Universiti Teknologi MARA, 40450 Shah Alam, Malaysia

*noriaho@yahoo.com

Abstract. In Malaysia, urban farming is viewed as a catalyst towards achieving the well-being of urban dwellers and natural environment. Urban farming is a strategy for Malaysia's food and economic security, and as one of the foci in the agriculture transformation whereby urban dwellers are encouraged to participate in this activity. Previous study proved that urban farming can help to address social problems of food security, urban poverty and high living cost, also provides leisure and recreation among urban dwellers. Thus, this study investigates the best urban farming practices suitable for urban setting, environment and culture of urban dwellers. Data collection was done via questionnaire survey to urban farmers of a selected community garden in Subang Jaya, Selangor. Meanwhile, on-site observations were carried out on gardening activities and the gardens' physical attributes. The study sample encompasses of 131 urban farmers of 22 community gardens in Subang Jaya. It was found that most of the community gardens practiced crops planting on the ground or soil base planting and dwellers in the lower income group with monthly low household income constitutes the majority (83.2%) of the respondents. Social and health benefits are the highest motivating factors for urban farmers. This study provides unprecedented insights on urban farming practices and motivations in a Malaysian setting.

1. Introduction

Food production and urbanization are two complementary mechanisms to the dynamic growth of cities. Global urban population expansion will lead to greater pressures on resources from increased competition, environment degradation and food insecurity, thus making it more difficult to sustain the environment and social relationships over time [1]. Recently, food scarcity is one of the global concerns where it is a basic need for human survival, important factor for community progression and crucial element of a nation's security and safety [2]. Lower food production and disturbance to food distribution chains in a country has resulted in rising food prices. Food becomes unaffordable to the public especially the lower income and underprivileged groups in urban areas. Their daily diet are affected and led to malnutrition and hunger. A global food crisis is expected in the year 2050. Hence, domestic production of foods is a strategic approach to reduce Malaysia's dependency for imported foods [3]. Urban farming is a prevalent movement worldwide and increasingly widespread in Malaysia. Although urban farming still at the infancy stage, Malaysia's government view this activity as a catalyst towards achieving the well-being of urban dwellers and natural environment. Urban farming was implemented in Malaysia since 1990's and somehow increasingly popular since 2011. However, there is still lack of knowledge about urban farming because newly exposure [4]. Urban farming is a strategy for Malaysia's food and economic security. It is also one of the foci in her agriculture transformation whereby urban and sub-urban populations are encouraged to grow vegetables and other crops to increase households' food security [3].

2. Literature Review

Rapid economic development over the past three decades has resulted in changes in food intake and growth in demand growth for agricultural products [5]. In Malaysia, smallholder farmers dominated vegetable cultivation at the urban fringes [6]. In National Agro-food Policy 2011 showed that Malaysia is a net food despite being blessed with fertile soil, abundant rainfall and suitable climate for producing food items [7]. Thus, urban farming can help address the problems of food security, urban poverty and high living cost among urban dwellers. It is among the peoples' initiatives to cope with living costs by reducing household expenditure. Urban farming also provides leisure and recreation through activities of crop growing and maintenance by the urban community for self-consumption and income generation [8, 9]. Illegal urban commercial agriculture farms and recreational urban farming have created conflicts between the farmers, residents and local government [6, 10]. Yet, urban farming can produce food for the urban growers and surplus produce can be supplied to local food banks [11]. Theoretically, urban farming had substantial impacts on urban system such as urban food security and poverty, ecology and health impacts.

Best practices are techniques, procedures or methods used to achieve desired goals that had consistently show superior results. Policy makers and practitioners use best practices as exemplary effective guides for policy making and practice improvement [12]. Best practices are also used as benchmarks to manage practice performance. The Department of Agriculture and Department of Standard Malaysia have produced a standard on agriculture practices namely Malaysian Standard MS 1784:2005 Crop Commodities – Good Agriculture Practice (GAP). This standard can be as suitable guide for urban farming best practices [13]. GAP described good agriculture practice for crop is sustainable crop production that is legally compliant, environmentally sound, socially acceptable and economically viable to ensure quality produce that is safe and suitable for utilization and consumption. The Guideline on Implementation of Green Neighbourhood Development Initiatives by the Department of Town and Country Planning Peninsular Malaysia can also be used as a guide for urban farming planning [14]. Producing better quality and affordable food is the aim of National Agro-Food Policy 2011-2020 [15]. Programs introduced under this policy intend to increase food production through optimization and sustainable land management, developing and upgrading agriculture infrastructure, and increasing the food quality and safety through the expansion and compliance of related standards.

However, in order to encourage people to do the best with excellent results requires identifying the motivational force, and understanding how the motivational forces behave with such concepts, goals and needs [16]. Practicing urban farming has been found to benefit the community, locality and country in many ways lead to increase the motivation of urban dwellers in urban farming participation. Urban farming can offer economic, environmental and health advantages that make it an appealing movement [17]. Furthermore, urban farming has potential to help in the economic revitalization of city through used of vacant land for economic opportunities [10].

3. Research Methodology

To ensure the objectives of the research can be achieved, the researcher conducted a mixed method approaches in data collection. The study population encompasses the practitioners of community gardens in Subang Jaya. The selection of study areas define based on the local authority initiatives, gardening program, strong community engagement and future land use in the community garden. Figure 1 shows a flow chart of the research method. Data collection was via questionnaire survey that collected among 131 respondents of urban farmers. The questionnaire explored on current practices of urban farming, knowledge in urban farming, access to resources, good agricultural practices,

IOP Conf. Series: Earth and Environmental Science **117** (2018) 012007 doi:10.1088/1755-1315/117/1/012007

motivating factors and impacts of urban farming towards urban dwellers. Prior to the survey, a pilot study was conducted to improve the validity and reliability of the survey questions. The research also used on-site observations of the study areas, i.e. the current urban farming practices and physical condition of the study areas. The observation data were field notes and site photos. This method aimed to identify and understand any relevant significant elements which may not be captured in the questionnaires. The use of questionnaire survey and observations provide for more reliable and valid findings. Descriptive statistics were used to analyse the survey data while critical characteristics description analysis was used for observation data.

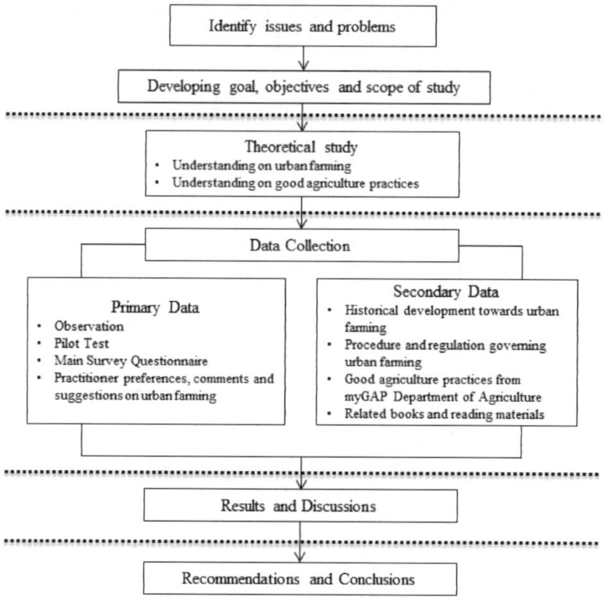

Figure 1. Research Methodology

4. Findings

4.1. Observation Survey

Observation was carried out using non-participant observation. The observation was done by looking at gardening activities up close and the gardens' physical attributes within 22 community garden areas in Subang Jaya. The garden setting in community garden was either within existing open spaces in neighbourhood areas, or within green spaces such as buffer zones and reserved land under the jurisdiction of the Subang Jaya local authority. The authority regulates the community garden projects through comprehensive rules and regulations regarding the gardens' development, operation and maintenance. It was found that most of the community gardens practiced crops planting on the ground or soil base planting. Vegetables planting pots were usually constructed using wood planks. Some of community gardens used raised beds in planting their crops. The plants were also sometimes grown in polybags and recycle containers. Some practitioners planted leafy vegetables in hydroponic units. The plant irrigation at community garden located in neighbourhood parks or green spaces that quite far away from residential areas are treated by water supply, rainwater or water from a nearby river. *Polyculture* was practiced in Subang Jaya community garden, whereby varieties of crops are planted in the same space. Varieties of leafy and fruity vegetables, root crops, herbs and fruit trees are planted. Generally, the gardens and surrounding area are clean. The existence of community garden added value to the open and green spaces by enhancing the landscape character, increasing the frequency and duration of the people visits, and creating a diversity of flora and fauna.

4.2. Questionnaire Survey

4.2.1. Socio-demographic Characteristics. The respondent demography is shown in Table 1. The respondents consist of 79 males and 52 females with the percentage of 60.3% and 39.7% respectively. The result also indicates that urban dwellers in the lower income group with monthly household income below RM5000.00 per month constitutes to majority (83.2 %) of the respondents. This finding implies support for a previous research finding [9] showed the urban poor mostly benefitting from urban farming through increased access to food, stabilizing household food security and malnutrition prevention.

Table 1. Socio-demographic Characteristics of Respondents

Characteristics	*f*	Percentage (%)
Gender		
Male	79	60.3
Female	52	39.7
Age		
18-30	11	8.4
31-50	58	44.3
51-60	44	33.6
>60	18	13.7
Race		
Malay	100	76.3
Chinese	18	13.7
Indian	9	6.9
Other	4	3.1
Household Income		
<RM3000	61	46.6
RM3001-RM5000	48	36.6
RM5001-RM10000	14	10.7
>RM10000	8	6.1

Table 2 shows the findings regarding the Subang Jaya community garden practitioners' opinion about their motivations for and perceived impacts from urban farming participating. Social and health benefits are the two top motivating factors.

Table 2. Motivating Factors and Impacts
(1=Strongly disagree, 2=Disagree, 3=Agree, 4=Agree, 5=Strongly Agree)

Motivating Factors	Mean
For a healthy lifestyle, leisure and recreation.	4.27
Enhance social bonding within the community.	4.24
The needs for fresh and healthy foods.	4.22
Supplement foods supply for the family.	4.15
Reduce the household financial burden on foods supply.	3.90
Enhance the image and aesthetic value of the neighbourhood.	3.90
One of the approaches in a sustainable urban environment.	3.86
Bring in cash income.	3.39

4.2.2. Best Practices of Urban Farming. This study also investigated the perception and understanding of practitioners of community garden regarding best practices of urban farming (aspects of urban farming planning, urban farming knowledge, access to knowledge and GAP). Parameters developed under this segment were to capture practitioner perception, opinion and understanding of urban farming practices that are sustainable and suitable for urban setting, environment and culture of urban people. The findings are given in Table 3. The urban farmers' access to resources and good agricultural practices were deemed to be good urban farming practices.

Table 3. Best Practices Analysis
(1=Strongly disagree, 2=Disagree, 3=Agree, 4=Agree, 5=Strongly Agree)

	Parameter	Mean	Finding
Urban Farming Planning			
1	Local authority has a clear policy on urban farming.	2.91	Disagree
2	Easy access to local plan (land use zoning, green space, availability of vacant land, regulatory information, etc.).	2.75	Disagree
3	Public owned vacant land easily access for gardening.	2.57	Disagree
Urban Farming Knowledge			
4	Authority or NGO's give technical support to improve knowledge.	3.07	Agree
5	Local urban farming information easily access on line.	3.00	Do not know
6	Easy to get training, consultation and extension services from relevant agencies.	2.96	Disagree
7	Enough awareness and promotion program to the community.	2.83	Disagree
Access to Resources			
8	Tenure and security of land is the key issue for practitioner.	3.74	Agree
9	Access to water is a major problem for farm management.	3.68	Agree
10	Access to input and material is a big concern for urban grower.	3.63	Agree
11	Government provides financial resources and/or others assistant with urban farming start-up, management and expansion.	3.34	Agree
Good Agriculture Practices			
12	Fungicide and insecticide being used to protect crops.	3.42	Agree
13	Applied precaution measure to avoid environmental degradation such as contamination of water resources with production inputs and soil erosion.	3.41	Agree
14	Integrated Pest Management (IPM) has been practice to control pest and diseases.	3.38	Agree
15	Synthetic fertilizers widely use to increase production as compare to organic fertilizer and compost.	3.37	Agree

5. Discussions and Findings

From demographic characteristics analysis, urban dwellers in the lower income group with monthly household income below RM5000.00 per month constitute the majority (83.2 %) of the respondents. Analysis on household income in motivating factors, urban farmers in Subang Jaya motivated in participated urban farming activity for social and health benefits. This finding is congruent with previous research that found urban farming helping to improve the diet quality of people and in fostering community social bounding and empowerment [18, 19]. Yet these findings contradict with a previous research finding [20] where the most common motivation factor for beginner cultivation was a need for food, means to access food without cash due to the needs for fresh and healthy foods. Two factors may contribute the findings contradiction. First, the implementation of the Subang Jaya community garden projects was in stages starting 2013 thus showing just three (3) years of operation. This short duration cannot prove the long term financial benefits of urban farming. Second, the average size of individual plots for planting is too small (4.0m²) for significant financial crop outcomes. Our study also showed that best urban farming practices were the availability of resources and good agricultural practices. Hence, there are better financial benefits and motivations if the Subang Jaya local authority were to provide more garden space and resources for the urban farmers so that they can have more surplus produce to sell.

6. Conclusions and Recommendations

The findings indicate that social and health benefits are the major motivating factors urban farming practitioners. It was also found that the local authority had put several fundamental initiatives through green initiatives and community gardening program to encourage urban dwellers to produce own food in community garden sites within the administrative boundary. Best practices are the availability of resources and good agricultural practices. Increased financial benefits to urban farmers can be obtained if the local authority increases the availability of more garden space and resources for urban farming. Further research may explore on the review of policy, guidelines or ordinance in authority that might impede the development and growth of urban farming. Besides that, providing infrastructure in community gardens leads to enhancing the participation of urban farming among urban dwellers.

Acknowledgements
The authors wish to thank Faculty of Architecture, Planning and Surveying, Universiti Teknologi MARA (UiTM) and Research Management Institute (RMI), UiTM for permission to conduct the research. Financial assistance for this research is funded by Minister of Higher Education (MOHE) under Fundamental Research Grant Scheme (FRGS): FRGS/1/2015/SS07/UITM/02/1.

References
[1] Chaipa I, and King B 1998 *Urban Agriculture in Zimbabwe, Nature, Significance and Prospects*. Techknikon, Pretoria.
[2] Smit J, Nasr J and Ratta A 2001 *Urban Agriculture - Food, Jobs and Sustainable Cities.* (Virginia: The Urban Agriculture Network, Inc.)
[3] Abd Rahman N, Idris K, Tajuddin Z and Abdullah F 2013 *Transformasi Pertanian Bandar*, (Serdang, Malaysia: Pusat Transformasi Komuniti Universiti (UCTC UPM))
[4] Yaakob FA, Assistant Director Urban Farming Division, Department of Agriculture Peninsular Malaysia (interviewed on 13th October 2014).
[5] Ariffin AS, Abas Z and Baluch N 2015 *Australian J. of Basic and Applied Sci. Issues and Challenges of Integrated Agro-food Supply Chain: An Overview of Malaysia Food Policy,* **9** 171
[6] Razak SA and Roff MNM 2007 *Status and Potential of Urban and Peri-urban Agriculture in Malaysia.* http://www.fftc.agnet.org/htmlarea_file/activities/20110719103448/paper-833732114.pdf
[7] Md Razak MI, Amir Hamzah ASM, Abas N, Idris R and Ibrahim Z 2013 *J. of Eco. and Dev. Stu. Sustaining Food Production for Food Security in Malaysia* **1** 2
[8] Mlozi MS 1997 *Environmentalist. Impacts of urban agriculture in Dar es Salaam, Tanzania* **17** 115
[9] Islam R, and Siwar C 2012 *Adv. in Env. Bio. The Analysis of Urban Agriculture Development in Malaysia* **6** 1068
[10] Hendrickson MK and Porth M 2012 *Urban Agriculture – Best Practices and Possibilities.* University of Missouri Extension.
[11] Ackerman K 2012 *The Eco. and Soc. Review. The Potential for Urban Agriculture* **45** 189
[12] Geertman S, and Stillwell J 2009 *Planning Support Systems Best Practice and New Methods.* (Netherland. Springer Science + Business Media B.V)
[13] Department of Agriculture 2007*Malaysia Farm Certification Scheme for Good Agricultural Practice (SALM) Standard* (Kuala Lumpur: Department of Agriculture (DOA))
[14] Jabatan Perancang Bandar dan Desa Semenanjung Malaysia 2013 *Panduan Pelaksanaan Inisiatif Pembangunan Kejiranan Hijau – Pembangunan Kebun Kejiranan.* (Kuala Lumpur: Jabatan Perancang Bandar dan Desa Semenanjung Malaysia)
[15] Ministry of Agriculture and Agro-based Industry 2011 *National Agrofood Policy 2011-2020.* (Putrajaya: Percetakan Watan Sdn. Bhd)
[16] Taboli H 2012 *Life Science Journal. Approach Based on Motivation Theories* **9** 556
[17] Mukherji N, and Morales A 2010 *American Planning Association. Practice Urban Agriculture, Chicago.*
[18] Armar-Klemesu M 1999 *Urban Agriculture and Food Security, Nutrition and Health.* Havana, Cuba.
[19] Mees C, and Stone E 2012 *Cities and the Environment (CATE). Zoned Out: The Potential of Urban Agriculture Planning to Turn Against its Roots* **5**1
[20] Warren E, Hawkesworth S, and Knai C 2015 *Food Policy. Investigating the Association between Urban Agriculture and Food Security, Dietary Diversity, and Nutritional Status: A Systematic Literature Review* **53** 54

Influential aspects of leader's Bourdieu capitals on Malaysian landscape architecture subordinates' creativity

R Zahari[1], M H Ariffin[2], N Othman[3]

[1]Centre of Postgraduate Studies, Faculty of Architecture, Planning and Surveying, Universiti Teknologi MARA, 40450, Shah Alam, Selangor, Malaysia
[2]Centre of Construction Management Studies, Faculty of Architecture, Planning and Surveying, Universiti Teknologi MARA, 40450, Shah Alam, Selangor, Malaysia
[3]Centre of Landscape Architecture Studies, Faculty of Architecture, Planning and Surveying, Universiti Teknologi MARA, 40450, Shah Alam, Selangor, Malaysia

*mhisham238@salam.uitm.edu.my

Abstract. Free Trade Agreements as implemented by Malaysian government calls out local businesses such as landscape architecture consultant firm to explore internationally and strengthen their performance to compete locally. Performance of landscape architecture firm as a design firm depends entirely on creativity of the subordinates in the firm. Past research has neglected studying the influence of a leader's capitals on subordinates' creativity, especially in Malaysian landscape architecture firms. The aim of this research is to investigate the influence of subordinates' perceptions of the leader's Bourdieu capitals towards promoting subordinate's creative behaviours in Malaysian Landscape Architecture firms. The sample chosen for this research are subordinates in registered landscape architecture firm. Data was collected using qualitative semi-structured interviews with 13 respondents and analysed using Qualitative Category Coding. Aspects of the leader's social capital (i.e. knowledge acquisition, problem solving, motivation boosting), human capital (guidance, demotivating leadership, experiential knowledge, knowledge acquisition), and emotional capital (chemistry with leader, respect, knowledge acquisition, trust, understanding, self-inflicted demotivation) that influence subordinates' creativity were uncovered from the data. The main finding is that the leader's capitals promote the subordinate landscape architects or assistant landscape architect to be more creative based on three main things, first is knowledge acquisition, motivation, and ability for the leader to influence through positive relationship. The finding contributes to a new way of understanding the leader's characteristics that influence subordinates' creativity.

1. Introduction

Malaysia has entered Free Trade Agreements with other countries so as to have reciprocating preferential access in each other's markets [1, 2]. This preferential access that includes creative service suppliers such as architects and engineers will lead to increase competition from foreign and domestic firms in a country's marketplace. Landscape architecture firm are firms that operate on continuous creativity as indication of their performance. Creativity is one of their competitive tools to compete with other firms in the industry. Institute of Landscape Architects Malaysia [3] has listed 10 Strategic Focus Area to improve the profession towards becoming more influential aspect in built environment industry. One of the strategic focuses is Human Capital Enhancement. In Human Capital Enhancement, the board suggested that the landscape architecture industry should focus more on the development of leadership and promoting creativity. The encouragement from Institute of Landscape

Architects Malaysia (ILAM) shows the importance of leadership and creativity enhancement for landscape architecture firms in Malaysia. To deal with global and economic conditions, firms around the world are encouraging employees to be creative for increased productivity [4, 5]. Creativity is the core competency of design occupations such as landscape architects, building architects etc. During the design processes, creativity is used to generate possible solutions to the design problem [6]. Andries and Czarnitzki [7] reported that subordinates also contribute to innovation and creativity performance of the firm. Hence, subordinate's creativity is significant in sustaining the firm's performance especially in creative-oriented-firm such as landscape architecture firm. The enhancement of the subordinate designers' creativity in the workplace is necessary for their firms to remain competitive in their respective industries. The role of leadership in promoting creativity has continued to be an object of research [8, 9, 10, 11]. The leader plays an important role to influence employee's creativity [12]. According to Zhang and Zhou [13], leader needs to be aware that their own leadership behaviour is capable to influence the development of employee's creativity. The research on leadership in promoting creativity mostly focussed on a basis of relationship between the leader and subordinate in the firm. However, there are few studies that emphasise the influencing role of a leader's intangible resources on his followers. Network diversity, human quality, and emotional intelligence are some of the intangible resources that leader have. Network or relationship diversity can give access to social resources such as support and signal credibility [14]. Bourdieu [15] identified diversity of relationship as one aspect of social capital. Social capital is one of the three main forms of capital. These capitals are useful in sustaining and improving of person's status in life. Capitals are sets of valuable resources and powers that a person accumulates through his life actions [15, 16]. Thus, a capital is a set of tangible and intangible resources that every person possesses. It can be referred to a person's social connection, a person's academic qualification, a person's work experiences, or a person's emotional resources.

2. Past Research on Capitals

Most past researches [17, 18, 19, 20, 21, 22] that studied the influence of Bourdieu's capitals tend to focus on a single capital in their research instead of using all Bourdieu's capitals. Past studies [19, 20] on the capitals' influences on creativity and innovation were mostly focused on social capital rather than human and emotional. Ooi, Hooy and Mat Som [23] combined social and human capitals, but did not incorporate emotional capital in their study. Their field of study also was not relating to creativity. Even though emotional theory has attracted a considerable amount of research in the recent decade, emotional capital has not been given enough attention in the management literature [22]. Past studies [17, 19, 20, 24, 25, 26, 27] in education and manufacturing indicated that social capital is proved to exert a clear influence on innovation and creativity. No studies have been carried out upon the creative industry and in Malaysian context. A knowledge gap regarding Bourdieu's capitals' influence on leadership exists because of the absence of a study using Bourdieu's capitals, and the minimal research on the capitals' influence on the creative work. Thus, this paper report the findings of the exploratory interviews undertaken in a research to investigate the influence of subordinates' perceptions of the leader's Bourdieu capitals (social, human and emotional) towards promoting subordinate's creative behaviours in Malaysian Landscape Architecture firms.

3. Research Methodology

This research is exploratory and uses a qualitative interview data methodology. The population comprises landscape architects or assistant landscape architects in registered landscape architecture consultant firms in Malaysia listed in the Institute of Landscape Architects Malaysia (ILAM) website directory (updated to 7 October 2015). The potential respondents are subordinates in landscape architecture firm working as landscape architect or assistant landscape architect with more than 3 years experiences in landscape architecture. The interviews were carried out by telephone calls to subordinate landscape architect and assistant landscape architect. The sampling technique used is purposive sampling. Fourteen firms from 18 contacted firms responded and agreed to participate in the telephone interviews. Only thirteen interviews were analysed (12 subordinate landscape architects and one assistant landscape architect) because one respondent was rejected for unsuitable demographics. The interviews were audio recorded and later transcribed for analysis. Qualitative Category Coding was used to generate the categorical findings.

4. Result From Exploratory Study

The respondents were landscape architects subordinate's minimum experiences of 3 years in the field except for a respondent who was an assistant landscape architect. Nine (69%) respondents were female while the rest were male. Half of the respondents (54%, 7 respondents) have 1 to 5 years' experiences in the field. The remaining respondents have 6 to 15 years work experiences. The academic background of the one respondent (8%) was from diploma in landscape architecture. The respondents have mostly bachelor of landscape architecture (77%. 10 respondents) as highest qualifications. Two respondents (15%) and one respondent have landscape architecture master degrees and diploma respectively. The respondents were coded from LA01 to LA13. They were located in Malaysia peninsular east coast region (LA01 & LA11), southern region (LA02 and LA03), central region (LA07, LA08, LA09, and LA13), northern region (LA05, LA06, and LA12) and lastly, Borneo region (LA04 and LA10). The responses on the influence of their leader's Bourdieu's capitals towards promoting their creativity differed according to the capitals (social, human and emotional). All respondents said that at least one or more of the capitals can influence their creativity. The transcript analysis generated thirteen meaning units that were then grouped into one of three capital categories. Table 1 describes the effect of capitals towards creativity as perceived by the respondents.

Table 1. Definition of category and meaning unit generated

Category	Category Definition	Meaning Unit	Meaning Unit Definition
Social Capital Creativity Influencer	Perception of subordinates towards their leader's durable social network of more or less institutionalised relationship of mutual acquaintance and recognition.	Knowledge Acquisition (SC)	The subordinate indirectly gaining knowledge from the social network of the leader.
		Problem Solving	The ability of subordinates to solve problems because of the indirect use of leader's social capital.
		Motivation Boosting	The increase of subordinate's motivations to be creative because the available potential use of the leader's social capital to achieve the subordinate design goals.
Human Capital Creativity Influencer	Knowledge, Institutionalised recognition, skills and experiences of a leader as perceived by his/her subordinates.	Guidance	The advice or information focused at resolving problems or difficulties using the leader's human capital.
		Demotivating Leadership	The subordinate's demotivation to become creative due to the design being overly dependent on input from the leader's human capital
		Experiential Knowledge	The design knowledge gained by the subordinate through sharing the leader's experiential knowledge.
		Knowledge Acquisition (HC)	The knowledge gained by the subordinate through sharing the leader's overt knowledge
Emotional Capital Creativity Influencer	A leader's emotional resources in the form of emotional support, expenditure of time, attention, care and concern that is available to his/her subordinates.	Chemistry with Leader	Good interaction between a leader and subordinate from the emotional capital that the leader has.
		Respect	A feeling of admiration for leader from subordinate caused by the leader's emotional capital.
		Knowledge Acquisition (EC)	The ease of the subordinate gaining knowledge from the leader due to the comfortable interactions between them.
		Trust	The subordinate's belief in the leader's reliability, truth, ability, or strength.
		Understanding	The subordinate's ability to understand the needs of his/her leader due to the comfortable interactions between them.
		Self-inflicted Demotivation	Demotivation arising from subordinate's acquiesce to leader's demands due to the comfortable interactions between them

Most respondents said that leader's social capital, human capital and emotional capital do influence their creativity. One common reason is knowledge acquisition. They believed the existence of their leader's capitals (social, human, and emotional) enable them to receive more knowledge from their leader. Respondents (LA01, LA02, LA03, LA04, LA05, LA06, LA08, LA09, and LA11) told that social capital that their leaders have helped them to acquire knowledge, to easily solve problems, and boost their motivation to be more creative. However, four respondents (LA07, LA10, LA12, and LA13) stated that their leader's social capital does not inspire or help them to be more creative. Most respondents (LA01, LA02, LA03, LA04, LA05, LA06, LA07, LA08, LA09, LA10, and LA11) except LA12 relate that leader's human capital influences their creativity in designing by guiding them, influencing them, give experiential knowledge and acquiring knowledge. LA02, LA03, LA07 and LA08 however mentioned that their leader's human capital limited their freedom to be creative because they need to always follow the leader's preferences, thus encouraging them to be overly dependent on the input from the leader. LA12 explained that in his/her office, the process started with individual brainstorming about the design and then regularly discussing with the principal who suggests many ways to make it more creative. There are six effects of leader's emotional capital towards subordinate's creativity as perceived by the respondents. They agreed that the leader's emotional capital leads them to have a good chemistry with their leader (LA01, LA02, LA07, and LA08). They also acquire respect, knowledge, trust, and understanding from the capital (LA03, LA04, LA08, LA09, LA06, LA11, LA12, and LA13). Yet, LA10 noted that the effect of her leader's emotional capital could over-pressure her to be creative to meet the expectation of the leader thus causing her self-inflicted demotivation. LA05 on the other hand believed that emotional capital does not influenced her directly through the relationship, but more on the expertise or skills and experiences of the leader that helped encourage her to be creative.

5. Discussion from the analysis

From the analysis, the leader's capitals influenced subordinate landscape architects or assistant landscape architect to be more creative based on three main things, first is knowledge acquisition, motivation, and ability for the leader to influence through positive relationship. These findings support Gu, Zhang, and Liu [27] that social capital helped to enable an individual to gain knowledge and information as well as triggering original ideas and creative thinking. One of the items described by respondents that can generate a positive relationship is trust. According to Zhou and George [28], subordinates or employees are encouraged to take risk to be creative because of the trust towards their leader. The trust gives them the confidence that the new ideas and directions that they proposed will be taken seriously by their leader [28]. Yoshida [29] also supported the important of trust through positive relationship in fostering employee's creativity. The positive relationship through emotional capital that the leader possesses is a significant factor in promoting creativity. The influence of the leader's human capital on the subordinate's creativity through giving guidance and knowledge from their accumulated skills and experiences is also consistent with Jaiswal and Dhar [30]. Transformational leadership which has elements of teaching and coaching can promote a follower's creativity [30]. However, leader's capitals also could cause demotivation through human and emotional capital aspect. Lengnick-Hall and Lengnick-Hall [31] stated in their research, 'wanting to give the expected answers' was one of the common barriers to creativity. The statement aligns with LA10's opinion on her leader's emotional capital. Thus, it supports the position that leader's capitals can also demotivate subordinate's creativity. This research has limitations. First, the findings of this qualitative study cannot be generalised to the population of landscape architecture firm's design subordinates. Secondly, the respondents gave information implying that the leader capitals had promoted the subordinates to be more creative, but does not specifically prove that the capitals change the creativity of the subordinates.

6. Conclusion

The leader's capitals influence on promoting subordinate landscape architects' creativity reflected on three main things; first is knowledge acquisition, second are motivation and demotivation, third is leader ability to influence (because of follower's trust, respect, and understanding towards the leader). This exploratory research contributes to knowledge relating to the influence of the leader's capitals upon subordinate's creativity by promoting them to be more creative.

Acknowledgements
The Authors would like to express their gratitude for the financial support from Malaysia MOHE (Ministry of Higher Education) under the Fundamental Research Grant Scheme (FRGS), 600-RMI/FRGS 5/3 (16/2014) given to this research which is supervised by Universiti Teknologi Mara (UiTM).

References

[1] MATRADE [Internet]. Matrade.gov.my 2017 [cited 14 August 2017]. Available from: http://www.matrade.gov.my/en/malaysian-exporters/going-global/understanding-free-trade-agreements
[2] Construction Industry Transformation Programme 2016-2020 – Driving Construction Excellence Together [Internet]. Citp.my 2017 [cited 14 August 2017]. Available from: http://www.citp.my/
[3] Institute of Landscape Architects Malaysia [Internet]. Institute of Landscape Architects Malaysia. 2017 [cited 25 April 2017]. Available from http://www.ilamalaysia.org/#!the-profession/c1jo0
[4] Hirani A 2014 Creativity of interior design Students: understanding the relationship between cognitive style, Personality Style, motivational orientation, and creativity of interior design students [dissertation on the internet]. [Minneapolis (United States of America)]: University of Minnesota; 2014 [cited 2017 Aug 12]. Available from: https://conservancy.umn.edu/bitstream/handle/11299/162641/Hirani_umn_0130E_14570.pdf?sequence=1
[5] Bai Y, Lin L and Li PP 2016 *J. of Business Research. How to enable employee creativity in a team context: a cross-level mediating process of transformational leadership* **69** 3240
[6] Crilly N 2015 *Desig. Studies. Fixation and creativity in concept development: the attitudes and practices of expert designers* **38** 54
[7] Andries P and Czarnitzki 2014 *Small Business Economics. Small firm innovation performance and employee involvement* **43** 21
[8] Amabile T M 1988 *A model of creativity and innovation in organizations Research in Organizational Behavior vol 10 in B M Staw and L L Cummings* (Greenwich: JAI Press Inc) pp123-167
[9] Sosik JJ, Kahai SS and Avolio BJ 1998 *Creativity Research Journal. Transformational leadership and dimensions of creativity: motivating idea generation in computer-mediated groups* **11** 111
[10] Politis D 2005 *Entrepreneurship Theory and Practice. The process of entrepreneurial learning: a conceptual framework* **29** 399
[11] Wang P and Rode JC 2010 *Human relations. Transformational leadership and follower creativity: the moderating effects of identification with leader and organizational climate* **63** 1105
[12] Sijbom RBL, Janssen O and Van YNW 2015 *European Journal of Work and Organizational Psychology. How to get radical creative ideas into a leader's mind? Leader's achievement goals and subordinates' voice of creative ideas* **24** 279
[13] Zhang X and Zhou J 2014 *Org. Behavior and Human Decision Processes. Empowering leadership, uncertainty avoidance, trust, and employee creativity: interaction effects and a mediating mechanism* **124** 150
[14] Venkataramani V, Zhou L, Wang M, Liao H and Shi J 2016 *Org. Behavior and Human Decision Processes. Social networks and employee voice: the influence of team members' and team leaders' social network positions on employee voice* **132** 37
[15] Bourdieu P 1986 The forms of capital *Handbook of Theory and Research for the Sociology of Education* ed J Richardson (New York: Greenwood Press)
[16] Bourdieu P 1984 *La Distinction: a social critique of the judgement of taste* (R Nice, Trans.) (Cambridge, Massachusetts: Harvard University Press)
[17] Santos-Rodrigues H, Dorrego PF and Jardon CF 2010 *The Int. Buss. & Econ. Research Journal. The influence of human capital on the innovativeness of firms* **9** 53

[18] Barczak G, Lassk F and Mulki J 2010 *Creativity and Innovation Management. Antecedents of team creativity: an examination of team emotional intelligence, team trust and collaborative culture* **19** 332

[19] Yu S H 2013 *Technological Forecasting and Social Change. Social capital, absorptive capability, and firm innovation* **80** 1261

[20] Liu CH 2013 *The Int. J. of Human Resource Management. The processes of social capital and employee creativity: empirical evidence from intraorganizational networks* **24** 3886

[21] McGuirk H, Lenihan H and Hart M 2015 *Research Policy. Measuring the impact of innovative human capital on small firms' propensity to innovate* **44** 965

[22] Liu CC, Chen SY and Liao CH 2015 *Chiao Da Management Review. The relationships among emotional capital, job satisfaction and organizational citizenship behavior: a cross-level analysis* **35** 1

[23] Ooi CA, Hooy CW, and Mat Som AP 2015 *Tour. Manage. Diversity in human and social capital: empirical evidence from asian tourism firms in corporate board composition* **48** 139

[24] Landry R, Amara N and Lamari M 2000 *Proc. 4th Int. Conf. on Technology Policy and Innovation. Does social capital determine innovation? to what extent?* (Curitiba, Brazil)

[25] Perez-Luno A, Cabello Medina C, Carmona Lavado A and Cuevas Rodriguez G 2011 *J. of Business Research. How social capital and knowledge affect innovation* **64** 1369

[26] Camps S and Marques P 2014 *Tech. Forecasting and Social Change. Exploring how social capital facilitates innovation: the role of innovation enablers* **88** 325

[27] Gu J, Zhang Y and Liu H 2014 *Thinking Skills and Creativity. Importance of social capital to student creativity within higher education in China* **12** 14

[28] Zhou J and George J M 2003 *The Leadership Quarterly. Awakening employee creativity: The role of leader emotional intelligence* **14** 545

[29] Yoshida DT, Sendjaya S, Hirst G and Cooper B 2014 *J. of Business Research. Does servant leadership foster creativity and innovation? A multi-level mediation study of identification and prototypicality* **67** 1395

[30] Jaiswal NK and Dhar RL 2015 *International J. of Hospitality Manage. Transformational leadership, innovation climate, creative self-efficacy and employee creativity: A multilevel study* **51** 30

[31] Lengnick-Hall ML and Lengnick-Hall CA 1999 *J. of Manage. Edu. Leadership jazz: an exercise in creativity* **23** 65

The influence of client brief and change order in construction project

N A A Mahat[1] and H Adnan[1]

[1]Faculty of Architecture, Planning and Surveying, Universiti Teknologi MARA, 40450 Shah Alam, Malaysia

*nooraisyahasyikin@yahoo.com

Abstract. Construction briefing is a statement of needs about intentions and projects objectives. Briefing process is the preliminary stage in the design process and successful briefing can achieve project delivery right on target time, cost and quality of project confidently. Although there are many efforts to approach client's requirement and needs for a project, it is still not collected adequately to make proper solutions in design. Thus, these may lead the client to include change orders during the construction phase. This paper is concerned toward the influence of client's briefing of a construction project that impact on the change order on the construction works. The research objective is to identify the influence of client's brief on change orders, therefore, the aims of the research is to reduce change orders in project delivery. This research adopted both qualitative and quantitative data collection methods which are content analysis and semi structure interview. The findings highlight factors contributing to change orders and the essential attributes of clients during the briefing stage that may help minimise them.

1. Introduction

The briefing is a process whereby client explains and notifies others of his goals, requirement and desires. The process is intended as a channel to convey decision and information of client requirements on the projects to the consultant. Literature revealed that briefing as a problem of formulation and solving process, which identifies the projects objectives and missions. According to [1] and [2], briefing process is crucial to the project's success delivery, as well as to the satisfaction of clients. Inadequate client briefs could result in inaccurate designs that fail to fulfil the clients' expectations. Literature revealed that inaccurate or inadequate brief is the main causes of time and cost overrun of construction projects [3] and
[4]. As most clients depend on their consultant team for assistance, it is vital that the team could interpret the client's requirements into a design, which stipulates technical characteristics, functional performance criteria, and quality standards.

.

2. Client Brief and Change Order

2.1Problem in the Client's Brief

The conception of construction briefs is a topical issue in the construction industry as unproductive preparation of briefs significantly rises the total cost of building. Lack of knowledge in briefing process triggered some clients incapable to describe their requirements that resulting in uncertain project objective [5] and [6]. Poor practices during briefing include inadequate involvement of projects participants and insufficient time for briefing. Completeness of information about the client's requirement is most pertinent to reduce the risk of miss-interpretation by the designer and delivery of the wrong kind of building to the client [6] and [5]. Conformity during briefing process is observed highly depending on client's attributes that translated into practices during briefing. Changing need by the client caused by the inability of clients to provide the required information during briefing due to lack resources during formulation of the brief [17]. This shortcoming can manifest itself in the ultimately poor performance of the resultant building. Due to this, client often instructs for change order to accommodate their requirement that resulting in cost overrun [18].Most of the variations were made through direct instructions by the architect to fulfil the client's requirements.

2.2 *Change Order*

Change orders are among the most significant sources of cost growth and disruptions to field productivity on building construction projects. There are also cases where the renovation works are carried out prior to moving-in by the user to the newly completed building to rectify the design of the building, to ensure that the building meets the client's requirement. Such renovation works and change order are a wasteful of resources in term of time and money thus reduces the level of client satisfaction on the project [16]. Hence, completeness of information in the client's brief is essential to avoid the risk of change order. In the Malaysian construction industry, [18] discovered the key factors that contributed to changes were due to belated clients' instructions, ambiguous briefs and requests from end users. The most frequent type of change met in constructing capital projects are changes to the original scope of work or those that a rise from unexpected conditions in the field. The nature and impact of change orders on overall project performance were investigated by [12], and the result notable that clients were identified as the most predominant sources of change orders due to unclear briefing and changing requirement.

3. The research

This paper deliberated the research findings carried out in Universiti Teknologi MARA (UiTM) which is involved in constructing projects for all its campuses nationwide. Seventeen (17) completed projects were examined including construction of academic buildings and mixed development projects. The aim of the research is to decrease the client influences contributing to change orders, thus the objective is to recognize the client's attributes during briefing which may affect change orders.

4. Research methodology

A triangulation approach is applied throughout the research activities, which began with multiple research method of qualitative and quantitative. A combine research approach engaged in this research is based upon the principle of triangulation in which it attempts to look at things from different angles. The investigations begin with the examination of content analysis. Content analysis is a quantitative method in nature that is perceived as hard data, which is reliable, countable, and purely based on evidences and records. The primary objectives of conducting content analysis are to identify and to measure the success level of project delivery in regard to time and cost achievement. It is also to analyze the number of change orders and to the grounds of change orders issued by the client. The primary data collected which involved an intensive examination the content of archival documents of the projects. Relevant documents and information such as contract documents and final accounts of the projects were inspected.

The second stage of data collection is through a semi-structured interview with client representatives Semi-structured interviews involve a series of open-ended questions based on the topic areas. The open-ended nature of the question defines the topic under investigation but provides opportunities for both interviewer and interviewee to discuss some topics in more detail. The objective of the interview is to investigate the briefing practices in UiTM, thus to identify problems associated with the briefing practice, which later influence the issuance of change order. Fifteen (15) project manager / client representative who was involved in the chosen projects were interviewed.

4.1 Value of change order

Majority of the project inspected in the study are found with low level of project achievement in term of cost in which, cost overrun occurred in most of the projects. As presented in the Fig. 1, only 6% of the projects with omission from projects cost, though, majority of the project suffered with cost overrun with highest group of cost overrun value is an additional of 10% of project cost. Cost overruns need a constructive analysis. Once the project is completed, a inquest meeting with significant individuals of the project team to discuss what went right and what went wrong is highly recommended.

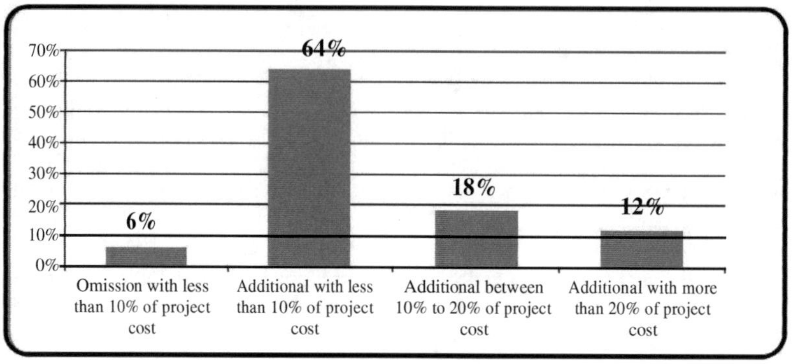

Figure 1. Value of Change Order

4.2 Cause of change order

The ranking of the ground of change order are based on the mean rank of the data. The higher the mean rank means the more significant the causes are. In this study, the author considers mean rank of 2.00 and above as the selected value for the significant cause of change order. Based on Table 1, Change in decision making is the highest rank, followed by Construction changes and site condition and natural incidents.

Table 1. Cause of Change Order

No.	Cause of Change Order	Mean Value	Rank
1.	Change in decision making	4.45	1
2.	Construction changes	3.55	2
3.	Site Condition and natural incidents	2.09	3
4.	Planning & design error	1.64	4
5	Local Authority Requirement	1.54	5

4.3 Sources of change orders

A change may be initiated from any source: a contractor's, clients or project consultant. Fig. 2 presents the origins or sources of change orders. It is found that majority (83%) of the change order is due to client request, in which the clients are recognise as the parties involved in the briefing process. This result narrates to the cause of change order, whereby, the pertinent cause of change order is due to the changes is decision making, decided by the client it-self.

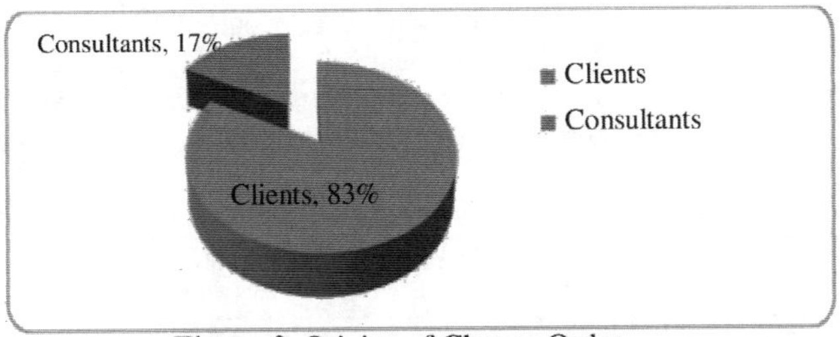

Figure 2. Origins of Change Order

4.4 Problems during briefing

Based from semi-structured interview, respondent revealed that inadequate management changes to client requirement as well as insufficient time allocated for briefing were the main problems during the briefing process. Though the participation of management is significant during the process, lesser efforts were given in incorporate managing of changes to meet the client requirement. This condition may be the cause of imprecise strategic brief as the process that deals with variations and uncertainty. Table 2 show the results of the problems during briefing process.

Table 2. Problem during briefing

No.	Problem During Briefing	Mean Value (N=15)
1.	Inadequate involvement of the relevant parties	2.83
2.	Insufficient time allocated for briefing	3.38
3.	Lack of communication	2.22
4.	Inadequate management changes to client requirement	4.12
5.	Lack of quality client representative	3.33
6.	Lack of commitment of client organization	2.83

4.5 Commitment of client organization

Table 3 signify the commitment level of management through briefing process. It is mystifying result when management set less commitment on the confirmation of scope of works, and confirmation of project cost. Based from the interview session, the confirmation on project cost is outside management control and it will only be decided at final phase of briefing. As the budget remained uncertainty, the probable of changing the design to meet the final cost will later influence the client to issued change order in the project.

Table 3. Commitment of Client Organization

No.	Client Organization Commitment in	Mean Value (N=15)
1.	Confirmation of the client objectives	4.33
2.	Confirmation of the scope of works (size, function, users of the project)	2.66
3.	Confirmation of the site location of the project	4.67
4.	Confirmation of the budget/estimated cost	2.22
5.	Confirmation on the project milestone	3.78
6.	Confirmation on project priority (cost, functionality, time)	3.67

4.5.1 Management efforts in managing brief development. Briefing process is iterative in nature that required shared understanding and commitment from the management. The effort in managing brief development is crucial to ensure the successful completion of project. In UiTM, the efforts in managing brief development were carried out by Bahagian Pembangunan. The result form the interview session shows that the efforts in organizing of client's project team scored highest (4.45) mean followed by effort on developing, documenting, and communicating clear brief (3.35). The results indicate that the management was very concern on the development of their project team, hence, the communication within their management is significant to clarifying the project requirement.

Table 4. Level of Management Efforts in Managing Brief Development

No.	Management Efforts in Managing Brief Development.	Mean Value (N=15)
1.	Organizing of the client's project team	4.45
2.	Allocating adequate time for briefing	1.64
3.	Planning for brief development	1.64
4.	Coordinating user group for brief development	2.09
5.	Developing, documenting and communicating clear brief	3.55
6.	Coordinating and monitoring of brief	2.09
7.	Communication within client organization	3.18
8.	Ensuring changes are evaluated and taken consideration	2.00

5. Conclusion

Briefing is the process of expressing the objectives and necessities of a project. The clients should understand of user needs and ensure that the final product meets project goals. The complexity of building projects requirements has increasing the difficulty of client organisations to prepare the client brief. Problems in briefing process range from various issues, hence, this will later lead to the issuance of change order in construction stage. The cause of change order not only significantly gives financial impact to the project, but also to the overall project performance. Project success is concern about the efficiency of the process that refers to completion on time, within allocated budget and builds quality as stipulated in contract. Therefore, clear project goal shall be define clearly at early stage of construction and accurate decision making during briefing are vital toward achieving success in construction projects.

References

[1] Shen, Q., LI, H., Chung, J. and HUI, P., 2004, 'A framework for identification and representation of client requirements in the briefing process', Taylor & Francis Ltd, UK

[2] Yu, A., Shen, Q., Kelly, J. and Hunter, K., 2007, 'An empirical study of the variables affecting construction project briefing/architectural programming', International Journal of Project Management, vol. 25, p.198–212.

[3] Clarke, A. 1999. A Practical Use of Key Success Factor to Improve the Effectiveness of Project Management. International Journal of Project Manager, 17, 139-145.

[4] Nguyen, L.D., Ogunlana, S., Quang, T., & Lam, D.T.X 2004. A Study on Project Succes Factors in Large Construction Projects in Vietnam. Engineering Construction and Architectural Management, 11(6), 404-413.

[5] Barrett, P.S and Stanley, C. 1999. Better Construction Briefing, Blackwell Science, Oxford.

[6] Blyth, A. and Worthington, J. 2001. Managing the Brief for Better Design, , London and New York, Spoon Press.

[7] Norizan, A. 2007. The Influence of Client Attributes on Project Success ; A Focus on the Briefing Stage. Unpublished PhD Thesis, Universiti Teknologi MARA, Shah Alam.

[8] Salisbury, F. 1998. Briefing Your Architect (Second Edition). Oxford : Reed Educational and Professional Publishing.

[9] Yu, A.T.W., Shen, Q.P., Kelly, J., and Hunter, K. 2005. Application of value Management in Project Briefing. Facilities 23(7/8), 330-342.

[10] Takim,R. and Akintoye,A. 2002. A Conceptual Model for Successful Construction Project Performance. Paper presented at Second International Post Graduate Research Conference in Built Human Environment, University of Salford, Salford, 11-12 April.

[11] Othman, A.A.E., Hassan, T. M., and Pasquire, C.L. 2004 Drivers for Dynamic Brief Development in Construction. Engineering, Construction and Architectural Management Vol. 11, Issue 4, pp. 248-258

[12] Abdul Rahman, H. Yahya, I.A., Mohammaed, O., & Munaaim, M.E.C 2006 Construction Cost Estimate and Monitoring : A Revisit of Quality Cost. Paper presented at the Quantity Surveying National Convention, Subang Jaya, Malaysia.

[13] Othman, A.A.E., Hassan, T. M., and Pasquire, C.L. 2006 The Brief Development Manager: As an IT Tool for Managing Dynamic Brief Development in Construction. International Journal of Construction Management, Vol. 6 No. 2, pp.31-47.

[14] Ndihokubwayo, R. and Haupt, T.C. 2008. Uncovering the Origins of Variation Orders. Paper presented at the 5th Post Graduate Conference on Construction Industry Development, Bloemfontein, South Africa

[15] Kuprenas, J.A. 2007. Design Phase Influences on Construction Change Order Costs. AACE International Transactions, 2007, 1-3.

[16] Clare and Brian S. 2008. Reviewing Change Order. AACE International Transactions, 2008, 1-5

[17] Fritz, E.H 1994, Change Order Management for Construction Project, PE CCE Cost Engineering Journal, Vol. 36, (3) March 1994.

[18] Norizan, A. 2011. Assessing Important Qualities of Design Team in Managing Client's Brief, Paper presented on IEEE Colloquium on Humanities, Science and Engineering, Penang, Malaysia.

Air quality assessment on human well-being in the vicinity of quarry site

W H W Ibrahim[1], E Marinie[2], J Yunus[2,3], N Asra[2] and K Mohd Sukor[1]

[1]Centre of Postgraduate Studies of Faculty of Architecture, Planning and Surveying, Universiti Teknologi MARA, 40450 Shah Alam, Malaysia
[2]Centre of Studies for Construction, Faculty of Architecture, Planning and Surveying,Universiti Teknologi MARA, 40450 Shah Alam, Malaysia
[3]Institute of Business Excellence, Universiti Teknologi MARA, 40450 Shah Alam, Malaysia

*wwanhasmirah@yahoo.com

Abstract. This study aims to investigate the variation of air pollutants associated with the quarry activities prior to classified distance from quarry site. Air pollutants were monitored with the use of instruments which are Rae System Multirae Lite Pumped (PGM-6208) to measure indoor air quality while TSI 8533 Dusttrack Drx Desktop Aerosol Monitor to measure outdoor air quality. Sampling will be replicated two times. The locations of quarry are at Bandar Saujana Putra and Taman Kajang Perdana 2, Selangor. The objectives of this study are to investigate the impact of quarry mining by preparing the suitable Indoor Air Quality Index and to prepare preventive measure for residential that caused from quarry mining activities. Both Qualitative and Quantitative approaches will be implemented in this study, which employed case study and interview survey. Both quarries identified previously will be the main case study. The Respondent's interviews are from Local Authority and Quarry Management Staff while questionnaire surveys from selected residences. Measurement method will be used to measure the Particle Matter (PM2.5) for indoor and outdoor in selected resident's area. However, this paper is primed to discuss the method used in this study. It is not only presents the beneficial information for future research on methodologies employed but also it is anticipated the benefit to environment which can increased residents' well-being in the vicinity of quarry sites.

1. Introduction

Quarry industry is one of the important industries in the development of country. For the economic development of the country, quarry industry continues to supply the raw materials for the construction, building and manufacturing. However, to sustain the supply and meet the development targets, both parties from industry and the government agencies must take all responsibility to control the consequences. Despite, the rapid growth of population and expanding of the housing development had forced the residential areas being built near the quarries [1]. This limitation of land issues had led to the high potential of more residential will developing near the quarry mining. The scenario is resulting is the level of Air Quality Index (AQI) for indoor environment become worsen. This is alarming for people's health and welfare, where 100% of a typical person's spend their time indoors. With regard to air quality, previous studies categorised five particles in the air quality index as suspended particle (PM2.5), suspended particle (PM1.0), carbon monoxide (CO), carbon dioxide (CO^2), and volatile organic compounds (VOCs).[2] Many researches had been utilized not only to understand this scenario but also analysed the facts and synthesis the problems with solutions in the end.

This paper is concentrating in research methodology particularly in procedure utilized for collecting and analysing data for air quality assessment in vicinity of quarry site.[2]This research will involve mixed method. Both qualitative and quantitative methods were applied.

First, interview session was conducted which involved participation from Local Authority and Quarry Management Staff. Secondly, the questionnaire surveys will be distributed to the selected residents to investigate the issues affects quarry activities. Thirdly, the measurement of air quality level will be conducted at the quarry neighborhood.

2. Literature Review

2.1. Analysis of document review on Quarrying

Documents were retrieved from reliable sources namely Science Direct, free access journal and Annual report of Department of Environment. The search criteria used the specific keyword "Quarrying in Malaysia and Indoor Air Quality (IAQ) in Malaysia" within years. There is an escalation of publication on quarrying-related research between 2007 and 2016. The pattern indicates an increased in the used of mixed methods approach compared to the single method. Table 1 shows the common research method applied e.g. Published interviews, case study, measurement and Environmental Impact Assessment (EIA) analysis. [3]

Table 1. Common research method utilised in the study

No	Research Approach	Descriptions
1.	Interview	This technique used by researcher to get involved and discuss the situation for a better understanding of the study.[4]
2.	Case Study	This enables researcher to closely examine the data within a specific context in order to explore and investigate the real-life phenomena.[5]
3.	Measurement Method	Techniques of real-time data collection from site, the parameters will be measured at the selected locations. For this study, PM10 concentrations will be measured at residential near the quarry mining area. The data obtained from each location will be plotted between distance and air quality level.[6]
4.	Environmental Impact Assessment (EIA)	Reports studied began from the year 1995 to 2002 and it contains the assessment, prevention and reducing method of the impact for environment. [7]

Table 2. The relationship between distances of residency to and the quarrying impact

Distance to Quarry Site (m)	Affected Respondent	Quarry Impact			
		Air Pollution	Explosion	Health Problem	Noise Pollution
<200	**6**	**6**			
401-600m	6	4	2		
601-800m	3		2	1	
801-1000m	3			2	1
>1000m	2	2			
Total		**12**	4	3	1

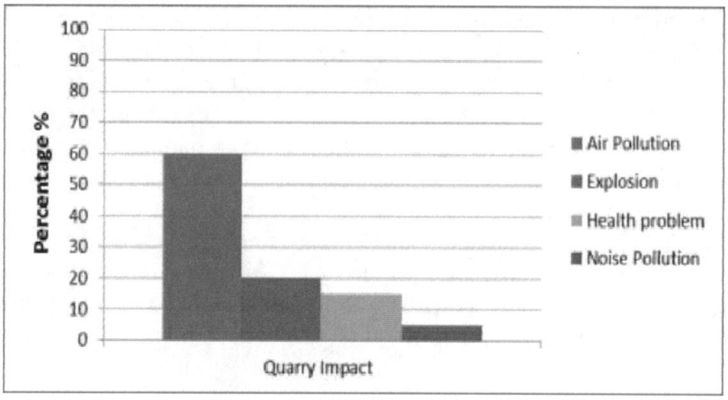

Figure 1. The impact of quarry

Table 2 shows the relationship between distances of residency and the quarrying impact of affected respondent. While, figure 1 shows the impact of quarry activities to the resident's. The most quarry impact affected the residents are air pollution, explosion, health problem and noise pollution. As results, the distance of residency to the quarrying area also played roles where neighbourhood within 200m of quarrying are most affected with air pollution.

3. Methodology

3.1 Sampling

Referring to Nartey, Nanor and Klake [8], the sampling method that had been practiced in their research where residents were randomly selected from two communities in Ghana based on the area located. 65 residents were selected for interview sessions. The scope of questions in the interview concerned to their social economic, perception on quarry activity and level of people health due to quarry mining activities. [8] While, cases for this studies were selected from residents at Bandar Saujana Putra and Taman Kajang Perdana 2 in Selangor within 1 kilometre (km) perimeter radius of the quarry mining. The characteristics of the houses within the distance of quarries in residential areas are shown in Table 3. Twenty (20) residents were chosen as the respondents from each case study. The scope of surveys were asked on impact of quarrying to resident, perceptions' about quarrying activities, resident's health experiences and mitigation plan to overcome the problems. The overall process of research in order to achieve the aim and objectives illustrated in the Figure 2.

Table 3. The Location of Selected Case Study

Selected Location	Type of House	No of Houses	Distance from Quarry To Residential Area
Bandar Saujana Putra, Kuala Langat, Selangor	Terraced House – Double Storey	10	200m 400m
Taman Perdana Kajang 2, Kajang , Selangor		10	600m 800m 1000m
Total	20		1 kilometres

IOP Conf. Series: Earth and Environmental Science **117** (2018) 012010 doi:10.1088/1755-1315/117/1/012010

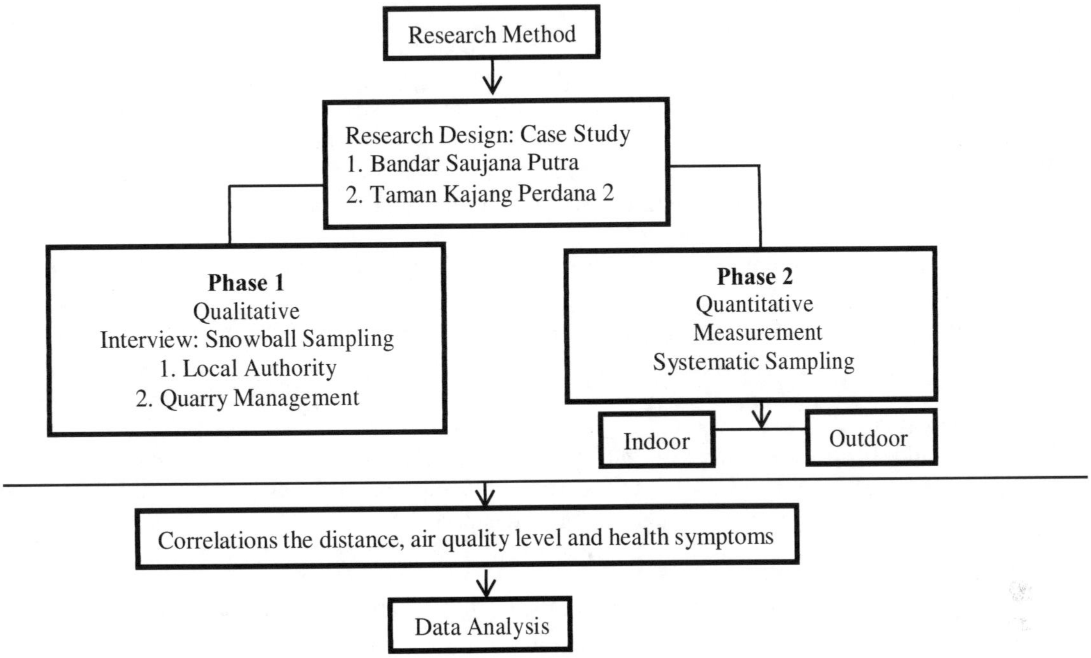

Figure 2. Flowchart of Research Method

3.2 Interview: Snowball Sampling

The first phase of data collection was the interview by using snowball sampling. Snowball sampling conducted to facilitate the identification of other members of population. The criterion for the sampling subject was only the Local Authority in Selangor. The interview was conducted to gain the information on current problems faced by the management of the quarry in their efforts to operate the activities of the quarry, the implementation of the existing guidelines and awareness of quarry management to the residential. Selected samples should be consisting of policy makers from government involvement to dust management. Due to the nature of qualitative research, the data was collected until saturation point is achieved. The interviews were designed in semi-structured interviews while process of data collection has conducted face to face interviews.

3.3 Measurement of AQI

The second phase of data collection is real-time measurement. It will involve the use of instruments that measure Air Quality Index (AQI). The measurement will be taken indoors, for houses located near the quarry. The parameter of air quality logged in this real-time measurements are PM 2.5 and PM10. Measurements will involve the level of oxygen (O^2) amount, carbon dioxide (CO^2) and any Particle Matter (PM2.5) in each house. According to the Air Pollutant Index (API) guide, published by the Department of Environment Malaysia (2000), [9] the air pollutants included in Malaysia's API are ozone (O^3), carbon monoxide (CO), nitrogen dioxide (NO^2), sulphur dioxide (SO^2) and suspended particulate matter of less than 2.5 microns in size (PM2.5) which are potentially harmful to human health. The instruments that will be used are Rae System Multirae Lite Pumped (PGM-6208) for measurement of indoor air quality and TSI 8533 Dusttrack Drx Desktop Aerosol Monitor for measurement of outdoor air quality [10] Different levels of air pollution signify conditions for alert, warning, emergency and significant harm. The correlations of the distance, air quality level and health symptoms will be analyse at the later stage to provide a stronger finding on impact of quarrying.

4. Conclusion

It is hoping, implementing the mixed methods, researcher would be able to identify and propose possible modification to overcome the impact of quarrying activities. This paper outlines the information for future researcher with regards to methodologies aspects and understanding the pattern of research methods on quarrying issues. The next step is to propose the preventive measure from the analysis of indoor air quality (IAQ) assessment based on the real-time measurements. In conclusion, it is hoping that the preventive measure will be beneficial to the quarry management as a guide to reduce the environment effect in surrounding areas. It also hopes it will aid the professional agency to overcome the issues of air pollution especially from the dust due to quarry mining. Finally, this study will be able to provide more knowledge and understanding on decreasing environmental quality by air pollution, which then aided in improving the comfortable condition of residents in their own houses.

Acknowledgements

The authors would like to express their gratitude for the financial support from Postgraduate of Centre of Postgraduate Studies of Faculty of Architecture, Planning and Surveying, Universiti Teknologi MARA (UiTM) in Shah Alam for the management and support of this research

References

[1] Tarmizi NI 2014 *Towards a Sustainable Quarry Industry In Malaysia* (Penambahbaikan Garis Panduan dalam Prosedur Penyerahan Lombong Kuari bagi Jabatan Teknikal), pp 123-12

[2] Chiang C, Lai C, Chou P, Li Y, and Tu, Y 1999 *ASIA - PACIFIC Conference on the Built Environment. The Study On The Comprehensive Indicators Of Indoor Environment Assessment For Occupants' Health.* (Taipei, Taiwan) pp 1–7.

[3] Lameed 2010 *African Journal of Environmental Science and Technology. Effect of quarrying activity on biodiversity: Case study* 740

[4] Omosanya KO and Ajibade OM 2012 *Journal of Applied Sciences. Environmental Impact of Quarrying on Otere Village, Odede, South Western Nigeria Ozean* **4** 75

[5] Adabanija MA and Oladunjoye M A 2014 *Journal of Geochemical Exploration. Geo-environmental Assessment of abandoned mines and quarries in South-western Nigeria* 148

[6] Olusegun O, Adeniyi and Adeola G 2009 *Ethiopian Journal of Environmental Studies and Management. Impact of Granite Quarrying on the Health of Workers and Nearby Residents in Abeokuta Ogun State, Nigeria* **2** 1

[7] Wahid R 2010 *Malaysian Journal of Environmental Management. Input Ekologi dalam Penilaian Impak Alam Sekitar (EIA)* 45

[8] Nartey VK, Nanor JN and Klake RK 2012 *Effects of Quarry Activities on Some Selected Communities in the Lower Manya Krobo District of the Eastern Region of Ghana* 362

[9] Department of Environment Malaysia 2000 *A Guide To Air Pollutant Index in Malaysia (API)* (Kuala Lumpur: Ministry of Science, Technology and the Environment)

[10] Ibrahim MH, Che Leh F, Adnan M, Mohd Isa NK 2016 *Malaysian Journal of Society and Space. Dust pollution in Malaysia: An investigation of the distribution of PM 10 in the city centre, sub-cities and suburbs of Ipoh, Perak* 104

Opportunities for corruption across Flood Disaster Management (FDM)

R Mohd Nordin[1*], E Latip[1], E M Ahmad Zawawi[1], Z Ismail[1]

[1]Faculty of Architecture, Planning & Surveying, Universiti Teknologi MARA, 40450 Shah Alam, Malaysia

*drrumaizah@salam.uitm.edu.my

Abstract. Flood is one of the major disasters in the world. Despite flood resulted in loss of life and damaged properties, it naturally imparts people to assist the victims that affected by the disaster. Malaysia has experienced many serious flooding events and proper flood disaster management need to be developed and adopted occasionally. Flood Disaster Management (FDM) seemed to be not working effectively especially during the Kelantan prodigious flood in December 2014. There were negative perceptions among victims and Malaysian citizens regarding the disaster management and government authorities in relation to corrupt practices. The FDM can be divided into four phases (i.e., prevention, preparedness, response and recovery) which undoubtedly corruption is perceived to exists in every phase. The aim of this study is to identify opportunities of corruption across FDM phases. The study presents a case study of Kelantan using the quantitative research approach which utilises questionnaire with government and private agencies. Further to that, this paper proved that opportunities for corruption may occur at every phase, undoubtedly response and recovery phase especially activities involving fund and donation are riskier. The findings are hoped to assist in developing an improved FDM in term of increased transparency.

1. Introduction

Disaster strikes quality, changes the lives of all it touches and its effects are felt long after the event [1]. Disaster has profound implications for large sector of the economy and political system of the affected region that may leads to political and social stability as well as international security and relations [2]. Despite disaster can be divided into three types such as natural, man-made, and hybrid disaster [1], natural disaster shows significant effects in term of loss of life, damage to property and environment, as well as the disruption of supply chains and markets [3].

Consequently, Malaysia has experienced various magnitudes of disaster with flood contributed to the highest number of occurrence. Looking at Malaysia's disaster and risk profile, Internationally Reported Losses of Disaster (1990-2014) by Centre for Research and Epidemiology of Disaster [4] states that flood is the most frequent disaster occurred (62.5%) compared to other disaster. Moreover, in term of mortality, flood contributed second place after storm with 24.1 % mortality and in term of economic issues, flood is ranked first place with 60% compared to earthquake, wildfire and storm. In addition, the 2014 year-end downpour and flood was the worst ever in Malaysian history, affecting more than half a million people with damages to public infrastructure alone was estimated at RM 2.851 billion [5].

Due to the fact that Malaysia has experienced many serious flooding events, the disaster management and relief have to be adopted and developed from time to time [6]. Building up an efficient and transparent disaster management is one of government agenda as primary approach to promote openness and reduce corruption problem [7]. In relation to corruption, Mahmud & Prowse [8] suggest that disaster management and relief sector has suffered from lower levels of corruption with 91% of individuals allegedly involved in corruption were elected representative with 54% is due to abuse of power whilst 37% of bribery. They further added that negligence to provide services and nepotism were most common within the pre-disaster intervention, whilst wage or asset stripping, bribery and the misuse of the resources were most prevalent in post-disaster intervention. Therefore, it is inevitable to determine the opportunities for corruption across FDM

In order to ensure an improved disaster management, transparency in decision making is important through preparing guidelines in advance so that they can be drawn upon at the time of a disaster without further delay [9]. Moreover, transparency and accountability initiatives seeks to improve the process through which actors and institution can affectively bring government to account and to effectively contribute for better development outcomes and socio-economic condition [10]. In relation to the latest scandal of misappropriation of state funds for flood victims, the Kelantan Menteri Besar has put forward that the state government is serious in curbing corruption in the distribution of flood relief funds for victim who were hit by the floods in 2014 [11]. Besides, China has been continuously marching toward transparency and openness in disaster management on the policy agenda since 2003 [12].

2. Perceived corruption activities in disaster management process

According to Islam et al. [13], the strategic framework on integrated flood disaster management includes four cyclic steps namely: (1) Preparedness before flood impact such as flood forecasting and warning; (2) readiness upon flood arrival; (3) emergency responses during flood impact and; (4) recovery and rehabilitation after flood impact. As for Malaysia, during the current 11th Malaysia Plan, the phases of disaster prevention, mitigation, preparedness response and recovery will be given emphasis to tackle problem of floods and other emerging hazards in a holistic manner through the explicit focus on strengthening disaster risk management, covering structural and non- structural measures [5].

Table 1 addresses the opportunities for corruption in disaster management process, particularly in flood disaster. During prevention phase, the perceived corruption activity was identified in organisational and community preparedness activity such as lack of transparency, lack accountability and corruption in sector related to organisation's core work that will negatively influence the beneficiaries' vulnerability to natural hazard/ disaster [14].

Table 1: Perceived corruption activities in disaster management process

Disaster Management Process	Prevention	Preparedness	Response	Recovery
Opportunities for corruption	• Organisational and Community Prevention Activity	• Improvisation central foundation of emergency management	• Donation • Temporary Supply • Sheltering and volunteers • Emergency Response Plan	• Transitioning and return to normal • Dealing with donation • Government welfare and emergency assistance

As for preparedness phase, improvisation of central foundation for emergency management is one of activity that allow non-transparency situation especially in the injection of large amounts of resources into poor economies, where institutions may have been damaged and destroyed, and can exaggerate power imbalance that leads to increase opportunities for corruption [15]. Moreover, donation, temporary supply, sheltering and volunteers, and Emergency Response Plan were found to create opportunities for corruption during response phase. Due to the fact that corruption is often not transparently reported, this situation gives problem to approvals for the donation [16]. Besides, the breakdown of accountability provides unscrupulous public official with a golden opportunity to siphon off money and supplies from government funds and donations [17]. According to Paterno [17], there are important people that put in charged for coordination of the distribution of relief good who will distribute relief pack in the basis of their favouritism and political patronage. Finally, during recovery phase, transitioning and return to normal and Government welfare and emergency assistance were detected with corruption occurrence. Recovery and reconstruction project rarely possess systematic consideration of whether such project will actually achieve the goals for which they implement [14]. Usually, a lot of money gets spent but one does not always know the reason. Extreme weather triggers an infusion of reconstruction cash of federal, state and local government, which corrupt official can misappropriate on the situation [18]. Furthermore, public officials are given a heightened opportunity to exploit victim desperation in order to advance their private interests [19].

3. Research Methodology

The research utilises quantitative data collection method by means of questionnaire survey. The questionnaire intended to identify the possible opportunities for corruption across disaster management phase. The questionnaire was designed to have 5 sections with a total of 22 questions. A 5-points Likert scale were utilised for the respondents to indicate their opinions. The target respondents were selected based on non-probability of judgement purposive sampling due to their expertise in the subject matters. There are five (5) departments (i.e., Department of Irrigation and Drainage (DID), Public Works Department (JKR), Fire & Rescue Department (BOMBA), Royal Malaysia Police (POLIS), Malaysia Civil Defence of Kelantan (APM KELANTAN), National Disaster Management Agency (NaDMA) and Malaysian Medical Relief Society (MERCY)) identified to be involved in flood management especially in Kelantan (as shown in Table 2) and the list of departments were obtained from official website of eBanjir Negeri Kelantan. The target respondents were based on the number of the workforces in the various departments inclusive of senior officers, executives and supporting staffs.

Table 2: Target respondents and response rate

Name of Department	Target respondents	Responses
Agensi Pengurusan Bencana Negara (NADMA)	50	5
Jabatan Pengairan dan Saliran Negeri Kelantan (JPS)	35	11
Jabatan Kerja Raya Negeri Kelantan (JKR)	30	13
Jabatan Bomba dan Penyelamat Negeri Kelantan	35	13
Ibu Pejabat Polis Negeri Kelantan	25	15
Jabatan Pertahanan Awam Kelantan (APM)	30	14
MERCY	27	15
TOTAL	**232**	86

The research focused with Kelantan since it is considered as the most severe state experienced flood disaster in Malaysia. In example, during 1967 flood, it is estimated that 84% of the villages in Kelantan
was affected with estimated damage of RM 30,000,000.00 [20]. Adding to that, in 2004, the water level of Sungai Kelantan at Tambatan DiRaja, which has a danger level of 25m raised to 29.70m [21] that caused 5,000 individuals sheltering in government reception centres [22]. Whereby, in 2014 flood, the Malaysia National Security Council confirmed that it was the worst in history with 202,000 victims were displaced [23].

Table 2 also presents the target respondents of 232 for the five departments and the response rate of 37.06 percent (86 responses) after one month of direct and postal questionnaire distribution (1st May

2017 until 21st May 2017). The response rate of 37.06 percent is appropriate due to postal questionnaire survey is around 20 to 30 percent [24]; [25]. In addition, a more recent survey done by Ismail [26] and Nordin [27] received 23.6 and 37.5 percent response rates.

4. Data Analysis and Discussion

Table 3 presents the respondent's position held in the departments. The result shows that most of the respondents (70 of 86 respondents or 81.4%) are senior officers and executives which indicates that they have the authority to give their opinion and statement on the issues. While the remaining 18.6% are the supporting staffs. On the other hand, table 4 shows the respondents' working experiences. Based on the results, 67% of the respondents have more than 5 years experience with 26 of them have more than 10 years experiences which can be considered as highly experienced as suggested by Masrum [28]. This implies that the respondents have reasonable knowledge on the issue.

Table 3: Respondents' position

Position	No. of Respondents	Percentages (%)
Senior Assistant Director	2	2.33
Operation officer	28	32.56
Administration officer	27	31.40
Training Officer	3	3.49
Financial	9	10.47
Engineer	1	1.16
Assistant Engineer	3	3.49
Secretary	3	3.49
Technician	2	2.33
Clerk	5	5.81
Volunteer	1	1.16
Intern	2	2.33
TOTAL	86	100

Table 4: Respondents' working experiences

Working Experiences	No. of Respondents	Percentage (%)
Less than 5 years	28	32.6
5 years and 10 years	32	37.1
10 years and above	26	30.2
TOTAL	86	100

Table 5: Implementation effectiveness of FDM

Phases	Mean	Rank
Response	3.67	1
Recovery	3.60	2
Preparedness	3.57	3
Prevention	3.57	4

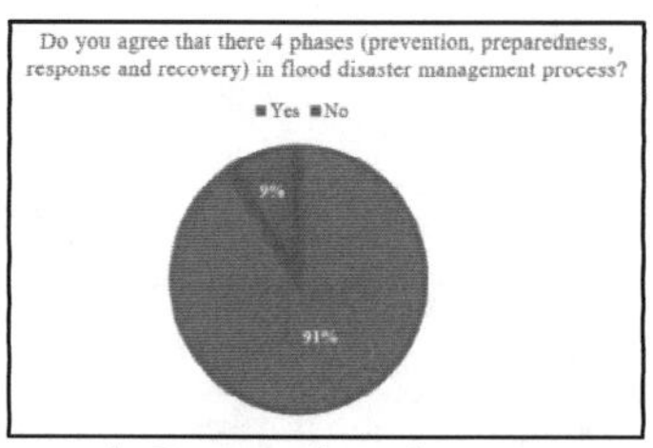

Figure 1: Respondents' agreement on the four phases of FDM process

Table 6: Implementation of FDM in Malaysia

Measures	Mean	Rank
FDM is important for your company	4.16	1
FDM is to manage and reduced the impact of hazard	4.14	2
FDM can help to overcome the economy environment and society in general	4.09	3

Figure 1 displays respondents' agreement on the four phases of FDM process with 91% agreed that FDM involve is prevention, preparedness, response and recovery which is in line with Islam et al. (2016) and Yassin [5]. However, based on Table 5, response phase was found to be more effective (M=3.67) compared to other phases. In addition to that, Table 6 shows the perception of respondents in term of implementation of FDM in Malaysia with 'FDM is important for your company' is ranked first (M=4.16), followed by 'FDM is to manage and reduced the impact of hazard' (M=4.14) and 'FDM can help to overcome the economy environment and society in general' (M=4.09). This verify the work of Islam et al. [13] that put forward a key to dealing with the disaster scenario is with a good planning and framework in disaster management with close collaboration between government and non-government agency. Hence, it can be deduced that the respondents are aware on the issue of FDM phases as well as the implementation of FDM.

Perceived corruption activities in flood disaster management process

Table 7 explains the perceived corruption activities across FDM process in Malaysia. During the prevention phase, 'individual and household activity' was found to be the most selected issue (M=3.52) followed by 'implement structural flood mitigation' (M=3.50) and 'implementation of complementary non-structural and non-engineering measure' (M=3.47). This is in line with Labadie [14] that put forward there is a lack of transparency and accountability that leads to corruption in community and organizational preparedness activities. For the preparedness phase, elements of 'emergency assessment activity', 'give public info media and disaster activity', 'improve central foundation of emergency management activity' and 'Prepared community of worship of disaster' were equally agreed by most of the respondents with mean = 3.20 and followed by 'expedient hazard mitigation' (M=3.15). This proved the work of Kompanek [15] that highlighted improvisation of central foundation for emergency management is one of activity that transpire corruption and non-transparency.

Table 7: Perceived corruption activities in flood disaster management process

Phases	Activities	Mean	Rank	Mean Score	Rank
Prevention	**Individual and household activity**	**3.52**	**1**	**3.09**	**4**
	Implement structural flood mitigation	3.50	2		
	Implementation of complementary non-structural and non-engineering measure	3.47	3		
Preparedness	**Emergency assessment**	**3.20**	**1**	**3.19**	**3**
	Give public info media and disaster	3.20	2		
	Improve central foundation of emergency management	3.20	3		
	Prepared community of worship of disaster	3.20	4		
	Expedient hazard mitigation	3.15	5		
Response	**Fund and donation**	**3.52**	**1**	**3.48**	**1**
	Hazard detection and Warning consideration	3.50	2		
	Sheltering	3.47	3		
	Temporary Supply	3.44	4		
Recovery	**Dealing with donation**	**3.45**	**1**	**3.32**	**2**
	Return to normal in social, economic and environment	3.38	2		
	Disaster recovery center	3.29	3		
	Post disaster policy	3.16	4		

Next, during response phase, 'fund and donation' was ranked first (M=3.52) as the perceived corruption issue. This is not surprising due to approvals for donations were not transparently done [16] which also leads to opportunity to siphon off money and supplies from government funds and donations by the officers [17]. The issue of donation also transpires in the recovery phase as 'dealing with donation' was also ranked first with mean value of 3.45 by the respondents. All in all, it can be seen from the results that among the four phases, response phase was perceived to be prone with corrupt activities (M=3.48), followed by recovery (M=3.32), preparedness (M=3.19) and lastly prevention (M=3.09). Despite the result proved that opportunities for corruption may occur at every phase, undoubtedly response and recovery phase especially activities involving fund and donation are riskier which has been proved by the scandal mentioned by Reduan [11].

5. Conclusion and recommendation

The study shows that effective FDM is vital in reducing the impact of hazard. As for that matter, Malaysian government have identified phases of flood disaster management as prevention, preparedness, response and recovery. The finding of this study revealed that most of respondent recognise and aware with this four phases in FDM process. However, majority of the respondents believed that response and recovery phase are the most effectively implemented in Malaysia. The study also indicates possible area and activities in FDM in Malaysia that may involve corrupt activities. Based on the findings, majority of the respondents believed that the response and recovery phase are positive toward corrupt activities with fund and donation is the higher possible activity intended toward corrupt activities. There may be risk of double funding and misuse. They also suggested that in recovery area, activity that dealing with donation contributed to corrupt activity. This showed that other than wage or assets stripping, bribery, and the misuse of resource were most prevalent in post-disaster intervention.

As for that matter, high transparency in FDM in Malaysia is believed to improve the disaster management process, particularly in term of sharing information including funding, budget utilization, beneficent, selection criteria, selection vendor and other prevent from the corruption. Moreover, the government should improve their enforcement by conducting more training to the local community in order to increase the public awareness, education and public participation in disaster management. Finally, the disaster management centre should have been positioned in a strong ministry which is perceived as having a low political profile and status. Also, links between the disaster management and local governments should be strengthened as teamwork is needed to produce effective coordination on disaster management.

Acknowledgments

The authors would like to acknowledge the contribution from the Universiti Teknologi MARA (UiTM) through supporting the research with LESTARI grant (600-IRMI/DANA5/3/ LESTARI (0154/2016)).

References

[1] Azmi NA, Hashim R and Zamhury N 2012 *IEEE Symposium on Humanities. Flash Floods and People Preparedness in Anticipating Disaster in Malaysia, Science and Engineering Research Kuala Lumpur* pp. 630-633

[2] Rodriquez JT, Vitoriano B, Montero J and Kecman V 2011 *OR Spectrum. A Disaster- Severity Assessment DSS Comparitive Analysis* pp. 451- 479

[3] W. E. Forum 2010 A New Private Public Partnership Model for Disaster Response," Engineering & Construction Resource Partnership

[4] C. f. R. a. E. o. D. (CRED) 2009 EM-DAT: The International Disaster Database (CRED, 2009) [Online]. Available: http://www.emdat.be/. [Accessed 10 February 2015].

[5] Yassin MM 2015 *Official Statement 3rd United Nations World Conference. Disaster Risk Reduction* (Sendai, Japan)

[6] Shafiai and Khalid 2016 *Management Malaysia International Reviews of Management and Marketing. Examining of issues on Flood Disaster*

[7] Bertot JC, Jaeger PT, and Grimes JM 2010 *11th Annual International Conference on Digital Goverment Research. ICTs, Social Media, and Goverment Transparency Iniatiative* (Mexico)

[8] Mahmud T and Prowse M 2012 *Global Environment Change. Corruption in Cyclone Preparedness and Relief Efforts in Coastal Bangladesh. Lessons for Climate Adption ?* pp. 933-943

[9] Bayntun C, Rockenschaub G and Murray V 2012 A qualitative analysis of the core Literature to Complement the WHO Toolkit for Assessing Health-system Capacity for crisis Management PLOS Current Disaster. Developing a Health System Approach to Disaster Management.

[10] Acosta AM 2013 *Development Policy Review. The Impact and Effectiveness of Accountability and Transparency Initiatives: The Governance of Natural Resources* **31** 89
[11] Reduan, 2017 Mb Vows Report Errant Agencies," New Straits Times: Prime News, Kota Bharu, 2016, March 17.
[12] Kang Y, Disaster Management in China in a Changing Era, Springer, 2015.
[13] Islam R, Kamaruddin R, Ahmad SA, Jan SJ and Anuar AR 2016 International Review of Management and Marketing. A Review on Mechanism of Flood Disaster Management in Asia **6** 29
[14] Labadie JR 2008 Disaster Preventation and Management. Auditing of post-disaster recovery and reconstruction activities 575
[15] Kompanek A 2017 Tackling Corruption in Disaster Relief Efforts," Center for International Private Enterprise (CIPE), 9 December 2013. [Online]. Available: http://www.cipe.org/. [Accessed 22 August 2017].
[16] Hees R, Ahlendorf ML and Debere S 2010 *Transparency International. Preventing Corruption in Humanitarian Operation, Berlin*
[17] Paterno B 2014 Learning From Disaster. Corruption and Environment Catastrophe, 7 November 2014.
[18] Sorensen J 2014 Why are Natural Disaster Breeding Ground for Corruption ? Northwestern Now, 03 March 2014.
[19] Chan N 2012 Impacts of Disasters and Disasters Risk Management in Malaysia The Case of Floods pp. 508-509
[20] Davies R Malaysia Floods – Kelantan Flooding Worst Recorded as Costs Rise to RM1 Billion Floodlist, 5 January 2015. [Online]. Available: http://floodlist.com/asia/malaysia-floods-kelantan-worst-recorded-costs. [Accessed 22 August 2017].
[21] Kent 2017 Flooding hits eastern Malaysia," BBC News, 13 December 2004. [Online]. Available: http://news.bbc.co.uk/2/hi/asia-pacific/4091169.stm. [Accessed 22 August 2017].
[22] Su-Lyn B 2017 Kelantan floods like Japan's 2011 tsunami, says NSC," Malay Mail Online, 15 January 2015. [Online]. Available: http://www.themalaymailonline.com/malaysia/article/kelantan-floods-like-japans-2011-tsunami-says-nsc#F2wJf5ozdTs7BKZp.97. [Accessed 22 August 2017].
[23] Akintoye A 2000 *Construction Management and Economics. Analysis of Factors Influencing Project Cost Estimating Practice*, pp. 77-89
[24] Black C, Akintoye A and Fitzgerald E *2012 International Journal of Project Management. An Analysis of Success Factors and Benefits of Partnering in Construction* 423
[25] Ismail K 2012 A Value for Money (VFM) Assessment Framework for Public Private Partnership (PPP) Approach (Universiti Teknologi MARA Malaysia, Shah Alam)
[26] Nordin RM 2015 A Framework of Transparency Initiative (TI) to Fight Corruption for Public Construction Projects. Unpublished PhD Thesis. (Universiti Teknologi MARA, Malaysia)
[27] Masrum MAN 2012 Developing a Predictive Contractor Satisfaction Model (CoSMo) for Construction Projects. PhD Thesis (Queensland University of Technology, Queensland, Australia)

The effects of green areas on air surface temperature of the Kuala Lumpur city using WRF-ARW modelling and Remote Sensing technique

N A Isa[1], W M N Wan Mohd[1], S A Salleh[1*] and M C G Ooi[2]

[1]Centre of Studies for Surveying Science and Geomatics, Faculty of Architecture, Planning and Surveying, Universiti Teknologi MARA, 40450 Shah Alam, Selangor, Malaysia
[2]Division of Environment, Faculty of Engineering, University of Nottingham Malaysia Campus, 43500 Semenyih, Selangor, Malaysia

*aekbal@salam.uitm.edu.my

Abstract. Matured trees contain high concentration of chlorophyll that encourages the process of photosynthesis. This process produces oxygen as a by-product and releases it into the atmosphere and helps in lowering the ambient temperature. This study attempts to analyse the effect of green area on air surface temperature of the Kuala Lumpur city. The air surface temperatures of two different dates which are, in March 2006 and March 2016 were simulated using the Weather Research and Forecasting (WRF) model. The green area in the city was extracted using the Normalized Difference Vegetation Index (NDVI) from two Landsat satellite images. The relationship between the air surface temperature and the green area were analysed using linear regression models. From the study, it was found that, the green area was significantly affecting the distribution of air temperature within the city. A strong negative correlation was identified through this study which indicated that higher NDVI values tend to have lower air surface temperature distribution within the focus study area. It was also found that, different urban setting in mixed built-up and vegetated areas resulted in different distributions of air surface temperature. Future studies should focus on analysing the air surface temperature within the area of mixed built-up and vegetated area.

1. Introduction

The benefits and advantages of urbanization are undeniable. However, excessive urbanization can become uncontrollable and resulted in environment degradation [1]. Rapid conversion of vegetated areas into urban areas increased the air surface temperature and modified the urban climate condition which leads to the formation of Urban Heat Island phenomenon [2-4]. Therefore, consideration on urban climate in urban planning is suggested by many previous studies [1, 5-8]. Many earlier studies have indicated that green vegetation has the capability to regulate the high temperature within urban areas [9-11]. Therefore, there is a need to study the response of green area towards urban cooling as one of the mitigation strategy in urban planning. Previous studies demonstrated that the Weather Research and Forecasting (WRF) model was able to simulate the air surface temperature well within tropical cities[12, 13]. Remote sensing and Geographical Information System (GIS) techniques are often employed by previous studies for spatial-based analysis regarding climates [1, 4, 14, 15].

Therefore, this study attempts to identify the relationship between the air surface temperature and the green vegetation by employing WRF modelling and remote sensing as well as GIS technique.

2. Study Area

This study focuses on the Kuala Lumpur city. This city has a tropical climate since it is located in lower latitude region (03°N, 101°E). The selection of this city as the study area is due to the rapid urbanization that has taken place in the last few decades. The variability of urban morphology and urban settings also made this city is suitable for this study. Kuala Lumpur is the capital of Malaysia and is the densest city in the country. Located in the west coast part of Peninsular Malaysia, this city is surrounded by satellite cities in Selangor State such as Batu Caves, Seri Kembangan, Ampang Jaya and Petaling Jaya.

3. Methodology

This study employed WRF-ARW model to simulate the air surface temperature and remote sensing technique to map the green area of Kuala Lumpur city as shown in Figure 1. The following subsections describe the steps involved to determine the relationship between green areas and air surface temperature of the Kuala Lumpur city. The steps involved were 1) Modelling the air surface temperature using WRF model, 2) Generating NDVI using remote sensing technique and 3) Determining the relationship between green areas and air surface temperature.

3.1. Air Surface Temperature Modelling Using Weather Research and Forecasting (WRF) model

In this study, the 2m air surface temperature was simulated using WRF Version 3.8 with Advanced Research WRF (ARW) core dynamic solver. The simulations were spun up for 24 hours for two different dates that coincide with the satellite images used in this study. Since the WRF-ARW simulation will only stabilize after 12 hours, this study runs the simulation for 48 hours for each year in order to simulate both dates well. Figure 1 shows the flow of the simulation employed in this study.

3.1.1. Domain Configurations. This study used four nested domains to simulate the 2m air temperature of Kuala Lumpur city with the grid size of 125km (30x30), 25km (51x51), 5km (131x131) and 1km (101x101). The dimension given in parenthesis is the number of grids in the easting and northing direction respectively. The area of the smallest domain (d04) covers the entire Klang Valley region where the Kuala Lumpur city is located as shown in Figure 2. NCEP FNL Operational Global Tropospheric Analysis data with a 1° by 1° resolution supplied by Global Data Assimilation System (GDAS) was utilized as the lateral boundary conditions every six hour interval. The common (default) WRF-ARW MODIS land use data available was used as to describe the terrestrial information of the Kuala Lumpur city.

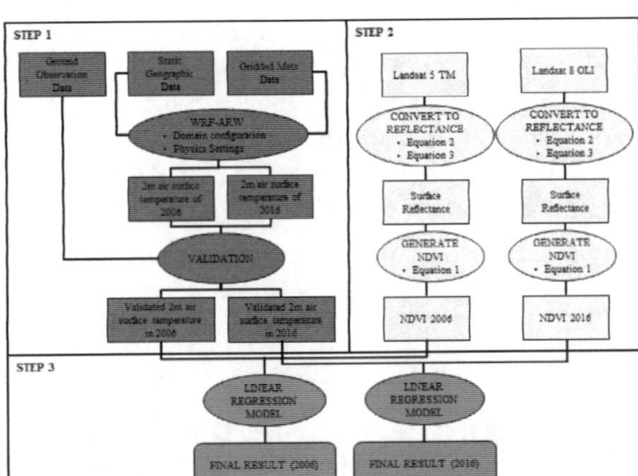

Figure 1: Overall methodology flow of the study

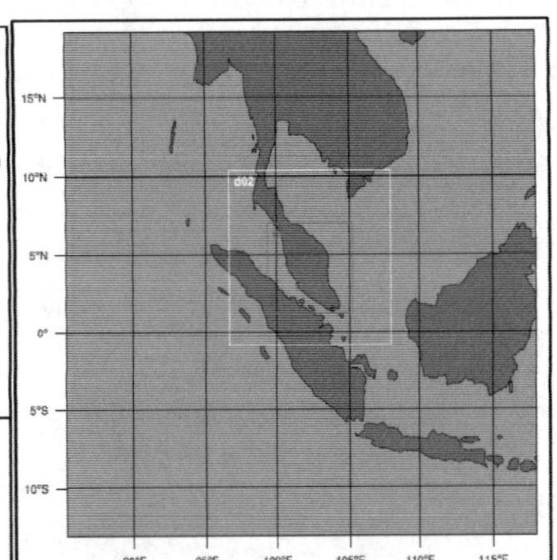

Figure 2: Domain configuration

3.1.2. Physics Options Settings. WRF Version 3.8 offers a variety of multiple physics options, range from simple to very sophisticated schemes as well as newly developed and well-tried schemes which can be combined together to suit the study area needs. Based on the previous studies conducted [13, 16], this study employed the physics options as described in Table 1.

Table 1. The physics options employed in the WRF-ARW model

Physics Option	Scheme
Microphysics	WRF Single-Moment 3-class scheme
Longwave Radiation	RRTM scheme
Shortwave Radiation	Dudhia scheme
Surface Layer	MM5 similarity
Land Surface	Noah Land Surface Model
Planetary Boundary Layer	Yonsei University scheme
Cumulus Parameterization	Kain-Fritsch scheme

3.1.3. WRF-ARW Outputs Preparation using GIS Technique. Since the WRF-ARW outputs were stored in binary format (NetCDF files), this study employed GIS platform to prepare the WRF-ARW outputs before being used in the analyses. The air surface temperatures were extracted from the binary files using a special tool developed in ArcGIS software. The layers were then converted into raster data format which are easily to be manipulated in any GIS platform. Since WRF-ARW simulated and offered the air surface temperature in Kelvin (K), this study converted the air temperature values into degree celcius (°C), a commonly used unit for temperature in urban climate studies.

3.1.4. Model Validation. The air surface temperature simulated using the WRF model were validated against the ground observations provided by the Malaysian Meteorological Department (MMD). The ground observations on both dates collected from three stations which located in Universiti Malaya (03° 07' N, 101° 39' E), Subang (3°07'N, 101°33'N) and KLIA Sepang (3°06'N, 101°39'E) were used as the reference. Based on diurnal comparisons, the RMSE and R^2 of both simulated models were calculated to identify how well the model simulated the air surface temperature of the Kuala Lumpur city.

3.2. Generation of Normalized Difference Vegetation Index (NDVI)

In order to extract the green areas of the Kuala Lumpur city, two satellite images from Landsat 5 TM and Landsat 8 OLI were utilized. The images were captured on two different dates; 2nd of March 2006 and 29th of March 2016. These dates were chosen due to very clear sky view which reduces the chance of atmospheric error to occur during image processing.

The generation of NDVI value using Landsat 5 TM and Landsat 8 OLI employed two different bands called Near Infra-red (NIR) and Red band. The NDVI values within the study area were extracted using commonly used equation as stated in Equation 1. However, in order to use the equation, the digital numbers (DN) of each band should be converted into surface reflectance. The NDVI outputs from this process were classified into 3 classes as suggested by [11].

$$NDVI = \frac{\rho NIR - \rho Red}{\rho NIR + \rho Red} \tag{1}$$

where ρNIR = surface reflectance of NIR band
ρRed = surface reflectance of Red band

3.2.1. Conversion to Reflectance for Landsat 5 TM. The conversion of DN values into surface reflectance using Landsat 5 TM image involved two steps. At first, the DN values should be converted into radiance using Equation 2. Then, the radiance values were converted into surface reflectance using Equation 3.

$$L_\lambda = \left(\frac{LMAX_\lambda - LMIN_\lambda}{QCALMAX - QCALMIN}\right) \times (QCAL - QCALMIN) + LMAX_\lambda \quad (2)$$

where L_λ = spectral radiance at the sensor aperture in watts
$LMAX_\lambda$ = spectral radiance that is scaled to QCALMAX in watts
$LMIN_\lambda$ = spectral radiance that is scaled to QCALMAX in watts
QCALMAX = the maximum quantized calibrated pixel value in DN
QCALMIN = the minimum quantized calibrated pixel value in DN

$$\rho\lambda = \frac{\pi \times L_\lambda \times d^2}{ESUN_\lambda \times \cos\Theta_s} \quad (3)$$

where $\rho\lambda$ = surface reflectance
d = Earth-Sun distance in astronomical units
$ESUN_\lambda$ = Mean solar exo-atmospheric irradiances
Θ_s = Solar zenith angle in degrees

3.2.2. Conversion to Reflectance for Landsat 8 OLI. The conversion to surface reflectance of Landsat 8 OLI was using a different equation due to different sensor's characteristics. The surface reflectance of Landsat 8 OLI can be obtained using Equation 4. However, the sun angle correction was calculated separately in Equation 5. The sun angle correction was performed after the bands involved were converted into surface reflectance.

$$\rho\lambda' = M_\rho \times Q_{CAL} + A_L \quad (4)$$

where $\rho\lambda'$ = surface reflectance without sun angle correction
$M\rho$ = band-specific multiplicative rescaling factor from metadata
Q_{CAL} = Quantized and calibrated standard product pixel value
A_ρ = band-specific additive rescaling factor from metadata

$$\rho\lambda = \frac{\rho\lambda'}{\cos(\Theta_{SZ})} = \frac{\rho\lambda'}{\sin(\Theta_{SE})} \quad (5)$$

where Θ_{SZ} = Local sun elevation angle
Θ_{SE} = Local solar zenith angle

3.2.3. Pixel Data Resampling. The original pixel size of Landsat 5TM and Landsat 8OLI was 30m by 30m which were not suitable to be compared to the simulated air surface temperature modelled by WRF-ARW. Therefore, resampling and aggregating the pixel size of the NDVI layers extracted from the satellite images were performed. The NDVI layers were resampled and aggregated into 1km by 1km resolution by assigning the mean NDVI value for each pixel.

3.3. Relationship between green areas and air surface temperature of the Kuala Lumpur City

The relationship between the green areas and air surface temperature of the Kuala Lumpur city was carried out using linear regression models. The study used pixel-to-pixel analysis to assess the effects of the green areas of air surface temperature within 1km by 1 km resolution. Through the linear

IOP Conf. Series: Earth and Environmental Science **117** (2018) 012012 doi:10.1088/1755-1315/117/1/012012

regression, the correlation between the green areas and air surface temperature were determined and presented using the value of goodness-of-fit for linear model (R^2).

4. Results and Analysis

4.1. Simulation Validation against Ground Observation

Based on this study, the RMSE of the model in 2006 were 1.9°C, 1.8°C and 0.8°C for Subang, Universiti Malaya and KLIA Sepang station respectively. Whereas, the RMSE of the model in 2016 were 1.9°C for Subang, Universiti Malaya and KLIA Sepang station respectively. Based on the RMSE tolerance, outstanding agreements between the prediction and ground observation were identified in this study for the year 2006 and 2016. Therefore, it can be concluded that the model employed in this study able to simulate the air surface temperature of the Kuala Lumpur city well. Figure 3 and Figure 4 illustrated the agreements on 2006 and 2016.

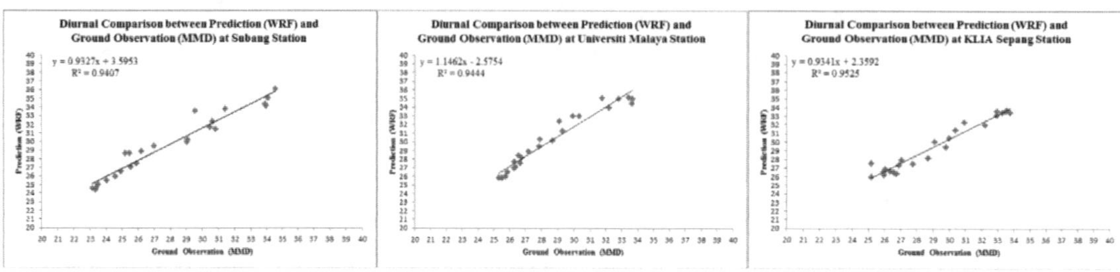

Figure 3: The agreement between simulated air surface temperature with ground observations in 2006

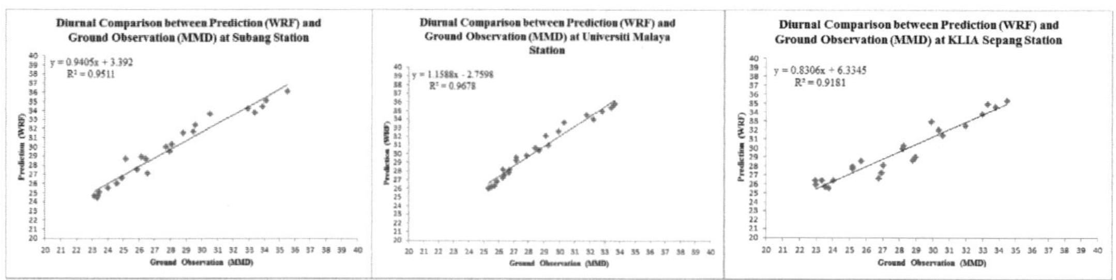

Figure 4: The agreement between simulated air surface temperature with ground observation in 2016

4.2. Air Surface Temperature of Kuala Lumpur City

The simulated air surface temperature distributions are shown in Figure 5 and Figure 6. From this study, it was found that, the air surface surface temperature of the Kuala Lumpur city was increased in the last decade. From the figures, it was also identified that the northwestern parts and some western parts of the Kuala Lumpur city are warmer than the others. This may be due to built-up coverage in the northwestern and western parts of the city which consist of mainly compact residential areas as compared to the other parts. Even though the eastern part is the busiest regions in Kuala Lumpur that includes the business areas and the city center, the areas were fragmented with green vegetation which makes the air surface temperature to be lower.

4.3. Green Area of Kuala Lumpur City

In Figure 7, the changes of green vegetation in Kuala Lumpur city can clearly be seen. In some parts, the green vegetation has become matured especially in Bukit Tunku and Mont Kiara with an increase of 5%. Also, it was found that, some vegetated areas have become mixed of built-up areas with vegetation portions. The conversion covered about 2.4% of the city areas. The decrease in 7.7% of cleared land and built-up areas was also identified in this study.

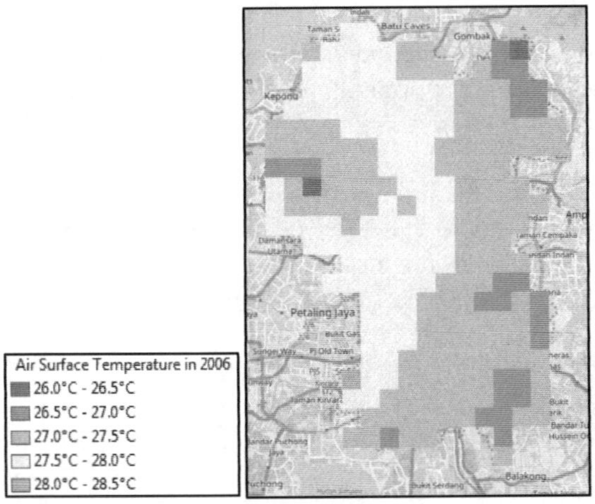

Figure 5: Air surface temperature distribution in 2006

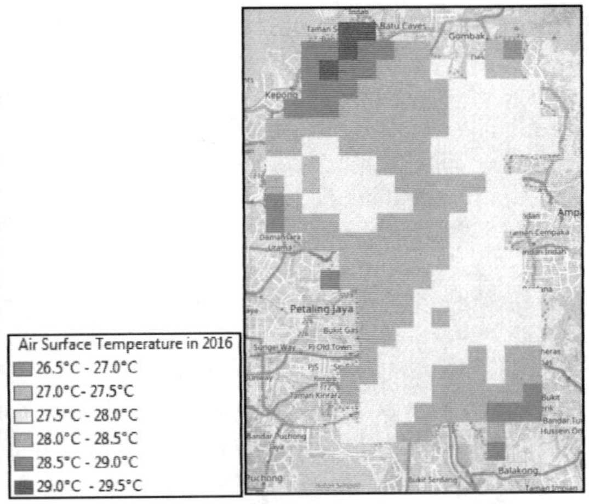

Figure 6: Air surface temperature distribution in 2016

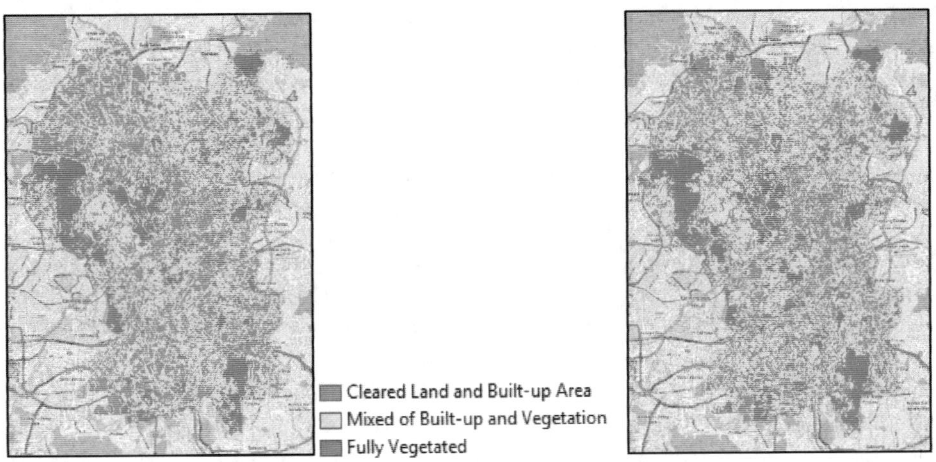

Figure 7: Green area coverage in 2006 and 2016

4.4. Relationship between Air Surface Temperature and Green Area in Kuala Lumpur City

This study has identified a strong correlation between the green areas and air surface temperature. In both years, the R^2 analyzed were 0.792 and 0.886 for 2006 and 2016 respectively. Figure 8 and Figure 9 show the correlation of green areas and the air surface temperature during the selected dates. A strong negative correlation indicates that the higher the NDVI values, the lower the temperature. Therefore, it can be suggested that, green vegetation has a positive impact towards the urban climate condition.

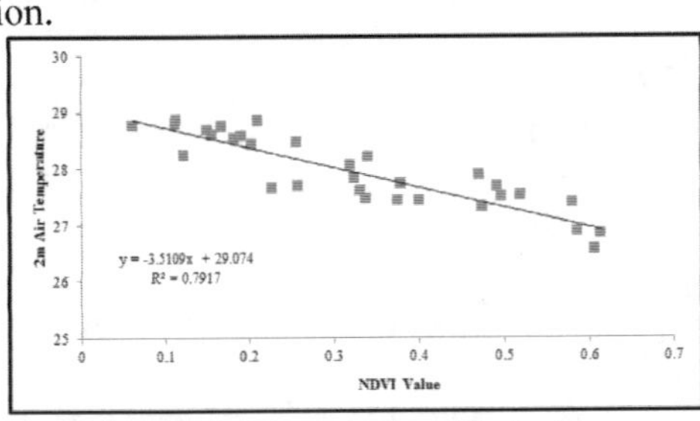

Figure 8: The relationship between simulated air surface temperature with green areas in 2006

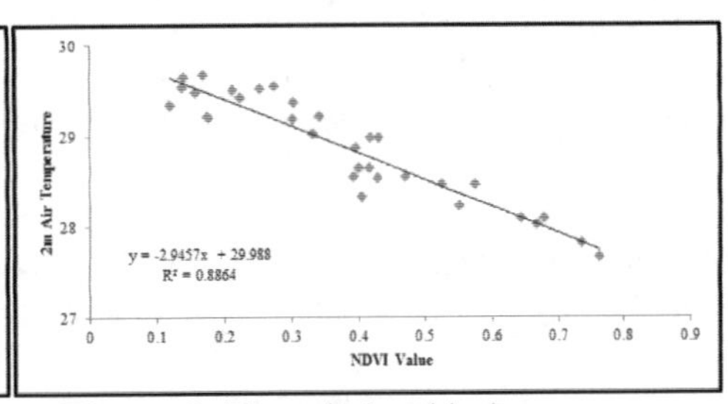

Figure 9: The relationship between simulated air surface temperature with green areas in 2016

5. Conclusion and Recommendation

From this study, it has found that green area has a significant effect on air surface temperature. Strong negative correlation was identified through linear regression performed in this study. Therefore, it can be concluded that the green area contributes to the decrease in air surface temperature of the Kuala Lumpur city and can be used as an urban cooling agent.

Spatially, the distribution of green vegetation helps in reducing the air surface temperature within the city. The regions with matured green vegetation tend to have lower air surface temperature distribution as compared to the regions with compact built-up features. However, the study has also identified that the increase of 5% matured green areas for the Kuala Lumpur cannot stop the city to be warmer as a whole. Future studies should quantify the response of green vegetation towards the urban climate for better understanding and specific mitigation measures can be constructed.

Northwestern and some western parts of the Kuala Lumpur city are warmer than the eastern parts. This may due to different urban morphology and settings between both parts. The western part of the city consists of compact residential area, whereas the eastern part of the city consists of the city centre and business areas which have been fragmented with vegetation portions. Future studies should quantify the difference in reactions towards climate by different urban morphology and settings.

Green vegetation is only one urban parameter that impacts the urban climate. Future studies should look into other urban parameters in order to analyse the behaviour of Kuala Lumpur climate in detailed. This study suggests that, future studies on mixed built-up and vegetation areas should be performed to quantitatively identify the built-up area and green area effects on urban climate.

Acknowledgements

The authors would like to express their gratitude to the Universiti Teknologi MARA (UiTM) and Ministry of Higher Education (MOHE) for funding this project under the Research Acculturation Grant Scheme (RAGS) 600-RMI/RAGS 5/3 (66/2015).

References

[1] Elsayed I 2012 *Middle-East Journal of Scientific Research. Mitigating of the Urban Heat Island of the City of Kuala Lumpur, Malaysia* **11** 1602

[2] Amanollahi J, Tzanis C, Ramli MF and Abdullah AM 2016 *Atmospheric Research. Urban Heat Evolution in a Tropical Area Utilizing Landsat Imagery* **167** 175

[3] Shahruddin A, Noorazuan MH, Takeuchi W and Noraziah A 2014 *Global Journal on Advances in Pure and Applied Science. The effects of Urban Heat Islands on HUman Comfort: A Case of Kalng Valley Malaysia.* **2** 1

[4] Buyadi SNA, Mohd WMNW and Misni A 2013 *AMER International Conference on Quality of Life. The impact of Landuse Changes on the Surface Temperature Distribution of Area Surrounding the National Botanic Garden, Shah Alam* (Langkawi Island: IOP Publishing) p. 516

[5] Ibrahim I, Samah AA and Fauzi R 2014 *Urban Planning and Local Governance. The Review for the Use of Urban Climatic Map in the Land Use Plan for Urban Planning* **3** 95

[6] Isa NA, Mohd WMNW, Salleh SA and Nasruddin NAM 2017 *Pertanika Journal of Social Sciences and Humanities. Urban Climatic Analysis Mapping of Kuala Lumpur City* **25** 291.

[7] Wong N-H, Jusuf SK, Katzscher L and Ng E 2015 *Urban Climatic Map Studies in Singapore. The Urban Climatic Map for Sustainable Urban Planning.* (United States of America: Routledge)

[8] Katzschner L and Burghardt R 2015 *Urban Climatic Map Studies in Vietnam. The Urban Climatic Map for Sustainable Urban Planning* (United States of America: Routledge)

[9] Morris KI 2016 *Computers, Environment and Urban Systems. Effect of Vegetation and Water Body on the Garden City Concept: An Evaluation Study Using a Newly Developed City, Putrajaya, Malaysia.* **58** 39

[10] Isa NA, Mohd WMNW and Salleh SA 2017 *International Conference Geomatic and Geospatial Technology. The Effects of Built-Up and Green Areas on the Land Surface Temperature of the Kuala Lumpur City* (Kuala Lumpur: ISPRS) p. 107.

[11] Buyadi SNA, Mohd WMNW and Misni A 2014 *8th International Symposium of the Digital Earth. Impact of Vegetation Growth on Urban Surface Temperature Distribution* (Sarawak, Malaysia)
[12] Ooi MCG, Chan A, Ashfold MJ, Morris KI, Oozeer MY and Salleh SA 2017 *Urban Climate. Numerical Study on Effect of Urban Heating on Local Climate During Calm Inter-monsoon Period in Greater Kuala Lumpur, Malaysia.* **20** 228
[13] Morris KI, Chan A, Salleh SA, Ooi MCG, Ozeer MY and Abakr YA 2016 *Urban Climate. Numerical Study on the Urbanization of Putrajaya and Its Interaction with the Local Climate, Over a Decade.* **16** 1
[14] Ren C, Lau KL, Yiu KP and Ng E 2013 *Cities. The Application of Urban Climatic Mapping to the Urban Planning of High-Density Cities: the Case of Kaohsiung, Taiwan.* **31** 1
[15] Eum J-H, Scherer D, Fehrenbach U, Koppel J and Woo J-H 2013 *Land Use Policy. Integrating Urban Climate Into Urban Master Plans Using Spatially Distributed Information - The Seoul Example* **34** 223
[16] Singh J, Yeo K, Hossieni R and Kalagnanam JR 2015 *14th EMS Annual Meeting and 10th European Conference on Applied Climatology (ECAC). Evaluation of WRF Model Seasonal Forecasts for Tropical Region of Singapore* (Prague, Czech Republic)

IOP Conf. Series: Earth and Environmental Science **117** (2018) 012013 doi:10.1088/1755-1315/117/1/012013

Satellite Image Analysis along the Kuala Selangor to Sabak Bernam Rural Tourism Routes

I Ibrahim[1], K Zakariya[2], N A Wahab[2]

[1]Department of Urban and Regional Planning, Kulliyyah of Architecture and Environmental Design, International Islamic University Malaysia, Gombak, Selangor, Malaysia
[2]Department of Landscape Architecture, Kulliyyah of Architecture and Environmental Design, International Islamic University Malaysia, Gombak, Selangor, Malaysia

*illyani_i@iium.edu.my

Abstract. This research focuses on the analysis of land cover map using satellite imagery along the rural routes. The aim of this research is to study the landscape features that can be seen by the tourists around the rural routes. The objectives of the study are twofold: (i) to analyse the land cover types along the rural routes and (ii) to create a tourist map along the rural routes. The method adopted was to use Supervised Classification by creating multiple polygons to ensure that each information is sufficient to create appropriate spectral signatures. The finding shows that 80% of the landscape features along the Point of Interest (POI) are paddy field. According to the analysis using the indicators criteria for choosing the rural routes, this research shows that this area has the potential to be part of a tourism area because it has many historical and cultural elements that can be exposed to tourists. Future research will be a factor analysis on the significance of the criteria to rural tourism attraction.

1. Introduction

Tourism is one of the largest and most rapidly expanding industries in Malaysia. Tourism has considered an effective catalyst of rural socio-economic development and revival (Sharpley, 2002). Tourism can be divided into several types of tourism: rural tourism, cultural tourism, tourist theme routes, religious tourism, and eco-tourism. Tourism corridor is a tourism mechanism that encourage a movement from a core of the township to the periphery of the boundary. The rural tourism activity will encourage regional economic development by giving tourists an experience of homestay, attractive landscape, and heritage. Thus, maintaining and upgrading the landscape scenery and attractions of the rural areas is necessary in order to enrich the tourist experience of the area.

The landscape character of a rural area plays an important role in attracting tourists. Preserving nature could contribute to the rural tourism at rural areas (Amir, Ghapar, Jamal, & Ahmad, 2015). The attractions should have a setting that can attract tourists to the area while offering them an experience that is unique and memorable. In rural tourism, the setting of the rural area (such as traditional villages, agriculture, local culture, and the natural scenery) becomes the main identity that set the rural area apart from the urban landscape. As mentioned by Rosley, Lamit, and Rahman (2013), a liveable village could be produced by an aesthetic landscape that could attract people to visit the village. The rural landscape in Malaysia has high potential because it is prominent with the natural landscape of forest and coastal areas with the agriculture such as paddy fields, coconut groves, rubber plantations and palm oil (Mizukami et al., 2012).

In addition, the scenic quality, sense of place, and tranquility of the rural area could be enhanced with natural landscape features such as landform, land cover, and cultural elements that could furnish as the landscape character (Jaal & Abdullah, 2012). The landscape character in Malaysia needs to be preserved because it carries (1) recreational, ecological, and natural values, (2) historical heritage, and (3) socio-cultural experience to the local and tourists (Clark & Hashim, 2010). The values that may attract the tourists to visit the rural area therefore need to be conserved and preserved by the local community with the support from the government and stakeholders.

Most of the current generations have never experienced the natural landscape character at their own country, especially those who grew up in the urban and suburban areas. Preserving the rural landscape may attract the local tourists especially the young generations who want to experience and learn about their culture and traditions. The landscape character could provide the traditional foods and fruits (Mizukami et al., 2012). The local plantation, for example, coconut trees, could produce variety of products from their leaves, trunk, fruits, and others, which could can be the source of income for the local people. The diminshing rural traditions have made these resources more valuable today than ever before.

The rural tourism could promote the local industry and get the community involved at the same time hence promote the country to the world (Amir et al., 2015). This advantage is also an opportunity for the local community to promote their local products. Visitors can experience the rural lifestyle and culture by looking at the local communities' daily lifestyle, such as agriculture activities, farming, and fishing as performed in the traditional ways. As mentioned by (Tudor, 2014), a landscape character can provide agriculture, fishery, cottage industry, and informal activities which can garner income to the local people and the country's economy. Table 1 shows the indicators criteria for choosing the rural routes according to the Rural Tourism Master Plan.

Table 1. Indicators Criteria for Choosing Rural Routes (Rural Tourism Master Plan, 2001)

Criteria	Indicator(s)
Quality of scenery	Mountains, lakes, rivers, coastline, panoramic views
Strength of attractions	Number of attractions x quality of each
Potential for recreational activity	Trekking trails, fishing locations, amenities, activities
Potential for product clustering	Combined number of attractions, amenities, activities
Range of accommodation options	Number of different types of accommodation category
Range of tourism amenities	Forest recreation area, agricultural parks
Distinctive culture and heritage	Number of museum, cultural events, historic links
General services and infrastructure	Dinning, shopping facilities, TICs, road quality, water
Quality of environment	Absence of litter, water pollution, attractive streetscape
Support from official agencies	Previous tourism investment, efforts to organize tourism
Active community organisation	Local development projects undertaken, sustained effort
Tourism marketing track-record	Promotional brochures, booking systems, publicity effort
Accessibility to main tourism centres	Within 3 hours travelling time
Proximity to international entry points	Within half a day's travelling time

In addition, the history and heritage of an area could be reflected by the image of traditional structures such as traditional houses, shops, bridges, religious institutions, and others that produce the sense of place at the rural area (Diti, Torreggiani, & Tassinari, 2015). The local and traditional built environment could help to enlighten the rural landscape as the rural tourism. Therefore, it could attract the tourists to visit the rural areas and experience the new view and sightseeing at the villages. Recent studies had done on the multi-decision analysis using remote sensing and GIS to designed the tourist route in each particular study area (Juodkiene, 2014; Zhang & Yang, 2014).

Thus, the objectives of the study are (i) to analyse the land cover types along the rural routes and (ii) to create a tourist map along the rural routes. Integrated method are used in order to fulfil the objectives, by using remote sensing technology to create a land use map and analysing the existing features available in the study area with the indicators explained in the literature review. This paper is organized as follows. Section 2 presents the related application of GIS in tourism planning. In section 3, the study area of this research is elaborated. Further section will elaborate the method used and the analysis of the result. Finally, our work for this paper is summarised in the last section.

2. Application of GIS in Tourism Planning

GIS mapping is one of the common tools used in built environment and tourism studies to document and analyse places. Reviews have attested that GIS mapping is an extensive technological tool used in tourism studies. A geospatial site selection method was used to select the sites of rural tourism by identifying criteria analysis (Calado et al., 2014). Another research adopted the GIS by establishing a tourist database along the thematic routes in Hajdu-Bihar country, Romania (Varga & Turi, 2014). It is important to distinguish the main rural tourism development resources in a particular area, for example, whether they are suitable for leisure (the picturesque landscape, forests, waters and other natural resources), and cultural heritage and recreational needs (towns having cultural heritage, villages, areas with valuable individual objects and living culture events) (Juodkiene, 2014). A different use of GIS (US AID, 2006) is to create a detailed GIS map that features accommodations, camping sites, biking trails, and other tourist attractions. One study used the analytic hierarchy process (AHP) analysis to catalogue rural buildings for restoration or reuse of the buildings by public organisation or individual (Cano, Garzón, & PJ, 2013). In one study on Cinque Terre, Italy, the GIS and CAD functions of (geo-referred) digital cartography data treatment were combined with attributed data to update and organise built heritage and the cultural landscape of the location (E.U. Program Culture, 2000) and a study using remote sensing and GIS had done in West Bengal to assess the spatial distribution of transport network accommodation facility that can improve the eco tourisms development in the study area (Bhaya & Chakrabarty, 2016).

3. Methodology

The study area is located in the boundary of Selangor along Jalan Kuala Selangor. Site visit was conducted along the Federal Route 5 from Kuala Selangor to Sabak Bernam for a total distance of about 60 km. This route is selected as the KS – SB route meets all the criteria for rural tourism route; villages, traditional houses, agriculture, natural resources, history, culture, local food and good accessibility. This route is located on the west coast of the state of Selangor, which connects to the state of Perak. Along the route, significant landscape features and points of interest were mapped using GPS tool and photographs. The GPS tool used was GPSmap 60CSx that important to ascertain the position of tourist attractions which were then transferred to a GIS desktop and then to the map. For each transect, the identified points will be categorised into four areas: landscape character, settlement, built environment and heritage.

As for the remote sensing image, it was downloaded from the USGS website (U.S. Geological Survey, 2017) dated March 29, 2016. This image was used due to its good quality of image, referring to its minimum cloud cover (3.61%) during the image was captured. This study is conducted using ArcGIS software. The supervised classification used was Maximum Likelihood Classification Algorithm for the land cover map generation. The spectral signatures was selected as training areas. Hundred numbers of polygon were selected as the training areas that can help to differentiate different signatures. Then, the computer algorithm used the selected spectral signatures to classify the whole image. Further process was conducted are to generate signature file and classification. Fig. 1 shows the process for supervised classification. Using the image, the indicators of criteria for choosing rural routes was used in order to look into the land cover as stated in Table 1 was used to verify the existing criteria for rural tourism route.

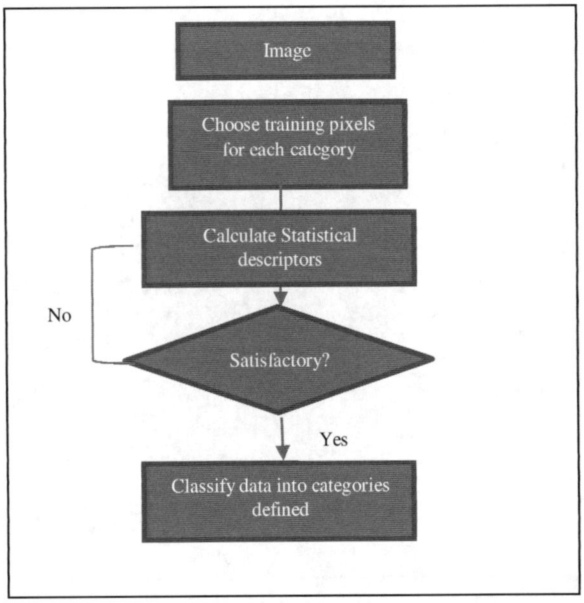

Figure 1. Processing image using supervised classification

4. Results

A land cover map of Landsat image that was captured on July 20, 2016, was analysed. Accuracy assessment was carried out and an overall accuracy of 93.33% was obtained. The overall Kappa coefficient is 0.9204. This number meets the 80% accuracy standard of the National Park Service (NPS) in the United States (Summer & Nordman, 2008), that showed that a good accuracy of classification was achieved for the land cover classification. The land cover types are divided into several categories: green area, forest area, bare land, buildings, roads, water bodies, and paddy field.

This map (Fig. 2) shows the land cover distribution along the Kuala Selangor route. This route is just located around 2 to 4 kilometres from the coastal area. The study area is covered with a linear route and is categorised into five segments of areas: Kuala Selangor, Tanjung Karang, Sekinchan, Sungai Besar, and Sabak Bernam.

Our visit on May 1, 2017, shows there are many attractions for tourism visits, mainly on the landscape feature, the rural houses, agricultural areas, local food, natural attractions, and historical landmarks. Most of the attractions are already highlighted by the Tourism Malaysia, such as the historical attractions on Bukit Malawati, Taman Alam Kuala Selangor, and the fireflies in Kampung Kuantan. All these major attractions are located in Kuala Selangor. Other attractions along the route are the fisherman village by the river, the coastal area in Tanjung Karang, the paddy fields and Malay traditional homestays in Sekinchan, the local food, the beach in Sungai Besar, and the rural Chinese in Sabak Bernam. As observed, most of these attractions are shaped by the geographical setting and landscape features of the area.

In the rural route of Kuala Selangor, the most obvious area is the Selangor River which has the potential to be a tourist attraction. Along the route, the river is covered with mangroves that are important to the sustainability of the area. Obviously the rural pattern is covered with green and forested areas. Many places for eat are also located close to the road.

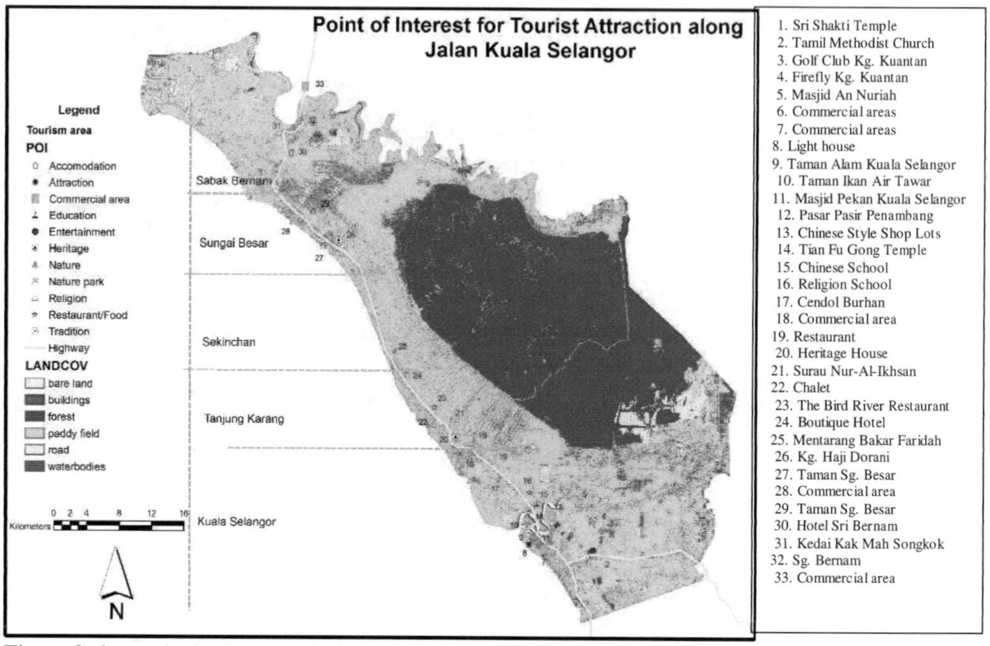

Figure 2 shows the land cover distribution in the study area.

Majority part of the areas (80%) are paddy fields, therefore information on the variety of paddy field season is an important element that needs to be informed to the tourists. For instance, according to the locals in the area, the paddy fields will give a green panoramic view during the month of October onwards after the new plots have been planted. As shown in the map, the nice view of green paddy fields are exactly on the right side of the main road to Sekinchan, if tourists were to travel north from Kuala Selangor. The views towards the paddy fields are mostly visible from the roadside. However, at spots where houses, institutions, and shops are built adjacent to the road, tourists would have to drive into the inner roads in order to access the larger view of the paddy fields. It should be noted that the rice transplanting season is in March and September every year. In addition, the harvesting season is in June and December every year. For those tourists who are interested to see a greenery view of paddy field, they can visit during the harvest season. Along the route, the view from the roadside is mostly towards the oil palm plantations which are located intermittently among the houses, towns, and other small-scale fruit orchards.

Using the criteria for selecting this rural routes as the discussed in the literature review (Table 1), below is the analysis for suitability of this area to be selected as a rural route tourism. First, the quality of scenery for the Kuala Selangor to Sabak Bernam route can be considered unique because in Malaysia, views of paddy fields are more synonymous to the northern region. In Selangor, this route is the only route that provides a scenic view of the paddy fields and a rural setting. Along the route at the right hand side there will be a paddy field scenery (Fig. 2). Some part of the area will need us to cross over the bridge, therefore we can see a nice view of rivers (Selangor River and Tengi River).

Secondly, the strength of attraction of this area is good because along the way from Kuala Selangor to Sabak Bernam, there are small stalls selling local food such as *mentarang* (a type of shellfish), *kerepek* (local chips made from banana and tapioca), and fresh seasonal fruits (mango, banana, and corns). Tourists who visit this area have the opportunity to see and experience not only the rural landscape but also the well-conserved natural attractions, which are the mangrove and the rivers. On top of that, the tourists can also enrich their experience by visiting Malawati and learning about its historical significance to Selangor, as well as walking to the highest point in the area to the light house at the top of the hill. The light house and Bukit Malawati are important in the state history they among the major shipping routes in East Asia. From the hill, we can see the stunning panorama of the rural surrounding area with a short trekking trail, thus a potential for recreational activities. Less than 2 kilometres at the left hand side from Kuala Selangor to Sabak Bernam, there is a coastal area that can serve as a spot for activities such as swimming and fishing. Lembaga Kemajuan Ikan Malaysia (LKIM) is also just beside the river. Fresh fish and seafood is part of the local resource of the area which has become an attraction to visitors from Kuala Lumpur. The geographical and environmental setting of this area shows a potential for product clustering due to several attractions in each particular area. For example, at Kuala Selangor, there are religious institutions that have become the local landmark (masjid and temples), and there are also river activities where visitors can observe fireflies at night. The local fish market and a historical golf club can serve as supporting attractions and as an extension from the route. The golf club, which aged more than 100 years, sports large trees of the same age. Many old and interesting pictures of the club owner can be seen here. In terms of accommodation, several lodging can be found along the way to Sabak Bernam. Hotels and homestays are located just near to the road thus are easily accessible. Many tourism amenities can be found here, such as parks and wetland, which can also provide an educational experience to the tourists. Many visitors or tourists can also pass by and enjoy stopping in Kuala Selangor. They can buy relatively cheap fresh fish, eat in the restaurant to enjoy the seafood, buy fruits, take a look at the fireflies, and get to feed eagles along their route.

In terms of general services and infrastructure, many restaurants (see Fig 2) can be seen along the route. In fact many restaurants can be seen in the commercial areas of each segment. In terms of shopping facilities, many small lots of traditional chips and fruits can be seen. For the section of the quality of environment, there are no obvious littering and water pollution, which implies that this particular rural area is away from industrial pollution. Last year (2006), Jelajah FT Roadpedia was launched by Jabatan Kerja Raya to encourage the people in Malaysia to experience using the federal road, including the Kuala Selangor to Sabak Bernam route. This promotional effort by the government showed the high potentials of the federal routes or the rural routes to be integrated and planned as a wider network of tourism routes in Malaysia. Essentially, attractions in the rural areas do not work individually, but as a system that integrates the landscape character, physical setting, culture and history, and activities that people can do while visiting. Tourists seek for a unique experience, therefore these aspects that make the rural areas attractive need to be sustained and managed.

5. Conclusions

This research is the preliminary findings on the potential of rural tourism route along Jalan Kuala Selangor, from Kuala Selangor to Sabak Bernam. The study has shown that this area has the potential to be part of a tourism area because it has many historical and cultural elements that can be exposed to tourists. The area can also be further planned for activities to have an integrated programmes on promotion for tourist attraction.

Acknowledgements

This study is funded by the Fundamental Research Grant Scheme (FRGS) FRGS/1/2016/WAB12/UIAM/02/1 from International Islamic University Malaysia (IIUM).

References

[1] Amir A F, Ghapar AA, Jamal SA and Ahmad, KN 2015 *Procedia - Social and Behavioral Sciences. Sustainable tourism development: A study on community resilience for rural tourism in Malaysia* **168** 116

[2] Bhaya S and Chakrabarty, A 2016 *Journal of Remote Sensing & GIS. A GIS Based Ecotourism Infrastructure Planning for Promotion of Tourism in Jungle Mahal of West Bengal* **5**

[3] Calado L, Rodrigues A, Silveira P and Dentinho T 2011 *European Journal of Tourism, Hospitality and Recreation. Rural tourism associated with agriculture as an economic alternative* **2** 155
[4] Cano M, Garzón E and PJ 2013 *Tourism and Hospitality. Tourism & Hospitality Preservation and Conservation of Rural Buildings as a Subject of Cultural Tourism : A Review Concerning the Application of New Technologies and Methodologies* **2**
[5] Clark I and Hashim H 2010 Exploring heritage value of a malay landscape **7** 5022
[6] Diti I, Torreggiani D and Tassinari P 2015 *Journal of Agricultural Engineering. Rural landscape and cultural routes: a multicriteria spatial classification method tested on an Italian case study* **46**
[7] E.U. Program Culture 2000 *From Smart History towards common European Heritage by a preservation model of CINQUE TERRE NATIONAL PARK – Italy.*
[8] Jaal Z and Abdullah J 2012 *Procedia - Social and Behavioral Science. User's Preferences of Highway Landscapes in Malaysia: A Review and Analysis of the Literature.*, **36** 265
[9] Juodkiene V 2014 *De Gruyter. GIS in Tourism Development using Spatial Modelling* **10** 98
[10] Mizukami MDG, Ribeiro LRDC, Sandseter EBH, Ali SM, Rostam K, Awang AH, Ahmad C 2012 . *Procedia - Social and Behavioral Sciences. Towards a Compatible Landscape in Malaysia: An Idea, Challenge and Imperatives* **15** 70
]11] Moradi N, Khoshnazar M, Aryanpour A and Namivandi MS *2014 Site Selection of Tourism Village in Mahabad Dam Shore Using AHP , GIS and SWOT Techniques*, **3** 9
[12] Rosley MSF, Lamit H and Rahman SRA 2013 *Procedia - Social and Behavioral Sciences. Perceiving the Aesthetic Value of the Rural Landscape Through Valid Indicators* **85** 318
[13] Sharpley R 2002 *Tourism Management. Rural tourism and the challenge of tourism diversification: The case of Cyprus* **23** 233
[14] Summer H and Nordman C 2008 *Accuracy Assessment: Abraham Lincoln Birthplace National Historic Site. Nature Serve* (Durham, North Carolina)
[15] Tudor C 2014 An Approach to Landscape Character Assessment, (October), 56
[16] U.S. Geological Survey. (2017). USGS Earth Explorer. Retrieved from https://earthexplorer.usgs.gov/
[17] US AID 2006 *Promoting Rural Tourism.*
[18] Varga OG and Turi Z 2014 *Agricultural Informatics 2013 International Conference. Application of GIS to Present Rural Tourism Values of Bihor-Hajdu- Bihar Euroregion*
[19] Zhang A and Yang Q 2014 Rural tourism spatial distribution based on multi-criteria decision analysis and GIS, (August)

IOP Conf. Series: Earth and Environmental Science **117** (2018) 012014 doi:10.1088/1755-1315/117/1/012014

Evaluating the interior thermal performance of mosques in the tropical environment

N I Nordin [1] and A Misni [2]

[1]Faculty of Architecture, Planning and Surveying, Universiti Teknologi MARA, Shah Alam, Selangor Malaysia.
[2]Faculty of Architecture, Planning and Surveying, Universiti Teknologi MARA, Puncak Alam, Selangor Malaysia

*ziksya257@gmail.com

Abstract. This study introduces the methodology applied in conducting data collection and data analysis. Data collection is the process of gathering and measuring information on targeted variables in an established systematic method. Qualitative and quantitative methods are combined in collecting data from government departments, site experiments and observation. Furthermore, analysing the indoor thermal performance data in the heritage and new mosques were used thermal monitoring tests, while validation will be made by meteorology data. Origin 8 version of the software is used to analyse all the data. Comparison techniques were applied to analyse several factors that influence the indoor thermal performance of mosques, namely building envelope include floor area, opening, and material used. Building orientation, location, surrounding vegetation and water elements are also recorded as supported building primary data. The comparison of primary data using these variables for four mosques include heritage and new buildings were revealed.

1. Introduction

Religious building such as mosque symbolizes the surrounding community's religious affiliation and it's cultural during it is built. The architecture and design of mosque emphasize three primary potential benefit includes social, economic and environmental. The mosque is an important religious institute and multi-functional building in Islamic societies to accommodate their daily education, spiritual and social activities. Better architecture and design of mosque can preserve the environment ecosystem, and improve the quality of ambient air and protection of natural resources as well as minimize the adverse impact with the surrounding environment. Lower operation and maintenance costs of the mosque will help on economic aspects. The building could improve the occupant's productivity and advance of life-cycle achievement. Moreover, it also can minimize the energy consumption of the mosque, while improving the quality of thermal comfort level and health of the users are in the category of potential social benefits. Heritage mosque building has a unique design and architecture and high value of art and technology. This heritage building is a precious national asset and should be preserved [14]. Heritage building is a historical retreat that plays an important role in society to convey the cultural and artistic heritage that exists in addition to modern development. In this study context, two types of building development are chosen; the old and new era mosque. In this context of study, the aged mosque is labelled as heritage category of building, while the ordinary mosques are new mosque. Both categories of the building have its context in architecture, construction and its surrounding environment. It's internal building architecture and exterior environmental context influences the interior thermal performances. Tropical climate with hot and humid atmosphere are the most demanding parts and as a primary parameter aspects to enhance through building design. This study analyses and evaluates the factors that can influence the indoor thermal performance of mosque in a tropical environment.

2. Literature Review

The early architecture era of a mosque in Malaysia is basic and straightforward in line with the simple and practical teachings of Islam. During the 15th century, the mosque architectural style and its construction materials were similar with the local traditional Malay houses, where the majority of the Islamic religious practitioners are Malays [14]. Since 19th century in the British colonial period mosque architecture in Malaya has changed, and influenced by British colonial style. Moreover, in early 20th century, the mosque architecture was further extended by the introduction of modernist style. Each heritage mosque has its own uniqueness and this clearly seen in an architectural point of view, settings and layout the cultural landscape and origin of materials. Identity is variations images of places, to distinguish each of it from another [7]. The component of heritage creates an identity in a place by the old existing buildings. The technical aspects of the importance of technology innovation and creativity are an important evidence of creation [11,13]. Mostly heritage mosques may tell their own history and certain old mosques still used as a royal mosque which has continued till today. Some of the historic building and monuments are still used as palaces and a place of worship or religious buildings; and some are preserved as a monumental heritage in such of statues and ruins [6].

Thermal performance of a building can be referred to as the process of modelling the heat transfer between the buildings and its surroundings. It calculates and estimates the indoor temperature variation, heating, cooling load and the duration of uncomfortable periods. Designing a building that responds to the natural environment can provide the desired level of comfort in the prevailing environment [3]. Health factors in sustainable building design are commonly associated with the comfort level. In architecture, comfort level can be achieved by application of passive design (e.g., natural lighting, natural ventilation and solar gain) and active design elements (e.g., renewable and energy saving gadgets). In contrast to active design, the passive design is sustainable to the environmental, social and cultural perspectives [1,5], which offers healthy condition, and the design does not depend on the mechanical system [10]. There are several criteria that support the sustainable design of heritage and modern mosque such as building orientation, natural ventilation, opening, building envelope, water element, and vegetation. Besides, building orientation is a significant design consideration, mainly with regard to solar radiation and wind flow through building envelope such as windows, building openings as well as external opaque walls and roofs.

The building envelope is the physical separator between the interior and exterior of a building [4]. Components of the envelope are typically: floors, walls, roofs, doors and fenestrations such as opening in the structure: windows, skylights, and clerestories. Envelope design on thermal performance is one of the most important components with respect to total heat gain of whole building and overall heat transfer coefficient which determines heat gain through the building envelope [2]. There are two types of parameters that strongly affect the thermal performance of a building, which is (i) climatic condition (i.e. solar radiation, cloud layer, ambient temperature, wind speed and relative humidity); and (ii) the design features (i.e. geometrical dimensions of building envelope, orientation, opening, shading devices and building material properties [8,9]. Local climate greatly affects the indoor thermal environment in buildings. In tropical climates, buildings are overheated during the day due to solar heat gain through the building envelope and solar penetration through windows [12].

3. Methodology

Three mosques were selected to be used as the case study. These mosques were located in Selangor and the locations are quite near to each other. The primary data on the actual sites were collected and recorded, while the secondary data will be used as supporting documents. The combination of both data sets can validate and ensure the accuracy of the results. The data are as follows:

3.1. Weather data.

Weather data will be used to validate the recorded data. This are obtained from the nearest Meteorology Department. The data includes temperature, relative humidity, wind speed and cloud covers in a same day of collected primary data. The nearby Meteorology Department involved are Subang Airport and Mardi Klang Stations.

3.2. Building construction plan

The Public Works Department provides building plan for the new mosque, and National Heritage Department, Kuala Lumpur provides a documented measured drawings of heritage mosques. The working and detail drawings of each mosque are used to understand the physical character of building in the beginning stage of the study. The study involved basic information on architecture, design, construction, material and any related factors or aspects of heritage and new mosques. The focus of studies are on the floor plan and all angle of building envelopes. Roof, walls and openings and material are the most critical parts to understand and study details. The floor plan of the prayer hall is studied to support the weather data recorded and on site and building observations.

3.3. Experimental survey

The experimental survey conducted in interior and exterior of the buildings are recorded a quantitative data that involve with the numbers, amounts, or quantities, of all variables. The experimental data are measured and recorded by equipment as shown in Figure 1. The recorded and measured data are the temperature, relative humidity and wind velocity. The main users use the space in the mosques as early as 5.00 am and their activities will finish at 10.00pm. As the state of art of the case study, hence the data collection exercise was conducted with a very careful consideration, because this is a noble and holy place for Muslim to do their prayers five times a day. The weather data are collected as early as 08.00 am till 19.00 pm in 30 minutes time interval. Four stations and checkpoints have been set for data recording and the focus area are in the exterior and interior of the main prayer halls of the mosques.

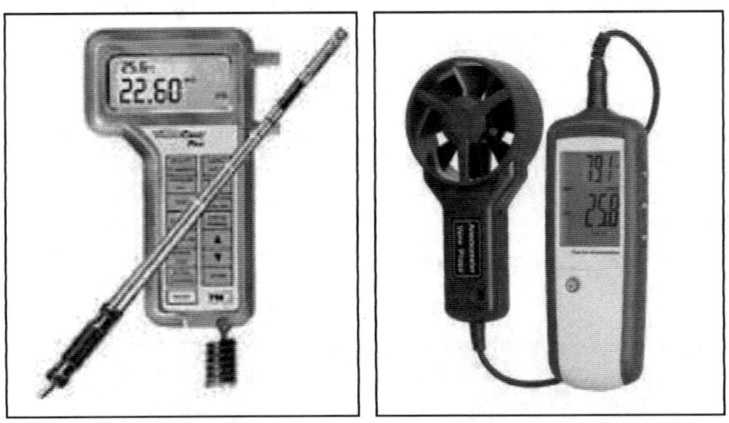

Figure 1. Tsi velocicals (left), and anemometer (right)

3.4. Observation

The in-situ observation is very important to collect the primary data in the real situation on site. The observation takes place at the physical building and their surroundings. The data need to collect by observation are building construction and building envelope elements, The other data is building orientation, material use and surrounding site elements and neighbourhood or site context of the study area. The comparison data from two categories of mosques are more specific and easy to analyse as well as to finalize the data findings. The primary data collection on site are illustrated and measured in detail along with floor plan, section and elevation drawings, supported by photographs and documented as a result of the study.

3.5. Analysis
All secondary and primary data were gathered using quantitative and qualitative analysis. The quantitative data from the experimental method are tested and analysed using Origin 8 software. The primary idea of the data analysis is to examine the differences of the interior thermal performance of each mosque. Origin 8 software can help to produce results in form of interactive scientific tabling and graphing. The results are formed in technical numbers, validated with the existing data and will be compared accurately for each mosque. While the data collected from observation/qualitative are set up and inventoried in tables to clarify the building criteria and it surrounding which influenced by the size, orientation, design and architecture of the building envelope of the mosques, and similarly, can influence its interior thermal performance. Both quantitative and qualitative results are analysed and supported by specific photographs and drawings.

4. Results and discussion
The selected mosques located in Klang, Port Klang and Shah Alam respectively. One of the mosque is 85 years old and this building is gazetted as National Heritage. The two new mosques age ranges are between 10 and 20 years old. The architecture of all mosques adopted few aspects of sustainable design in tropical climates such as a general building direction toward north and south, applied passive lighting and ventilation system by designing suitable type of openings in it building envelope; walls and roof. These green design aspects potentially can affect the interior thermal quality of the buildings. The architecture style and design of the building envelopes and interior spaces of each mosque is varied but the size and volume of interior of worship halls are relatively the same. The worship hall is a primary space of the mosque was used to evaluate its indoor thermal performance.

4.1 Mosque buildings and spaces
The Royal Mosque of Sultan Suleiman was built on 1932, categorised as historical building. This mosque was designed by Leofric Kesteven, a British architect. On Friday, the Sultan of Selangor always visits this mosque for Friday prayer. The management office and the mosque areas are administered by Islamic Religious Department of Selangor (JAIS), while for daily social activities was conducted by mosque committee members. The hall of this mosque can hold more than 1000 congregation at a time of prayer. The Al-Munawarah mosque is a two-storey mosque that was built on 1998. This building signifies the strength and dignity of Islam as official religion to the local residents of Section 27, Shah Alam. The prayer hall is huge, it can accommodate 3900 people at one prayer time. Total site area is almost 6-acre. Other facilities provided; conference rooms, committee members' rooms, etc. The second new mosque, Kg. Raja Uda mosque was built on 2007, located at 3.8 acres site. The size of the worship hall is medium, can hold 3500 people at one prayer time. Other facilities provided around and near this mosque are a public multipurpose hall, meeting rooms, and rooms for committee members.

Figure 2. Three mosques as case study models; Royal Mosque of Sultan Suleiman (above left), Al-Munawarah (above right), and Kampung Raja Uda (below)

4.2. Mosque construction and materials

The construction styles and materials used in all mosques were marginally diverse. The heritage mosque was built during British administration in Malaya in 1932 was influenced by Western Art Deco and Neoclassical cathedral style architecture. The floor plan of the mosque is designed with a simple geometric Islamic calligraphy in a proportion symmetry shape. The roof of the mosque is decorated by a semicircle dome fixed at the top of the worship hall to provide a high volume space, the design of the mosque is reinforced with the pillars, while small domes are added as classical symbols for the mosque buildings. The new mosque's design practised current style of architecture in Malaysia, modern contemporary style. The royal Mosque of Sultan Sulaiman and Al-Munawarah Mosques built in two-storey, while Kg. Raja Uda is a single storey mosque. The primary structures and foundations, and walls material and construction of all mosques were similar, used reinforced concrete and bricks, but different applied roof construction and materials. The heritage mosque covers it roof by concrete material and dome while new mosques used metal decking as roof covers.

4.3. Mosque orientation and passive ventilation

In a tropical climate, building position on site is the essential criteria for reaching a comfort level in interior spaces. The location of the mosque is referred to the location of the sun path and the direction of the wind. In this case study, the orientation of the three mosques was similar to each other in the direction of the Qibla and for having a square plan for prayer halls. The Royal Mosque of Sultan Suleiman neighbourhood contexts are the main roads, open areas, and Alam Shah Palace. Adjacent to the east side facing the open areas (a garden and a parking lot) recorded moderate increase from the morning and achieved the highest wind speed of approximately 7.1m/s at 17.00 hours and the north side facing the new building. Al Munawarah mosque is closed to a housing area at Section 27. The general wind speed adjacent to the south of an open area reached the highest level at the average of 7.7–10.2m/s in the morning and gradually fell in fluctuating patterns from noon until evening. On average, wind speed around the Kg Raja Uda Mosque was approximately 0-4.2m/s for south direction all day, while on the east side, the mosque recorded the wind pattern reaching the lowest speed of approximately 1m/s at 19.30 hours.

4.4. Mosque envelope

The heritage mosque and new mosques of Kg. Raja Uda has almost similar areas of building envelope; walls areas are around 915m^2, roofs cover areas are about 2135m^2 and 750m^2 base floor areas. During mid-day, the heritage mosque received high temperatures adjacent to the worship walls. The high temperature contributed by dome roof cover located above the worship hall. This is because it was decorated with stained glass and decorative grilles which allow direct solar radiation. Glass windows on the wall allow direct sunlight penetration and potentially raise the interior temperature. Therefore, the worship hall in the heritage mosque increases its temperature during mid-day in the range of 1.5–1.8°C. This temperature is higher compared to the interior space of new mosques. Solar reflectance from light colour walls or roofs for all mosques contributed to reducing gain entering the worship hall in the mosque. All wall used light-coloured paint and different colours for roof-covering. Interior ceiling heights are between 9 and 10 metres, this high volume of interior spaces can help to maintain a low temperature.

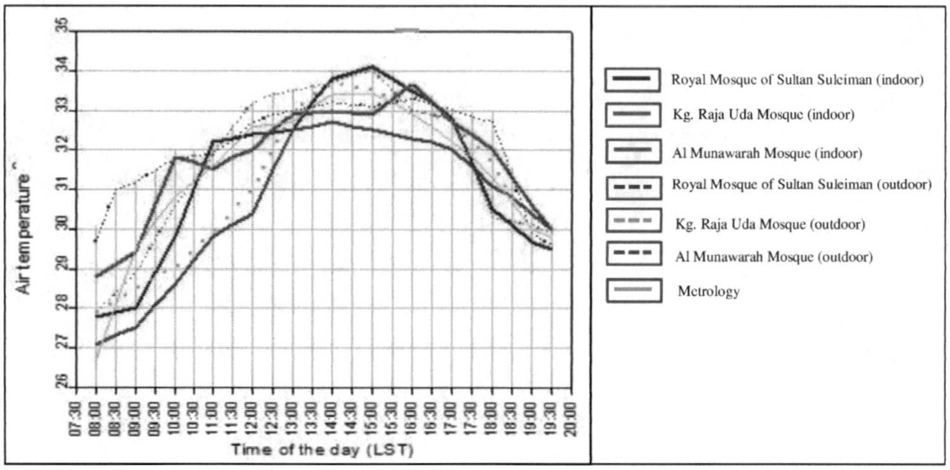

Figure 3. Temperature adjacent to the South wall of the three mosques.

Figure 4. Wind speed adjacent to the South wall of the three mosques.

4.5. Glass and another opening

Each mosque practice passive lighting systems during daytime, and natural ventilation all the time. Opening systems located in building envelopes such as doors, windows, art ornaments, top hung windows, and roof and floor openings are widely applied to encourage cross ventilation and stack effect in interior spaces of the mosque. Cool air from the exterior will circulate and replacing the hot temperature inside the mosque especially in the prayer hall. Each mosque also has different total areas of openings, include glasses, and doors. The first of new mosque, Al Munawarah has the most number of glass surface areas in the building envelope. Oppositely, Kg. Raja Uda as a second new mosque category has the least total of glass surface areas. While heritage mosque has medium areas of glass surfaces which primarily located in all wall sides at the first level of the mosque. The other openings such as doors and art ornaments are in the all walls angle located at the lower level except the west.

Ornament openings with Islamic calligraphy designs also can encourage passive ventilation in heritage mosque. The surrounding areas such as open space or garden may allow wind flow and circulate around building envelope. Balcony in upper level can act as shading device to cover the glass window surfaces from the penetrating of solar radiation during mid-day.

5. Conclusion

The method of the study briefly clarified the approach taken in primary data collection and analysing the interior thermal performance of two categories of mosque buildings; heritage and new mosques. The thermal data were recorded, tested and listed using technical equipment. Both categories of mosques have similar technology in construction and building orientation towards in the direction of the Qibla. However, both categories of mosques have different ages, building materials, total glass areas, and numbers for openings (ornaments). Two stages of collecting primary data; observation and experimental were gathered and analysed accurately. Finally, this study summarised that the heritage mosque has practised and applied the green aspects in design but the surrounding climate of hot and humid together with urban heat island effects produces high interior temperature as much as 1.8°C. State Islamic Department should collaborate with National Heritage Department for the better green design practice in order to achieve good thermal comfort quality in interior spaces of heritage mosque.

References

[1] Abu-Ghazzeh TM 1999 *Habitat International. Vernacular architecture education in the Islamic society of Saudi Arabia: Towards the development of an authentic contemporary built environment* **21** 229
[2] Al-Tamimi NA and Syed Fadzil SF 2010 Journal of Science & Technology. *Experimental and Simulation Study for Thermal Performance Analysis in Residential Buildings in Hot-Humid Climate (Comparative Study)* **15** 17
[3] Baker NV 1987 *Passive and low energy building design for tropical island climates. Pall Mall, London: The Commonwealth Secretariat.*
[4] Cleveland CJ, and Morris CG 2009 *Building envelopergy. Expanded Edition.* (Burlington: Elsevier)
[5] Eben Saleh MA 1999 *Building and Environment. Reviving traditional design in modern Saudi Arabia for social cohesion and crime prevention purposes* **34** 649
[6] Harun SN 2005 *Amalan Kerja Pemuliharaan Bangunan Bersejarah Di Malaysia*. Ph.D Thesis. Univesiti Sains Malaysia.
[7] Idid SZA 1996 *Pemeliharaan Warisan Rupa Bandar: Panduan mengenali Warisan Rupa Bandar berasaskan Inventori Bangunan Warisan Malaysia. Kuala Lumpur: Badan Warisan Malaysia.*
[8] Misni A 2013 *Procedia-Social and Behavioral Sciences. Modifying the outdoor temperature around single-family residences: The influence of landscaping* **105** 664
[9] Nordin NI and Misni A 2017 *Environment-Behaviour Proceedings Journal. A Comparative Study on the Indoor Thermal Performance of New and Old Mosques* **2** 23
[10] Moujalled B, Cantin R and Guarracino G 2008 *Energy and Buildings. Comparison of thermal comfort algorithms in naturally ventilated office building* **40** 2215
[11] Orbasli A 2008 *Architectural conservation principle and practice* (USA: Blackwell Science)
[12] Rajapaksha HN, Okumiy M 2003 *Renewable Energy. A ventilated courtyard as a passive cooling strategy in the warm humid tropics* **28** 1755
[13] Young G 1991 *Australia Planner. Authenticity in Cultural Conservation* **29** 3
[14] Baharudin NA 2016 *International Journal of Built Environment and Sustainability. Architectural Style of Da'wah Mosque in Malaysia: from Vernacular to Modern Structure* **3** 70

IOP Conf. Series: Earth and Environmental Science **117** (2018) 012015 doi:10.1088/1755-1315/117/1/012015

Comparison of WRF local and nonlocal boundary layer Physics in Greater Kuala Lumpur, Malaysia

M C G Ooi[1,2*], A Chan[1], S Kumarenthiran[3], K I Morris[4,5], M Y Oozeer[1], M A Islam[1] and S A Salleh[6]

[1] Department of Civil Engineering, University of Nottingham Malaysia Campus, Malaysia
[2] Department of Atmospheric Science, National Central University, Chung-Li, Taiwan
[3] Climate & Weather Model Development Section, Malaysian Meteorological Department, Malaysia
[4] Center of Excellence for Sustainable Innovation and Research Initiative (CESIRI), Port Harcourt, Rivers State, Nigeria
[5] Department of Mechanical Engineering, Ashesi University College, Berekuso, Ghana
[6] Faculty of Architecture, Planning and Surveying, Universiti Teknologi MARA, Malaysia

*chelgee.ooi@gmail.com/kebx2oce@nottingham.edu.my

Abstract. The urban boundary layer (UBL) is the internal advection layer of atmosphere above urban region which determines the exchanges of momentum, water and other atmospheric constituents between the urban land surface and the free troposphere. This paper tested the performance of three planetary boundary layer (PBL) physics schemes of Weather Research and Forecast (WRF) software to ensure the appropriate representation of vertical structure of UBL in Greater Kuala Lumpur (GKL). Comparison was conducted on the performance of respective PBL schemes to generate vertical and near-surface weather profile and rainfall. Mellor-Yamada-Janjić (MYJ) local PBL scheme coupled with Eta MM5 surface layer scheme was found to predict the near-surface temperature and wind profile and mixing height better than the nonlocal schemes during the intermonsoonal period with least influences of the synoptic background weather.

1. Introduction

Urban boundary layer (UBL) is the vertical atmospheric structure formed due to urban land use that contribute to the dryness, turbulent instability and increment of mixing height [1]. The urban-induced vertical air movements essentially determine the development and the structure of the boundary layer where fluxes of heat, moisture and component exchange between land surface and atmosphere. However, such eddies are unable to be resolved using conventional mesoscale numerical model and have to be parameterized through the planetary boundary layer (PBL) model. In Weather and Research Forecast (WRF) software, the PBL physics resolve the subgrid-scale vertical interaction of surface heat, moisture and air constituent flux such as nonlocal transport, entrainment near PBL top, transitions between stable (SBL) and convective boundary layer (CBL) [2]. The PBL scheme determines the lapse rate, vertical turbulence fluxes and radiation admittance of the vertical layer under the free atmosphere layer (500 - 3000 m from Earth surface). Hence, it can potentially affect the heat exchanges fluxes between the surface layer and the boundary layer to determine the urban-induced heating. The boundary layer turbulence problem is assumed with the gradient transport theory (K-theory) and is generally closed with two methods. They are (a) the local closure approach that estimates the turbulent fluxes at each point from the mean properties of its adjacent points and (b) the nonlocal treatment which determines the fluxes from the depth of boundary layer.

Comparative studies in the south-central USA have confirmed that local schemes produce closer agreement of vertical profile and near-surface variables under stable condition [3,4]. Nevertheless, it is prone to overestimation of the moisture content during summer. Nonlocal schemes excel in reconstructing the PBL height and estimating the lower boundary lapse rate which is pertinent to the prediction of pollution distribution [5]. Notwithstanding the inconsistent and site-specific performance of each boundary layer scheme, the PBL schemes were widely tested in the midlatitude region [6,7], but little information is available for applicability in tropical regions. This has therefore rendered the site-specific investigation necessary to adopt more appropriate scheme for detailed UBL study within the region.

This paper tested the performance of the boundary layer physics to ensure the applicability and weakness of the schemes to represent the transportation and dynamics of boundary layer structure in Greater Kuala Lumpur (GKL), Malaysia. From which, the comparison result can be referenced for future UBL study in GKL or region with similar background weather and urbanization conditions. Section 2 first discussed the parameterization method of the selected PBL schemes as well as the experiment design. In Section 3, each PBL physics was evaluated before settling on the better performing scheme to reproduce the turbulent transportation and mixing within the UBL.

2. Experiment design

Table 1. WRF domain and physics settings

Parameter	Value/ Option	Parameter	Value/ Option
Analyzed period (1-day spin up)	19 days (2nd April/Oct 2003 0800 MYT – 20th April/Oct 2003 0800 MYT)	**Boundary layer schemes**	**MYJ**: MYJ PBL with Eta MM5 SUR scheme; **YSU**: YSU PBL with MM5 SUR scheme; **ACM2**: ACM2 PBL with MM5 SUR scheme
Domains setting	4 domains (27km, 9km, 3km, 1km)		
Vertical levels	37 vertical eta levels. 15 level for bottom 12 km		
Lateral boundary data	ERA-interim 6-hourly reanalysis model level data		
Land use map	Updated with 3 urbanization level (Low and high density residential and commercial/industrial) [8]	**Microphysics scheme**	**Lin**: Purdue Lin single moment scheme
Land surface model (LSM)	Noah LSM coupled with single-layer urban canopy model (UCM) with calibrated urban parameters [8]	**Cumulus scheme**	**BMJ**: Betts-Miller-Janjić scheme for 1st and 2nd domains only

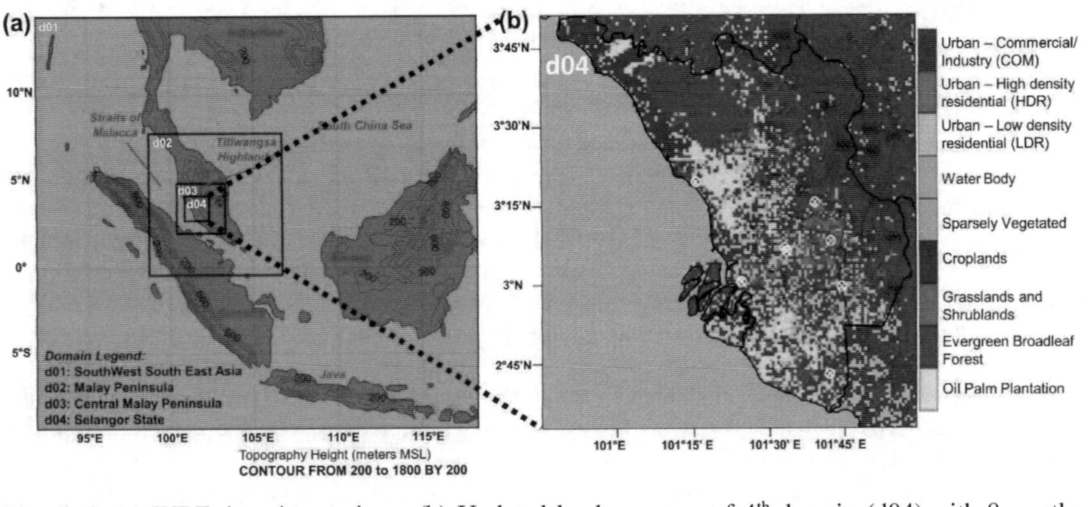

Figure 1. (a) WRF domains settings, (b) Updated land use map of 4th domain (d04) with 8 weather stations (ST), 1 rain gauge (ST6-Sepang) and 1 sounding station (ST6).

Three widely applied PBL schemes, namely the local scheme, Mellor-Yamada-Janjić (MYJ) and two nonlocal schemes, Yonsei University scheme (YSU) and Asymmetric Convective Model (ACM2) were chosen for comparison. Firstly, MYJ is an implementation of Mellor-Yamada PBL parameterization into numerical weather prediction field [9]. Additional prognostic equation for TKE is introduced to predict its generation, transportation and dissipation and the entrainment near PBL top [2]. This scheme is suitable for stable and slightly unstable flows because it is able to ensure the conservation of boundary layer [3]. The YSU scheme [10], on the other hand, is an improved version of the nonlocal parameterization Middle Range Forecast. It combines specified profile for diffusivity and counter-gradient term in the same equation. The explicit treatment of entrainment process is also included at the top of PBL. The scheme is therefore suitable to predict unstable flows by addressing the nonlocal fluxes when rigorous eddies circulation approaches the depth of the boundary layer [3,4]. The ACM2 scheme [11] is a hybrid scheme combining both local eddy diffusion and nonlocal transport scheme from ACM1. The ratio of nonlocal fluxes to total fluxes near-surface is introduced as a stability indicator for the smooth transition between SBL and CBL [2,5]. In WRF, each PBL scheme is bonded to associated surface layer (SUR) scheme.

The intermonsoonal periods, April and October with weaker synoptic forcing were studied to emphasize on the local climate, especially urban heating and its influence on topographic flow [12]. Due to the variability of atmospheric boundary layer to the background weather condition, this paper tested the PBL scheme performance in 2003 with least influence of rainfall anomaly [8]. The domain information for the simulation was given in Figure 1a with other input data and physics settings in Table 1. Simulation cases adopting MYJ, YSU and ACM2 PBL schemes were respectively run for the two mentioned intermonsoonal periods (April and October). Abilities to predict the vertical profile and boundary layer height were tested with measured data (marked in Figure 1b) to ensure the schemes reproduce the vertical flux dynamics and later the near-surface profile and distribution of precipitation. The performance of each PBL scheme was evaluated with statistical error indices, including mean absolute error (MAE), root mean square error (RMSE) and fractional absolute error (FAE).

3. Result and discussion

3.1 Vertical profile

The performances of each PBL schemes in October and April were evaluated with sounding data (ST6 marked in Figure 1b) during the morning (0800 MYT) and evening (2000 MYT) transitional hours in Figure 2. The temperature and humidity profiles were similar for all schemes with less than ±0.4 K and ±0.6 g/kg bias respectively with the largest deviation observed around the typical mixing layer height of UBL (1-1.5 km) [12]. The positive bias between the schemes at 0800 MYT was significantly greater in October (1.2±0.8 ms^{-1}) than in April (0.6±0.3 ms^{-1}). The boundary layer was relatively calm and invariant in April compared to October. It was therefore arrived that the performance of the schemes deviated greatly under influence of strong synoptic flow. The lower modeled wind speed suggested that the vertical mixing was suppressed within the boundary layer in Figure 2c,f. The insufficient mixing hence explained the negative bias of the modeled temperature and mixing ratio around the mixing height. The tabulated error statistics in Table 2 showed that October was better estimated with the nonlocal schemes while the weak wind condition during April was better estimated with MYJ. This clearly revealed the respective strength of PBL schemes. ACM2 with the physical representation of local and nonlocal eddy diffusion movement returned the best compliance among the nonlocal schemes under convective synoptic condition [11]. The MYJ is known to estimate vertical quantities through prognostic eddy diffusivity of adjacent grids and hence simulated cases under calm condition better [9]. Although MYJ has induced insufficient mixing under large-scale boundary condition in October [3], it showed an edge in solving cases of stable and localized boundary conditions well in the calm April.

3.2 Planetary boundary height

The vertical tendency of variables is principally dependent on the boundary layer height (PBLH), especially for the nonlocal schemes [10,11]. It was noticeable that the modeled PBLH was less than 1000 m and well below the level derived for the SEA region (PBLH = 1250 ± 670 m [13]). The lapse rate method ($\Delta\theta/\Delta z \leq 0.002\ Km^{-1}$) was adopted to standardize the PBLH of each schemes [3,14]. In Figure 2g,h, the unified derivation approach increased the modeled PBLH and associated well among the schemes (< ±100 m) for the daytime PBLH (0900 – 1900 MYT). The PBLH of each scheme was evaluated against the measured PBLH extracted from sounding (ST6) with hit rate and MAE (hit rate threshold <100 m), lower value was desirable. All the schemes predicted the PBLH reasonably well during the transition period (0800 and 2000 MYT). In Table 2, the adjusted PBLH demonstrated that MYJ was able to capture the entrainment of the free atmosphere into the boundary layer. Among the nonlocal schemes, the ACM2 produced a better mixed UBL with the highest modeled PBLH peak during the day. It was also noticeable that the hotter month (April) produces a generally higher mixing layer height during the day due to enhance convective activities nearing summer.

3.3 Near-surface profile

The ability to reproduce the near-surface condition depends on the accompanying SUR scheme of each PBL scheme [4,5], hence, diurnal data from the 8 ground weather stations (marked in Figure 1b) were analyzed for respective SUR schemes. The near-surface condition predicted by each scheme has smaller bias between each other compared to the vertical profile. The bias level recorded 0.4±0.2 °C T2, 2.1±1.2% RH2 and 0.5±0.2 ms^{-1} wspd10. YSU and ACM2 cases delayed the response of peak temperature and relative humidity up to two hours. MYJ case, nevertheless produced closer diurnal profile especially in October. All schemes overestimated the wind strength in the day [7] as shown in Figure 2k, with MYJ case produced more reasonable agreement. In Table 2, MYJ recorded overall the lowest statistical error for near-surface condition. This indicated that Eta similarity theory SUR of MYJ scheme predicted T2 and RH2 better. The good performance was attributed to the determination of heat transfer coefficient according to the roughness length of the surfaces [15]. The large deviation of T2 during the peak hour was attributed to the application of MYJ that generate insufficient mixing in daytime (Section 3.2). Due to the incompetency of the nonlocal scheme, the MYJ remained as the optimum choice to model the near-surface condition. A large bias for T2 and RH2 around 1700 – 2000 MYT during both intermonsoon seasons was potentially related to the concurrent precipitation events.

3.4 Rainfall amount

The precipitation amount simulated by full grid microphysics (RAINNC) and subgrid cumulus (RAINC) drivers were summed for total precipitation to evaluate against the rain gauge data at Sepang (ST6). All schemes, notably MYJ has successfully captured the initiation time of the rain at 1500 - 1600 MYT in both months and 0500 MYT in October in Figure 2m,n. A good agreement of rainfall amount was also produced by MYJ model. This could be explained as the heterogeneous land sea mask in MC shelters the GKL region from large scale turbulence movement and created a relatively calm weather condition compared to the rest of the Malay Peninsula [16]. The nonlocal scheme was considered less suitable for application unless critical Richardson number that determines the regime SBL and CBL is re-iterated for further adaptation. Conclusively, MYJ remained a more suitable candidate of scheme for the current study under relative stable weather condition, resonating with previous findings [4].

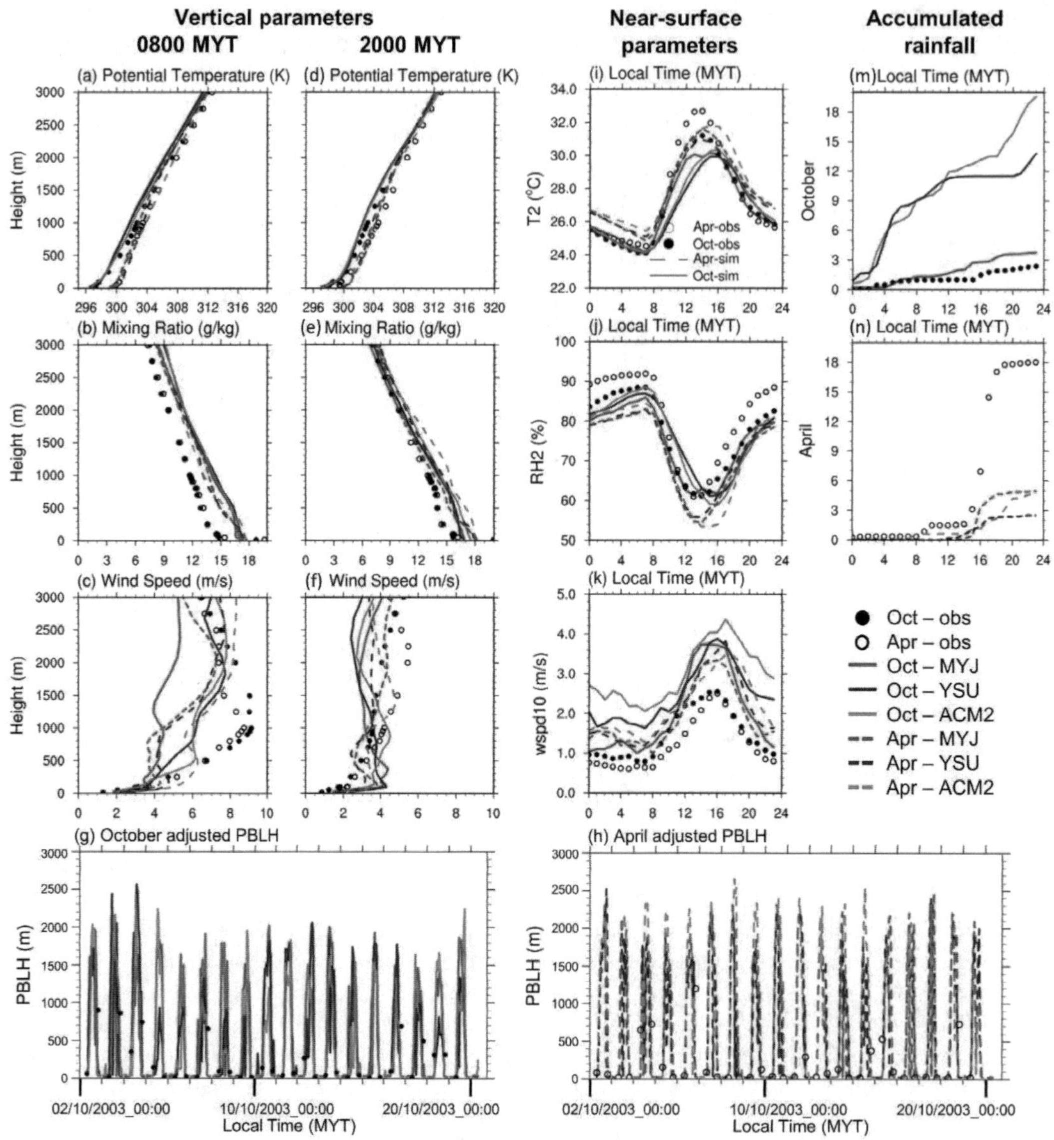

Figure 2. Evaluation of (a)(d) potential temperature, (b)(e) mixing ratio, (c)(f) wind speed, (g)(h) PBLH against sounding data at 0800 MYT (Column 1) and 2000 MYT (Column 2). Evaluation of hourly averaged of (i) T2, (j) RH2, (k) wspd10, (m)(n) accumulated rainfall with ground weather stations. Marker for observation while lines for simulation result.

Table 2. Error indices deviation of PBL schemes for vertical variables, PBLH and near-surface variables. (†) FAE are dimensionless, (*) multiplication of 100 for vertical potential temperature, T2 and RH2.

Schemes	**MYJ**		**YSU**		**ACM2**		**MYJ**		**YSU**		**ACM2**		**MYJ**		**YSU**		**ACM2**	
	Vertical parameters																	
	Vertical potential temperature (K)						*Vertical mixing ratio (kg/kg)*						*Vertical wind speed (ms^{-1})*					
Time	8am	8pm	8am	8pm	8am	8pm	8am	8pm	8am	8am	8pm	8am	8am	8pm	8am	8am	8pm	8am
Error indices	**October**																	
MAE	0.64	0.66	0.63	0.45	0.42	0.29	1.58	1.41	2.05	1.40	2.02	1.54	2.39	1.66	1.16	1.47	1.00	1.45
RMSE	0.74	0.73	0.73	0.54	0.46	0.34	1.83	1.61	2.18	1.56	2.13	1.73	2.81	2.22	1.49	1.99	1.45	1.65
FAE*†	0.21	0.21	0.21	0.15	0.14	0.09	0.13	0.12	0.18	0.12	0.18	0.14	0.41	0.29	0.20	0.24	0.15	0.23
	April																	
MAE	0.35	0.69	0.43	0.75	0.31	0.65	0.73	0.85	0.81	1.07	1.22	1.39	0.89	0.73	1.16	0.90	1.14	0.79
RMSE	0.43	0.77	0.48	0.83	0.37	0.71	1.03	1.12	1.07	1.37	1.34	1.77	1.06	0.83	1.42	1.01	1.23	0.91
FAE†	0.12	0.22	0.14	0.24	0.10	0.21	0.06	0.07	0.07	0.08	0.11	0.10	0.26	0.20	0.35	0.23	0.33	0.21

	Near-surface parameters										***PBLH***		
Schemes	**MYJ**	**YSU**	**ACM2**	**MYJ**	**YSU**	**ACM2**	**MYJ**	**YSU**	**ACM2**		**MYJ**	**YSU**	**ACM2**
Month	*2-meter temperature, T2 (°C)*			*2-meter relative humidity, RH2 (%)*			*10-meter wind speed, wspd10 (ms^{-1})*				*Adjusted PBLH (m)*		
Error indices	**October**												
Average		27.09			77.08			1.47		Hit Rate%	66.7	73.0	72.2
MAE	1.43	1.72	1.67	7.35	8.18	8.45	1.05	1.39	1.77				
RMSE	1.82	2.19	2.13	9.27	10.50	10.65	1.45	1.86	2.25	MAE	15.50	16.97	21.95
FAE*†	5.30	6.37	6.16	9.93	10.99	11.57	6.56	7.41	8.23				
	April												
		27.49			81.12			1.22		Hit Rate%	72.2	72.2	77.8
MAE	1.76	1.77	1.83	10.73	10.22	10.36	1.02	1.16	1.26				
RMSE	2.24	2.20	2.33	13.19	12.46	13.20	1.38	1.55	1.64	MAE	11.36	12.34	18.06
FAE*†	6.36	6.42	6.61	14.35	13.57	14.19	7.12	7.56	7.99				

4. Conclusion

In general, all schemes in prediction of vertical profile, near-surface profile and precipitation amount were more uncertain in the October which experienced a stronger synoptic forcing. The PBL schemes also gave larger variation for the vertical profile and rainfall amount during the convective period. This was attributed to the relatively significant influence of the synoptic condition on these PBL processes. It was also notable that the nonlocal scheme, namely ACM2 gave a closer agreement to the vertical profile in October. However, the remaining model evaluations of vertical and precipitation profile clearly inclined toward the MYJ local scheme. Similarly, the local scheme simulated shallower PBLH but closer agreement with measured data after normalization. The allocation of thermal roughness length to respective land use also improved the near-surface condition prediction for Eta MM5 SUR scheme. Therefore, the strong coupling of the MYJ PBL scheme and Eta MM5 SUR scheme prompted the future analysis on the effect of urbanization to be conducted with the combination.

Acknowledgments

The authors would like to thank University of Nottingham Malaysia Campus for awarding a research scholarship and the access to the University of Nottingham High Performance Computing Facility. We are also grateful for the RAGS funding provided by MOHE under 600-RMI/RAGS 5/3 (66/2015).

References

[1] Cooper D I and Eichinger W E 1994 *J. Geophys. Res. Structure of atmosphere in an urban planetary boundary layer from lidar and radiosonde observations* **99** 37
[2] Skamarock W C, Klemp J B, Gill D O, Barker D M, Duda M G, Wang W and Powers J G

2008 *A description of the Advanced Research WRF version 3* (Boulder, Colorado, USA)
[3] Hu X-M, Nielsen-Gammon J W and Zhang F 2010 *J. Appl. Meteorol. Climatol. Evaluation of three planetary boundary layer schemes in the WRF model* **49** 31
[4] Shin H H and Hong S-Y 2011 *Boundary-Layer Meteorol. Intercomparison of Planetary Boundary-Layer Parametrizations in the WRF Model for a Single Day from CASES-99* **139** 61
[5] Xie B, Fung J C H, Chan A and Lau A 2012 *J. Geophys. Res. Atmos. Evaluation of nonlocal and local planetary boundary layer schemes in the WRF model* **117** 1
[6] Shin H H and Dudhia J 2016 *Mon. Weather Rev. Evaluation of PBL parameterizations in WRF at sub-kilometer resolution : Turbulence statistics in the convective boundary layer* **144** 61
[7] Tateo A, Miglietta M M, Fedele F, Menegotto M, Monaco A and Bellotti R 2017 *Adv. Sci. Res. Ensemble using different Planetary Boundary Layer schemes in WRF model for wind speed and direction prediction over Apulia region* **14** 95
[8] Ooi M C G, Chan A, Ashfold M J, Morris K I, Oozeer M Y and Salleh S A 2017 *Urban Clim. Numerical study on effect of urban heating on local climate during calm inter-monsoon period in greater Kuala Lumpur, Malaysia* **20** 28
[9] Janjic Z I 1994 *Mon. Weather Rev. The step-mountain Eta coordinate model: Further developments of convection viscous sublayer, turbulence closure schemes* **122** 27
[10] Hong S-Y and Pan H-L 1996 *Mon. Weather Rev. Nonlocal boundary layer vertical diffusion in a medium range forecast model* **124** 22
[11] Pleim J E 2007 *J. Appl. Meteorol. Climatol. A Combined Local and Nonlocal Closure Model for the Atmospheric Boundary Layer. Part II: Application and Evaluation in a Mesoscale Meteorological Model* **46** 396
[12] Li X X, Koh T Y, Entekhabi D, Roth M, Panda J and Norford L K 2013 *J. Geophys. Res. Atmos. A multi-resolution ensemble study of a tropical urban environment and its interactions with the background regional atmosphere* **118** 04
[13] Von Engeln A and Teixeira J 2013 *J. Clim. A planetary boundary layer height climatology derived from ECMWF reanalysis data* **26** 75
[14] LeMone M a., Tewari M, Chen F and Dudhia J 2013 *Mon. Weather Rev. Objectively-determined fair-weather CBL depths in the ARW-WRF model and their comparison to CASES-97 observations* **141** 30
[15] Chen F and Zhang Y 2009 *Geophys. Res. Lett.* On the coupling strength between the land surface and the atmosphere: From viewpoint of surface exchange coefficients **36** 1
[16] Chang C P, Wang Z, McBride J and Liu C H 2005 *J. Clim. Annual cycle of Southeast Asia – Maritime continent rainfall and the asymmetric monsoon transition* **18** 287

IOP Conf. Series: Earth and Environmental Science **117** (2018) 012016 doi:10.1088/1755-1315/117/1/012016

An observational study of defensible space in the neighbourhood park

M A Marzukhi[1], M A Afiq[1], S Ahmad Zaki[1] and O H L Ling[1]

[1]Faculty of Architecture, Planning and Surveying, Universiti Teknologi MARA, Selangor Branch, Puncak Alam, Malaysia

**marlyana@salam.uitm.edu.my*

Abstract. The planning of neighborhood park is important to provide space for interaction, leisure, and recreation among residents in any neighbourhood area. However, on an almost daily basis, newspapers report inappropriate incidents such as snatch theft, robbery and street attack that occurred in the neighborhood park. These cases reflect the significance of physical planning and design of neighborhood park that directly affect the safety and comfort of the users. Thus, this study attempts to engage with the defensible space concept in ensuring the security elements be applied in the planning of the recreational area. This study adopts a qualitative method form of research that is retrofitted to an observational study. The observational study is significant for revealing the condition of a neighbourhood park in the 'real-world,' in which direct observation is conducted on Taman Tasik Puchong Perdana. The observer focused on four elements or variables of defensible space concept including the provision of facilities in the neighborhood park, territoriality, surveillance, image and milieu. The findings revealed that the planning of Taman Tasik Puchong Perdana does not deliberate the defensible space elements, which may contribute to the crime activities in the park. In these circumstances, the planning of neighbourhood park needs to include proposals for the implementation of defensible space in response to the challenges underpinned by crime problems. Besides, the awareness among the residents needs to be emphasized with the support from local authorities and other organizations to manage and sustain the safety environment in the neighborhood park.

1. Introduction

Open space or recreational park is an important aspect in contributing to the ecological balance in the physical development of an area. People need a comfortable, safe and secure park for leisure and recreational activities that can promote a better quality of life. The existence of recreational park requires the integration of aspects such as appropriate location, safety, good facilities, and cleanliness, in accordance with local culture and the aspirations of the local population. In fact, the design of spaces and recreational areas that follow the principles, standard and guidelines is an important element to provide a sustainable recreational area that will directly affect the safety and comfort of the community [1]. However, almost daily in the news and media reports on crimes such as attacks on pedestrians, parking lots, and public transportation, street crime, violence against individuals, rapes, and thefts occurred in parks [2]. These problems lead to the question: Is the recreational park nowadays increasingly insecure due to this problems? This is because of current crime trends more concentrated in the recreation park especially in the neighbourhood area. Furthermore, robbery and snatch thefts in the recreational area are at an alarming rate from time to time as a result of fewer safety features that caused discomfort among the residents to visit the neighborhood park.

2. Literature Review

According to Section 2 (1) of the Town and Country Planning Act 1976 (Act 172), open space is define as "any land whether enclosed or not which is laid out or reserved for laying out wholly or partly as a public garden, park, sports, and recreation ground, pleasure ground, walk or as a public place" [3]. The open space is an area that has components of public facilities such as surau, community centre, community hall, playground and other facilities [4]. Moreover, open space is important to provide green spaces and nature particularly in an urban environment settings which allows the public to access and carry out activity that aims for recreational activities. This includes either active or passive activities such as "walkways, sidewalks, and bicycle lanes - and the areas devoted to public parks, squares, green recreational areas, public playgrounds and open areas of public facilities" [5, 6]. According to *Manual Garis Panduan Dan Piawaian Perancangan Negeri Selangor* (2010), there are seven (7) hierarchies of open spaces in Selangor which includes play lot, playground, neighbourhood park, community park, town park, regional park and national park [7].

2.1 Neighborhood Park

Neighbourhood Park is one of the hierarchies of open space in the residential area. The provision of the park is important for recreational activities, sports and social interaction particularly for 3,000 local population in the vicinity [7]. The location of the open space area is within a radius of 1.5 kilometres from the residential units. The provision of the park is vital to create a sense of belonging and neighborliness among the residents which further can also create a safe environment within the neighborhood area [8]. The *Manual Garis Panduan dan Piawaian Perancangan Negeri Selangor* (2010) has set out certain criteria and requirements of neighborhood park as listed in Table 1 below.

Table 1. Hierarchy and Function of the Neighborhood Park

Hierarchy	Area	Population Size	Distance	Facilities
Neighborhood Park	1.2 hec. (3.0 ac.)	Every 3,000 population; and Each increase 3,000 population requires 2 hec. (5 ac.)	In the radius of 1.5 km	Children's playground; Football field; Courts of badminton, tennis, takraw, etc; Jogging area; Parks and shelter; Place for rest; Public phones; Toilet; and Parking: ▪ 35 car parks for the first 2 hec. (5 ac.). ▪ Additional 1 car park / 1500mp (16,145.8 kps) remaining base area. ▪ Additional 30% motorcycle park. Additional bus parking according to the type of activity and site requirements.

Source: Manual Garis Panduan dan Piawaian Perancangan Negeri Selangor (2010)

Furthermore, the physical design of the neighborhood park is important to provide comfort and safety for the users to combat crime problems within the neighborhood area [1]. Therefore, the planning of neighborhood park which focuses on physical design for crime prevention is an important aspects, particularly through the concept of defensible space. At this point of reasoning, observation of this defensible space is critical to illuminate its application in practice that may affect society as a whole. In this sense, to better understand defensible space, this article turns to the next section.

2.2 Defensible Space

The concept of defensible space is introduced by Newman [9] to restructure and rearrange the physical layout in order to create social strength among the communities and to prevent crime in the residential area. Defensible space is defined as;

> "....model for residential environments which inhabits crime by creating the physical expression of a social fabric which defends itself....an environment in which latent territoriality and sense of community in the inhabitants can be translated into responsibility for ensuring a safe, productive and well-maintained living space" [9].

Besides residential area, this study suggested to adapt the defensible space concept as one of the key features in planning of the open spaces and recreational areas. The suggestions includes the preparation of layout design through the arrangement of spaces in the neighborhood park. Further, defensible space entails an involvement and participation of residents or the communities in the area to reduce the crime activities and eliminate the presence of offenders. It is important to plan and develop a conducive open space and recreational area which further can create a sense of belonging among the community. Furthermore, a proper planning of open space will lead the local people to appreciate and value the area. This will also create a safe neighbourhood park that will give comfort and guarantees the safety of the users. In this sense, the planning of neighbourhood park is more liveable and can be managed by the residents within their locality [10, 11]. There are four elements that create the defensible space; territoriality, surveillance, image and milieu [9]. The territoriality is one of the defensible space elements that identify the boundaries of the area either it is private or public space. It is important to determine this space to create a sense of security in an area. Further, the surveillance elements are related to the supervision of the area as one of the ways to prevent crime to occur. Lastly, the image of the recreational area is vital to reflect a good image of the area which further will provide secured feelings to the community or the users of the recreational area. Therefore, the planning of neighborhood park should emphasize and implement these variables to create a park which is defensible by physical design to ensure the safety of the public. Therefore, the insight of the defensible space in relation to the research methodology selected is elaborated in the section that follows.

3. Research Methodology

This study used an observational study to investigate the 'real-life' condition of the neighborhood park. The researcher or an observer records the elements in a systematic manner by using rating scales in which the variables are identified from the literature. Through observation, the data collected is often more appropriate and reliable due to the direct observation either from a distance or near from the study area. The observation focuses on the physical elements such as the provision of facilities, safety elements and condition of environment that contribute to the movement of the offenders in the study area. The case study selected is Taman Tasik Puchong Perdana in the District of Petaling which is under the administration of Municipal Council of Subang Jaya. The study area is surrounded by mixed development of land uses such as residential, commercial and public facilities. The total size of the study area is 7.70 acres, in which 3.19 acres is covered with open space, whereas 4.51 acres is covered with lake area [12].

3.1 Observational Analysis

The observation was conducted on the 5th of May, 2017 during day and night, where an observer was focused on the physical elements in Taman Tasik Puchong Perdana. The observation focused on four variables of defensible space concept including the provision of facilities in the neighborhood park (refer Figure 1):

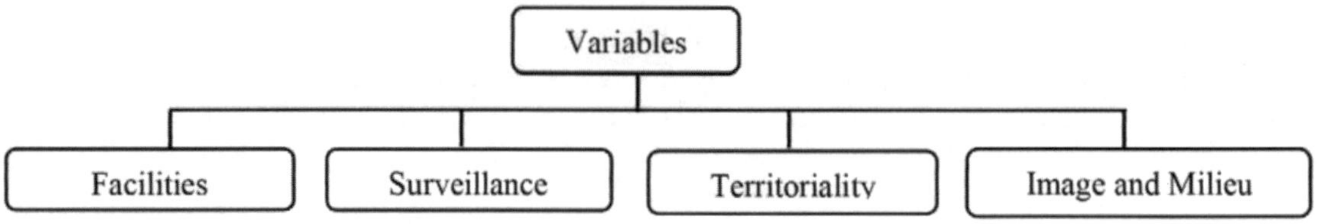

Figure 1. Variables for Observational Analysis

The observation was conducted systematically and recorded neatly to obtain correct results. The checklist was used as shown in **Table 2** below. Also, photographs were taken for each of the elements observed.

Table 2. Checklist of the Observational Analysis on the Defensible Space in Taman Tasik Puchong Perdana

Aspect	Elements	Condition
Facilities	Children's Playground	Good
	Jogging Track	Good
	Park and Shelter	Not Good
	Football Fields	Not Good
	Sports Courts	Not Good
	Toilet	Not Good
	Parking	Not Good
Surveillance	Lighting	Not Good
	CCTV	Not Good
Territoriality	Fence / Landscape / Bollard	Good
Image and Milieu	Lake	Not Good
	Landscape	Not Good
	Cleanliness	Good

4. Discussion and Findings

The findings focused on the provisions of facilities in the neighborhood park and four variables of the defensible space; territoriality, surveillance, image and milieu. Accordingly, the results show that the implementation of defensible space elements were not applied in Taman Tasik Puchong Perdana.

i. Facilities

The first variable observed is the condition and provision of facilities in the neighborhood park. From the observation, the provision of facilities still did not fulfill the characteristics and requirements for the planning of neighborhood park as shown in Table 1. This is due to the lack of provision of facilities such as football field, badminton court, takraw court and futsal court in the study area.

Figure 2. Condition of Children's Playground

ii. Territoriality

The observation analysis shows that the environmental aspect of the study area is not safe for the public to engage in the recreational activities. This is because the area did not have strong element of territoriality to aid the local community to defend their neighborhood park against crime. In fact, the area is exposed to crime activities particularly snatch theft due to no physical barriers provided between the recreational area and non-recreational area. Thus, it creates a weak territoriality of Taman Tasik Puchong Perdana. Physical barrier can be designed and provide in the form of landscaping or fencing which can control the entrance and exit of the area.

Figure 3. No Barrier around the Study Area

iii. Surveillence

Surveillance, which is divided into formal and informal categories is an important variable of defensible space to prevent crime. The analysis found that there are lack of informal surveillance such as community programs and awareness among the local residents which further led to the lack of public participation and social interaction in the park. The community programs and activities are important to attract public to join and participate in the activities conducted in the neighborhood park, thus, become one of the surveillance agents. Moreover, the analysis also shows the lack of formal surveillance such as lighting, CCTV, organized security patrol, and guard post provided in the Taman Tasik Puchong Perdana. Lighting is one of the crime deterrents which can act as natural surveillance for the community around the study area. However, the analysis found that the lighting condition in the area were poorly maintained and not well-functioned. Thus, it makes the study area obscured and at risk to offenders. Also, there is no regular patrol by the security patrol or neighborhood watch which further contribute to the crime activities in the study area.

iv. Image and Milieu

Image plays an important part in ensuring the neighborhood park have a good appearance and not vulnerable to the crime activities. Nonetheless, the finding shows that Taman Tasik Puchong Perdana is exposed to the offenders due to the lack of safety elements which can prevent from crime activities. For instance, the form of landscapes within the area are not well-maintained. In fact, many trees are blocking the views that can also become the hidden place for the offenders. Furthermore, there are no provision of proper landscapes within the study area that can be used as barriers to prevent the offenders to enter and exit the neighborhood park.

Figure 4. Landscape in the study Area

The children's playground is also located out of sight and do not have barrier around it which can lead to the crime activities such as kidnapping. Moreover, the image of this neighborhood park do not have sense of welcoming that can attract the public to engage with the recreation activities. Milieu is also one of the defensible space variables that is related to the design and placement of buildings surrounding the study area which can create an area prone to crime. Moreover, part of the neighborhood park is visible from the main road and can be accessed directly without proper barricade, which is at Jalan Persimpangan Puchong Perdana. Thus, this shows that the element of milieu is not fully applied in this neighborhood park.

5. Conclusions and Recommendations

The findings indicate that the crimes activities that occurred in the Taman Tasik Puchong Perdana are due to the lack of implementation of defensible space in the neighborhood park. In this sense, the factors of lighting, safety features, surveillance and maintenance often contributes to the crime in the study area. Therefore, to prevent the problems and issues that led to the crime in the neighborhood park, the planning of recreational area should be well-planned with the implementation of defensible space concept as an approach to deterring the crimes activities. Security control in open spaces and recreational areas must be one of the key initiatives in the form of social interaction of a community environment that is safe, and harmonious. Also, the facilities should be provided in all aspects so that the study area will become a recreational area that is fully equipped, comfortable and secure for visitors.

Acknowledgments

The authors would like to thank Universiti Teknologi MARA (UiTM), Malaysia for partly funding the study by given the BESTARI grant (600-IRMI/MYRA S/3/BESTARI (008/2017)). The authors would also like to thank all the government departments, organizations, and respondents who had contributed to this study.

References

[1] Ibrahim W 1991 *Perancangan Bandar: Aspek Fizikal dan Kawalan Pembangunan.* Kuala Lumpur. Dewan Bahasa Dan Pustaka, Kuala Lumpur.

[2] Lee Lam Thye, Deputy Chairman of the Malaysian Crime Prevention Foundation (MCPF) 2014 *World Town Planning Day Public Lecture and Exhibition.* Kuala Lumpur.

[3] Town and Country Planning Act 1976 2009 Laws of Malaysia. Kuala Lumpur: Percetakan Nasional Malaysia Berhad.

[4] Francis M 2003 *Urban Open Space.* Island Press, London.

[5] Thompson 2008 Design Principle –Parks and Open Space. *Healthy Spaces & Places. Assessed on June 2016, www.healthyplaces.org.au*

[6] United Nations Human Settlements Programme 2015 Adequate Open Public Space in Cities. *UNHabitat for a better future. Assessed on May 2016, www.unhabitat.org*

[7] Jabatan Perancangan Bandar dan Desa Semenanjung Malaysia 2010 *Garis Panduan dan Piawaian Perancangan Negeri Selangor.* Kuala Lumpur: Jabatan Perancang Bandar dan Desa Semenanjung Malaysia.

[8] Chow KY 2014 Taman Macallum, Taman Kejiranan Usaha MPPP. *Buletin Mutiara*, 16th January 2014.

[9] Oscar N 1996 *Creating Defensible Space*. U.S. Department of Housing and Urban Development Office of Policy Development and Research – Institute For Community Design Analysis.

[10] Mohit MA and Hannan MHE 2012 *Social and Behavioral Sciences. A Study of Crime Potentials in Taman Melati Terrace Housing in Kuala Lumpur: Issues and Challenges* **42** 271

[11] Mair JS and Mair M 2003 *Health.Violence Prevention and Control through Environmental Modifications* **24** 209

[12] Rancangan Tempatan Majlis Perbandaran Subang Jaya (Pengubahan 1) 2020. Majlis Perbandaran Subang Jaya, Selangor.

A perspective study on green cleaning for Malaysian public hospital

N A Ramli [1,2], E M Zawawi [1], N R M Arif[1], N S Mahbob [2], Z Sulaiman [2] and N N Zainol [3]

[1] Faculty of Surveying, Planning and Architecture, UiTM Shah Alam, Shah Alam Malaysia
[2] Lee Kong Chian Faculty of Engineering and Science, Universiti Tunku Abdul Rahman, Sungai Long, Cheras, Malaysia
[3]Department of Urban Studies, Imperia Institute of Technology, Malaysia

[a]aqlimaramli@gmail.com,[b]emmamarinie@salam.uitm.edu.my,[c]norri550@salam.uitm.edu.my, [d]syima@utar.edu.my, [e]zamharira@utar.edu.my, [f]nurulnadiah@imperia.edu.my

Abstract: Cleaning being a major contributor to the operations and maintenance expenditure and also Indoor Environmental Quality (IEQ) issues. Improper and ineffective cleaning can harm the environment and poses greatest risk to health. The use of traditional cleaning products presents a variety of human health and ecological concerns; and may contribute to poor IEQ. As an effort to reduce the issue of operations and maintenance costs and IEQ issues in a building, it is important to establish a green cleaning programme to ensure that the buildings are cleaned in a green way. Numbers of scholars has pointed out the factors which had prevented the green cleaning implementation in hospital buildings. Nonetheless, the significance of these factors has yet to be practically explored in the Malaysian context. Hence, the aim of the paper is to identify the most critical factor that prevents the implementation of green cleaning in Malaysian hospital building. A questionnaire survey and personal communication (i.e. interview) was conducted which involved two groups of respondents. They are the hospital maintenance staff (*Cleansing Service Department*) and cleaning contractors. Frequency and criticality index calculations have been used to rank these factors according to the level of importance. The result showed that an "unclear components and requirements of green cleaning" indicated as the most critical factor that prevent the implementation of green cleaning in Malaysian hospital building. In the concern for a successful implementation of green cleaning, it is hope that the findings of these studies can be enlightenment to the cleaning contractors as well as the hospital maintenance management team in Malaysia.

1. Introduction

In the beginning of 1990's, Malaysian Healthcare Industry has been developing sharply and steadily. Within the year 2011 to 2015, under the 10th Malaysian Plan, the healthcare industry has been recognized as one of the National Key Economic Areas (NKEA). Across the years, the Malaysian Healthcare Industries has been constantly prudent with the Malaysian Healthcare System but had undergone a revolutionary alteration [1]. Nonetheless, in the present day of the healthcare industry, it is time to re-evaluate the existing healthcare system. These are due to the endless increase in cost, the worsening profitability, resource inadequacy and the dissemination of diseases.

With a building normal life expectancy of about 100 years [2], a building goes through several phases in its lifespan beginning with its creation and erection, maintenance and management right through to destruction and removal. Throughout these phases, maintenance and management of the building plays a sizeable aspect and should be one of the core business decisions in any management itinerary [2]. In spite of its considerable impact on the management expenditures, maintenance and management aspect have always been largely ignored throughout the phases of the building [3]. It is acknowledged that generally the phases of a building plays primary elements in influencing the environmental aspect as well as the human aspect and more likely to increase with population growth and the fluctuations of economic issues. In order to minimize the impact of the latter, a sustainability model has to be adopted into the entire phases of the building especially the maintenance and management aspect [3].

As mentioned on the above, maintenance and management of the building consists of sizeable aspect including cleaning. Cleaning, ranked as the most functional and crucial of the building facility maintenance within the maintenance and management aspect in comparison to other maintenance operations of a hospital building such as lighting, Mechanical & Electrical services, plumbing inclusive of sanitary services, air-conditioning services etc [24]. According to Ministry of Health (MOH) official website, all the maintenance and management services are controlled by Hospital Operation Section (HOS) team under Engineering Service Division. The HOS itself comprise of six services, namely; Facility Engineering Maintenance Services (FEMS), Biomedical Engineering Maintenance Services (BEMS), Linen and Laundry Services (LLS), Healthcare Waste Management Services (HWMS) and Cleansing Services (CLS). The cleaning activity has been allocated under CLS, which supervise the cleaning of wards, clinics and other areas in the hospital building include operating theaters, laboratories, pharmacies and the surrounding area.

Numerous counts of cleaning services carried out per day with a primary portion of the facilities management expenses which facilitated about one-third (30%-33%) of overall maintenance and management finance provision [18]. These are an indication of the significant impact on the maintenance and management cost and the integral function of cleaning within the healthcare building. In comparison to the other maintenance operations (mainly M&E, plumbing, security services etc.) within the breakdown of the maintenance and management estimated costing [4], cleaning services (inclusive of the cost for machinery & tools, chemical elements and largely the cost of resources such as manpower etc.) has been clearly expressed as the most costly.

Indoor Environmental Quality (IEQ), a standard should be taken into account when introducing sustainability into a building. Cleaning service being a primary factor to the maintenance and management aspect, inappropriate and inefficient cleaning can impair the environment and poses risk to human health [26]. Intrinsically, it would have a substantial impression to the IEQ issues [25]. In trying to keep to the traditional chemical materials to administer the cleaning processes, it will only bring forth a poor IEQ reading [26 and 27] by introducing concerns about the well-being and environment which in turn will ultimately influence the building performance.

To implement a green cleaning programme with the determination to lessen the maintenance and management expenses as well as the IEQ issues, hospital buildings should be cleaned effectively and efficiently in a green manner. There are few factors that prevent the execution of a green cleaning and this study has established five main factors. These factors are imperative to ensure the effectiveness of the Green Cleaning Programme that is to be introduced to Malaysia. As green cleaning is new here and without sufficient researched and practices, it is fundamental to organize an analysis to establish the criteria and adopt it in accordance to the order of prominence and significance.

2. Hospital Operations and Maintenance

Regardless of how remarkable the design of the hospital building, it will only continue perform if the building is maintain and operated appropriately. Furthermore, being amongst the major vulnerability buildings, hospital building has a sizeable influence on the surroundings what with its 24/7 usage. Henceforth, the hospital building must be maintained and operated on the sensitivity towards safety & health, comfort and others.

In many kinds of building, in order to ascertain a building is maintained and operated in an effective manner, the maintenance and management aspect has to be a continuous activity. Much like other building, a hospital building entails a proper maintenance to ensure the facilities and buildings are in excellent state. As such, the requisite for green maintenance practice in hospital building [4]. The ecological value is continually affected through maintenance by the reduction in waste, contaminants and other sources; which in return influences the financial operations with the reduction in the consumption of water and energy utilization and ultimately cost with an inadvertent effects to people's security & safety, solace, health and productivity [6, 7, 8, and 9]. With this, an introduction to sustainable maintenance, it can be characterized as maintenance procedure that attained the system merit of the existing body of conceding the aptitude for attaining the merit for the forthcoming body [5].

Notwithstanding the various writings and findings on hospital that is accessible, a disparity existed on the nature of the way the hospital is administered, upheld and worked [4]. Many kinds of maintenance services throughout the hospital maintenance and management aspect it encompasses cleaning services. Cleaning services can be considered as an uppermost building maintenance services throughout the maintenance and management aspect in relation to building maintenance services such as Mechanical & Electrical services, plumbing inclusive of sanitary services, air-conditioning services etc. [10]. Therefore, it is essential to conduct an analysis on the cleaning facet of the hospital buildings considering that it is a primary portion of the management expenses [11].

3. Green Cleaning

The desire to preserve the surroundings of a hospital building with minimal pathogen (bacterium, virus or other microorganisms) level which is imperative for circumventing difficulties during the convalescence process and care of patients is the reason the cleaning of hospital buildings are implemented. In comforting the patients and their families, promoting healthful, safety and an appealing place by means of the hygienic surface give the imprint of quality upkeep minus the health threat [12]. Whilst cleaning is needed in all economic sectors, the healthcare industry requires it to operate as a double purpose of surface cleanliness mainly, contamination hindrances and control. Intrinsically, hospital needs a rigorous and regular cleaning with a wide variety of merchandise, methods and measures. Consistently, [12], hospital area may be characterized by two distinct factors in facilitating environmental cleaning:

- Non-medical areas – wherein an area that does not affect the residence hospital; this comprises areas of public access i.e. lobbies/waiting rooms, offices, corridors, elevators, stair areas as well as common areas.
- Medical areas – wherein an area that is directly associated with the residence hospital; this encompasses residence/patients sections (inclusive of nurse stations), medical rooms, lavatories, outpatient rooms, diagnostic and treatment areas.

IOP Conf. Series: Earth and Environmental Science **117** (2018) 012017 doi:10.1088/1755-1315/117/1/012017

Cleaning to protect health without harming the environment is delineated as Green Cleaning. It is achieved by using processes and materials by ensuring that the cleaning is performed with the interest of all the building residences' well-being and the environmental health [14]. The objective is the reduction of the consumption of chemicals substances, water and energy. Minimizing environmental and human-health impacts are the aim of green cleaning, whilst maintaining and/or refining the effectiveness of the cleaning procedures [15]. Green cleaning is linked to the notion of sustainability [14, 15, 16, and 17]. Implementation of green cleaning will have an encouraging correlation between the economy, society and environment. When participating a green cleaning practice, it is not sufficient by simply insinuating the usage of an environmental friendly cleaning product in a hospital building without taking into account the measures and techniques requirements. Inappropriate and negligence in the usage of the cleaning product will not only lead to ineffectiveness in cleaning but may also cause more harm to the environment [18]. A variety of chemical substances have been applied in the cleaning practice (removing dust and killing microorganism) which routinely performed in hospital building. Increasing evidence which showed the ingredients of the chemical substance give unintended negative impacts on the environment and occupants' health. Patients and staff are more exposed to cleaning chemicals which can cause dermatitis, endocrine and neurologic effects, cancer, asthma, and other respiratory disorders [28 and 29]. Cleaning chemicals may contribute to the pollution of outdoor air and water supplies, damage to ecosystems, bioaccumulation in animals and plants, and ozone depletion [28]. Reference to [19] and [20], the green cleaning provisos should tackle the types of supplies and machineries/tools used, the efficiency of using the product, coupled with a suitable coaching and staffing levels, as well as supervision and resident accountability in order to create the healthiest building achievable. Hence, the green cleaning requisite should be distinct from ensuring that the cleaning processes of a building are conducted in an effective and efficient manner.

The importance of the green cleaning procedures was highly supported by the aforementioned reviews of hospital buildings. Contrariwise in Malaysia, the specific standard or regulations has not been established for green cleaning, although there are a few researchers that recommend the Malaysian Authorities need to emphasize and enforced green cleaning in their guiding principle [4, and 21]. Green Building Index (GBI) rating tool is a recognizable green rating tool for buildings in promoting the endurance of systems and processes in the built environment. Unfortunately, there are no specific requirements emphasized on the GBI rating tools for the green cleaning. A general reference to green cleaning in the GBI rating tools are limited to the usage of environmentally friendly materials. However, it is not restricted to that aspect but also incorporates all janitorial services inclusive of methods, measures, training and schemes against the protection for the health of the residents in its design whilst also not causing more harm to the environmental [21]. Consequently, it is imperative to comprehend the idea of green cleaning and its inputs to a sustainable advancement.

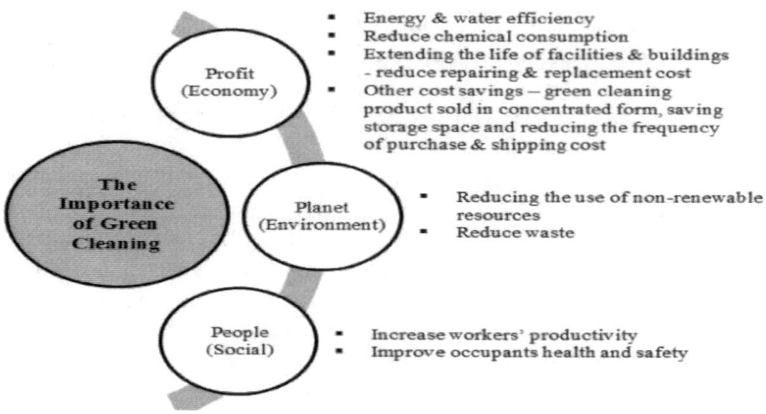

Figure 1. The importance of green cleaning implementation for sustainable development *(Sources [4])*

4. The Study

Figure 2 below shows the steps taken to achieve this study.

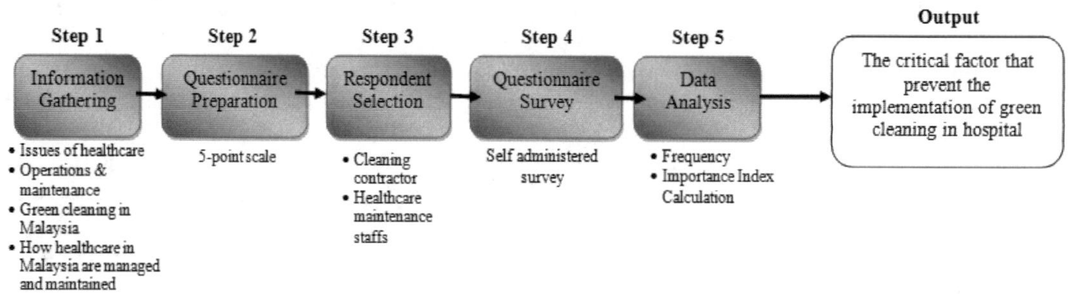

Figure 2. The study's research diagram

Step 1: Information Gathering

The researcher gained all the information about issues of hospital and green cleaning from primary sources (interview) and secondary sources (articles, journals etc.). The interview sessions have been conducted purposely to get the view of the practitioners regarding the factors and most critical factor that prevents the green cleaning implementation in Malaysian hospital building.

Step 2: Questionnaire Preparation

The questionnaire designed for this research employed closed ended questions i.e. the Likert scale of five ordered responses levels. The Likert scale is frequently seen as 5-point scale starting from "strongly disagree" until "strongly agree", on the other with "neutral" in the middle. The five scales chosen is to make sure that the respondents who are unable to decide if they "agree" or "disagree" will be able to state their "not sure" or "neutral" views (neither agree nor disagree) regarding the statement given.

Step 3: Respondent Selection
This study encompassing two groups of respondents, namely; hospital maintenance staff from MOH Cleansing Service Department (head of department and operation engineer) and cleaning service providers (manager and supervisor). They were chosen due to the role they play in the cleaning services in public hospital. The hospital maintenance staff (*Cleansing Service Department*) monitor and coordinate the cleaning services which has been conducted by the cleaning service provider. The scope of cleaning for hospital covers cleaning of wards, clinics and other areas in the public hospital building include operating theaters, laboratories, pharmacies and the surrounding area. The hospital maintenance staff involved are from 14 general public hospital (700beds and above) and has operating for over 15years. The general public hospital as Table 1 below;

Table 1. General public hospital in Malaysia

Hospital Kuala Lumpur	Hospital Sultanah Aminah
Hospital Melaka	Hospital Sultanah Bahiyah
Hospital Permai, Johor Bahru	Hospital Sultanah Nur Zahirah
Hospital Pulau Pinang	Hospital Tengku Ampuan Afzan
Hospital Raja Perempuan Zainab II	Hospital Tengku Ampuan Rahmah
Hospital Selayang	Hospital Tuanku Jaafar, Seremban
Hospital Sultan Ismail, Johor Bahru	Hospital Raja Permaisuri Bainun

Cleaning service has been outsourced and awarded to three (3) cleaning service providers for three (3) different regions (northern, central and southern). All these 3 providers involved in this study. The cleaning service providers for Malaysia hospital building namely;

- Radicare (M) Sdn. Bhd.: Wilayah Persekutuan Kuala Lumpur, Selangor, Pahang, Terengganu, and Kelantan.
- Edgenta Mediserve Sdn. Bhd.: Perlis, Pulau Pinang, Kedah and Perak
- Medivest Sdn. Bhd.: Negeri Sembilan, Melaka, and Johor

Step 4: Questionnaire Survey
A self-administered survey was carried out involving 30 public hospital maintenance staffs (*Cleansing Service Department*) and 30 from selected cleaning contractors. The respondents were given a month to complete the survey. After a month, the number of returned questionnaire is 52 where 25 of the respondents were from hospital maintenance staff and 27 from cleaning contractors.

Step 5: Data Analysis
The data were then analysed using Frequency and Importance Index Calculation.

Based on the data gathered, the study has indicated that there are five major factors that hinder the green cleaning implementation for Malaysian hospital buidling. The factors are; (i) cost and financial problems; (ii) lack of knowledge and skills of cleaning service provider; (iii) availability of product; (iv) lack of awareness; and (v) absence of green cleaning components and requirements. The result from the questionnaire survey has been shown in Table 2 below.

Table 2. The result of questionnaire survey

Factors	SD (1)	D (2)	N (3)	A (4)	SA (5)	Criticality Index	Ranking
Lack of knowledge and skills	0	0	7	8	37	0.9154	2
Cost and financial problems	0	0	8	10	34	0.9000	3
Absence of green cleaning components and requirements	0	0	0	4	48	0.9846	1
Availability of product	0	3	5	8	36	0.8961	4
Lack of awareness	0	0	16	31	5	0.7192	5

SD – Strongly Disagree D – Disagree N – Neutral
A – Agree SA – Strongly Agree

According to the result above, the factor about "absence of green cleaning components and requirements" is the most critical factor that hinders the implementation of green cleaning in Malaysian hospital building. This result is extremely supported by a study by [22, 23] which an absence of the requirements is among the key factors of a project failure. Green cleaning program considered as a project and the absence of green cleaning components and requirements, building owners are unable to grasp as to what cleaning contractors are supposed to furnish with their services for an effective and efficient cleaning. With this mindset, the owner or building management would have set the specifications to which the cleaning service providers have to abide by. As a result, green cleaning agenda could not be successfully implemented in a building.

5. Conclusion

As green building in Malaysia which is one of the sustainability approaches gave impact to the total human well being, green cleaning were also play an important role to provide clean indoor environment and save costs as well. A complete green cleaning programme will produce a good return on investment because of the green cleaning benefits that can be achieved. Green cleaning programme is one of the approaches for energy saving and reduce the energy cost up to 10%, and also saves water usage up to 70% [16]. Besides, green cleaning generates a healthy indoor environment which enhancing worker's productivity up to 5-7%, enhance attendance by more than 11%, decreases illness by 20-30% [16]. Inefficient cleaning practice because of the absence of requirement cause the hospital missed the chances to receive these profits. Consequently, the study extremely emphasis that it is vital to study the green cleaning practice for Malaysian hospital building, which consists of processes and procedures that should be incorporated in green cleaning practice.

[1] Castro JM 2009 Health Care in Malaysia. Available at http://www.expatforum.com/articles/health/health-care-in-malaysia.html. Accessed September 26, 2012.

[2] Hashim AR 2006 *Maintenance Management And Services (Case Study: Perkeso Building's In Penisular Of Malaysia)* Master Thesis (Universiti Teknologi Malaysia, Malaysia)

[3] Izran, Nurul, Shardy, Neo and Nur 2014 *Theoretical and Empirical Researches in Urban Management. Critical Factors that Lead to Green Building Operations and Maintenance Problems in Malaysia* **9** 68

[4] Nurul, Izran, Maizan, Neo, and Abdul 2015 *Jurnal Teknologi. Green Cleaning: An Essential Aspect of Malaysian Green Buildings* **10** 65
[5] Khamidi MF, Ashola OA and Idrus A 2010 *Malaysian Construction Research Journal. Building Maintenance: A Path towards Sustainability.* **7** 47
[6] Zakaria H, Arifin K., Ahmad S, Aiyub K and Fisal Z 2011 *Journal of Techno-Social. PengurusanFasilitiDalamPenyelenggaraanBangunan: Amalan Kualiti, Keselamatan dan Kesihatan* 23
[7] Olanrewaju AL 2008 Journal of Building Appraisal. Building Maintenance Management in Malaysia **4** 207
[8] Training Toolkit on Community Asset Management for Engineers 2003 Overview of Maintenance.Retrieved fromhttp://home.wmin.ac.uk/MLprojects/CAMweb/CAM1/Intro.htm.
[9] Sakina MA, Fassman E, Wilkinson S, and Adi Irfan CA 2012 Journal of Legal Affairs and Dispute Resolution in Engineering and Construction Management Practice to Achieve Energy Efficiency Performance (Case Study –Green vs. Conventional Office Building in Malaysia)
[10] Nik Mat NEM, Kamaruzzaman SN and Pitt M 2011 International Building Control Conference. Assessing the Maintenance Aspect of Facilities Management through a Performance Measurement System: A Malaysian Case Study
[11] Klungseth NJ and Olsson NOE 2013 *Journal of Facilities. Norwegian Cleaning Research: An Overview and Categorization* **31** 290
[12] Singapore Environmental Cleaning Guideline (2017). Environmental Cleaning Guidelines for Healthcare Settings.
[13] Cleaning Standard for South Australian Healthcare Facilities 2014 Available at www.sahealth.sa.gov.au/infectionprevention. Accessed July 1, 2017.
[14] Young J, Schwinghammer K, Steen E and Zaffrann D 2010 Clean Sweep: How A New Approach to Cleaning Commercial Buildings In The Twin Cities Can Protect Our Health And The Environment while Securing Jobs and Saving Money. Retrieved from http://www.bluegreenalliance.org/news/publications/document/CleanSweep.pdf.
[15] Kohls J 2010 Why Green Cleaning Makes Cents for Sustainable Facilities. Retrieved from http://www.newequipment.com/Resource.ashx?sn=makingcents.
[16] Betco Green Cleaning Workbook 200. Sustainable Green Cleaning: Cleaning for Health and the Environment. Retrieved from http://www.betco.com/SiteCollectionDocuments/Training%20Workbooks/Green%20Cleaning%20Workbook.pdf.
[17] Corbett-Shramo J, Wagner D and Esbensen P 2011 Sustainability Guide: Global Green Cleaning. IFMA Foundation.
[18] Berry M A 2011 Commentary - The Green Movement and Science. Retrieved from http://www.ciriscience.org/a_286COMMENTARY_The_Green_Movement_and Science
[19] Green Seal 2013 Green Seal Environmental Standard for Cleaning Services (pp. 18). Washington, DC: Green Seal, Inc.
[20] CIMS 2011 Cleaning Industry Management Standard-Green Building. Lincolnwood, IL: ISSA.
[21] Atifi N 2012 Clean Up, Green Up. *Business Today, 12,* 30-35.

[22] Kile JF, Little D and Shah S 2005 *The Importance of Effective Requirements Management in Offshore Software Development Projects* (Doctor of Professional Studies in Computing), Pace University, United States.

[23] More NT, Sapre BS and Chawan PM 2011 *International Journal of Internet Computing (IJIC).* An Insight into the Importance of Requirements Engineering **1** 34

[24] Kamaruzzaman S.-N, Myeda NE and Pitt M 2013 Performance Levels of High-Rise Private Office Buildings Maintenance Management in Malaysia. *Maintenance and Reliability* **15** 111

[25] IDAHO 2006 H2E 10 Step Guide to Green Cleaning Implementation (pp. 8). Lyme, NH: IDAHO Department of Environmental Quality.

[26] GGHC 2008 Green Guide for Health Care.

[27] Rumchev, Spickett, Bulsara, Phillips and Stick 2004 *Thorax. Association of domestic exposure to volatile organic compounds with asthma in young children* **59** 745

[28] Environmental Protection Agency. (n.d.). *Green cleaning pollution prevention calculator.* Retrieved May 23, 2011 from: http://www.fedcenter.gov/janitor/

[29] Quan X, Joseph A and Jelen M 2011 *Green Cleaning in Healthcare: Current Practices and Questions for Future Research.* Retrieved from Effect of Employee Training on the Perceived Organisational Performance: A Case Study of the Print-Media Industry

The satisfaction level of indigenous people towards planned housing development. Case study: Orang Asli Kg. Sejagong, Sri Medan, Yong Peng Johor

S M Kamaruddin[1], N S Zaini[1], A Misni[1], P Ahmad[1]
[1]Faculty of Architecture, Planning and Surveying, Universiti Teknologi MARA, 40450 Shah Alam, Selangor, Malaysia

*sitim065@salam.uitm.edu.my

Abstract. Orang Asli possesses a unique relationship with their traditional land. Indigenous people have their thoughts on development, life values, visions, needs, and priorities. The government through Department of Orang Asli Development or Jabatan Kebajikan Orang Asli (JAKOA) implemented various development programs such as structured settlement program, physical improvements such as Regrouping Plan (RPS), Village Restructuring project (PSK), New Village Plan (RKB) and Orang Asli settlement customary Land Survey (village) / Land Acquisition. Orang Asli Community. However, their sense of satisfaction towards the development implemented is rarely evaluated. This paper aims to highlight Orang Asli's thoughts and level of satisfaction towards the planned housing development of their settlement. A mixed methodology was employed using survey and interview to record their thoughts. The findings suggest that the Orang Asli Community's sense of satisfaction towards housing development has improved. However, their needs to adhere to their social traditions and norms such as living adjacent to relatives, surrounded by nature within a low-density environment must be respected.

1. Introduction

According to [1], there are greater than 370 million indigenous people spread through 70 countries around the world. Indigenous peoples or Orang Asli in Malaysia are the holders of knowledge systems, specific languages, beliefs and own precious information of practices. Orang Asli possesses a unique relationship with their traditional land. Indigenous people have their thoughts on development, life values, visions, needs, and priorities. In Peninsular Malaysia, the Orang Asli comprise of subgroups such as Negrito, Senoi, and Malay-Proto. To protect the rights of Orang Asli and improve their living standards, the Government as early as the 1950's had established the Department of Orang Asli Affairs or Jabatan Hal Ehwal Orang Asli (JHEOA). This department later became the Department of Orang Asli Development or Jabatan Kebajikan Orang Asli (JAKOA) governed under the Ministry of Rural and Regional Development. The government implemented various planned development programs such as structured settlement program, physical improvements such as Regrouping Plan (RPS), Village Restructuring project (PSK), New Village Plan (RKB) and Orang Asli settlement customary Land Survey (village) / Land Acquisition. Orang Asli Community. The government has applied particular development programs for the indigenous, the Orang Asli, to improve their dwelling. Within the Orang Asli community, JAKOA identified the communities settlement hierarchy and divided this according to urban, rural and semi-urban. In 2010, the Census indicated a total of 876 settlements with six settlements were in towns (0.7%), 547 villages lived in rural (62.4%), and 323 villages live at the edge of the wooded area (36.9%). A study by researchers on Weighted Quality Life Index for Orang Asli (IKHOAw) [2] found that in almost all respects the Malaysian aborigines' quality of life was very satisfactory; it is only less so with respect to electricity and water supplies, motorised vehicle ownership and housing quality [2].This paper aims to highlight Orang Asli's thoughts and level of satisfaction towards the planned housing development of their settlement. The findings would enable policy makers to determine the extent of impact the development has towards the community.

2. The Pattern and Design of Orang Asli Settlement

According to [3], currently there are three Orang Asli planned settlement layouts in Malaysia, i.e., cluster, linear and mixed (Figure 1-3). The pattern of every settlement is unique. However, there are similarities of each settlement, i.e., the community consists of the same ethnic family who is related often geographically located in proximity to the river and surrounded by orchards, farm, and woodland.

2.1 Cluster Layout

The houses are clustered together and located adjacent to the river. The layout maintains an informal setting with good accessibility. An example of this form of settlement can be found in Chinggung Village, Tanjung Malim, Perak.

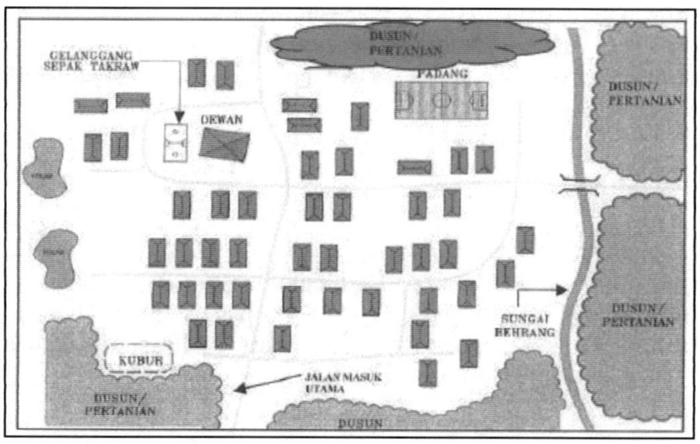

Figure 1. The Illustration Clustered Settlement Pattern In Chinggung Village, Tanjung Malim, Perak [2]

2.1.1 Linear Settlement. Based on [3] the linear or parallel settlement is usually formed along the river and road. The pattern of settlement was built to facilitate them to water and food supplies and accessibility. An example of a settlement which has this type of pattern is within the Semai race in Ulu Geruntum Village, Gopeng, Perak

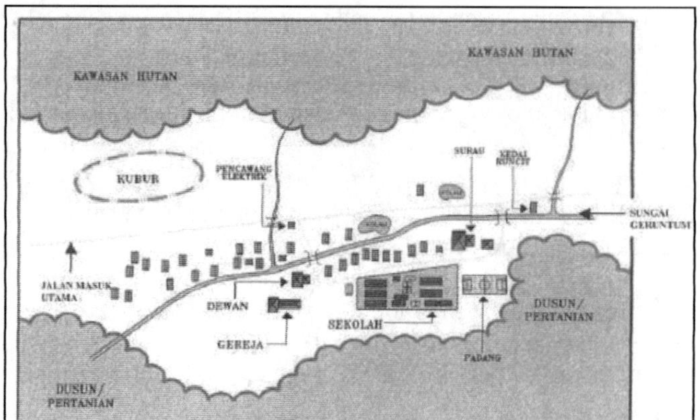

Figure 2.The Illustration Linear Settlement Pattern in Chinggung Village, Tanjung Malim, Perak [2]

IOP Conf. Series: Earth and Environmental Science **117** (2018) 012018 doi:10.1088/1755-1315/117/1/012018

2.1.2 A Mixed Development. According to [3], a mixed settlement is a mixture of two forms of agreement parallel and cluster settlement. This settlement will shape when they are many humans inhabit in one vicinity. Typically, it is populated by a few households as a collection. Each house is separated using physical barrier or plants. An example in Bukit Manchong Village, Hulu Selangor.

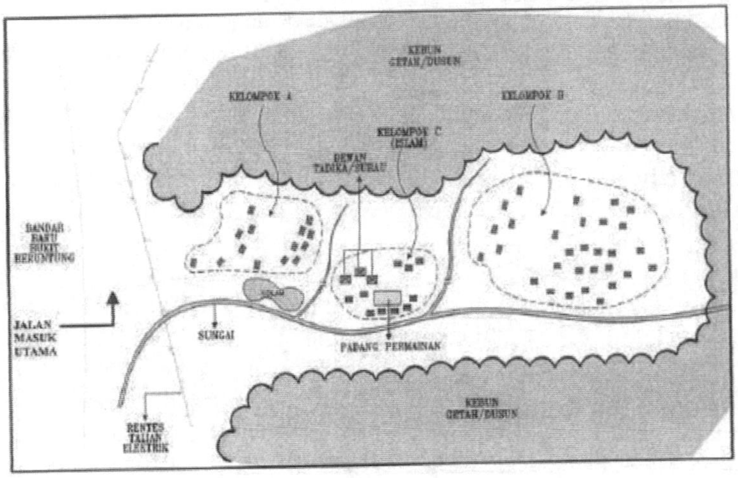

Figure 3. The Illustration Mixed Settlement Pattern in Chinggung Village, Tanjung Malim, Perak [2]

2.1.3 The Design of Traditional Orang Asli Housing. According to [3], the maximum traditional Orang Asli housing design emphasizes natural elements suitable to the climate and ventilation. They use palm leaves, bamboo, rattan, and timber to make a home which is easily available from their natural surroundings. Their traditional house size is 25' x 20'. There are three basic indoor components, i.e., living, bedroom, and kitchen. The homes are typically near the homes of different spouses and children. The strategy would facilitate social interplay and close supervision by the head of the family. According to [3], housing is built taking into consideration proximity to timber and water resources as the Orang Asli will not construct a house at the hill or valley slit region because this may disrupt the natural spirits. Based on their beliefs a person who desires to build a house must select a clean area. The Orang Asli also considers the significance of dreams in choosing a site for residence. The housing design for Orang Asli is similar to the conventional Malay house in which there are three specific regions with specific features. There are six components of a traditional Orang Asli home, i.e., the site, supporting poles, floors, walls, windows, and roofs.

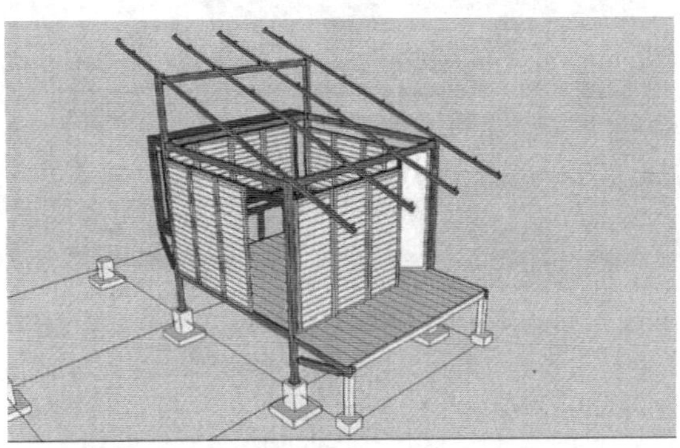

Figure 4. Illustration of Design of Traditional Orang Asli House.
Source: http://www.starproperty.my/index.php/articles/property-news/new-homes-for-orang-asli-families

2.2 Study Area

The Orang Asli settlement of this case study is Kg Sejagong, Mukim Sri Medan, Yong Peng, Johor. The area covers 230.51km2. (56,960acres). It is one of the earliest Orang Asli village under the structured planned development project implemented by JAKOA under the Ministry of Rural and Regional Development. It is under the jurisdiction of the Yong Peng Municipal Council, District of Batu Pahat, Johor. The earliest settlers of Hutan Maokil near Kg. Sejagong were Orang Asli families. According to the Tok Batin, the pioneer Orang Asli family who settled in this village was Jamilah binti Katan and her husband Khamis bin Samin in 1972. Their generations and family still reside in this area. The settlement layout is linear with access from the PLUS Highway via Jalan Kampung Parit Warijo-Jalan Kampung Parit Singgahan. The distance between the main roads to the study area is about 20km, and it takes 30 minutes by car to the study area. The Malay Settlement near Kampung Sejagong is Parit Warijo and Parit Jayus, The land use is also dominated by some commercial consisting of formal and informal business activities. The market area in Sri Medan attracts residents from Kampung Sejagong, Parit Warijo, Parit Baru and Parit Haji Noor.

2.3 Data Collection

A mixed methodology approach was employed. The interview with the Tok Batin considered an expert in Orang Asli matters, forms the qualitative data while questionnaires record the respondent's opinions on the subject matter. The questionnaire lists three aspects designed to record respondents' opinion on their current level of housing development satisfaction and how they felt two years ago. The main questions asked were related to "Home space for family activities," "Sense of Safety," and "Housing layout." This research was conducted on 26th and 27th May 2017. The enumerator distributed 52 questionnaires to all 52 heads of household in Kg. Sejagong and the return rate was 100%. Although there are only 27 houses within the settlement, it is possible for two related families to be living under one roof. The enumerator was of the same ethnicity and speaks Bahasa Orang Asli which expedited the process of data collection.

2.3.1 Inventory of Housing Development. The interview with the Tok Batin revealed that within the study area, the renovation was carried out on 21 out of a total of 27 houses with funding assistance from the government. The aim for renovation works was to improve the living conditions. All of the houses were originally built by the older generations of the Orang Asli while, the younger generation undertook the renovation. The structure of the housing is similar to a typical Malay house (refer to Photo 2.4 and Photo 2.5). The original building material was wood, but some homes were renovated using bricks. The original roof made of Nipah has been replaced with zink roofing while maintaining a pitched roof design.

Photo 1. Renovated traditional Orang Asli house in the study area

2.3.2 Interview. The expert interview was conducted directly involving officers from Department of Orang Asli Affairs (JAKOA) responsible for the development of Orang Asli. The data Triangulation of the qualitative data This step would ensure the validity of the information about this group of community.

2.3.3 Survey. A survey was conducted among the Orang Asli heads of household to obtain information regarding the residents' satisfaction towards the physical development. Their response was recorded using a questionnaire. The researchers used a numerical value for each of the scale designed to indicate the respondents level of satisfaction to a certain aspect of physical development. The questions incorporated psychological aspects which interpreted a place in which the occupants feel safe within a peaceful atmosphere merely than the physical conditions of the house [4]. A home can also be interpreted as a place of shelter from external surroundings and is a place for rest and leisure. The internal home space provision for family activities, an overall sense of safety and external housing layout are the three components that were considered by the respondents to evaluate their level of satisfaction towards the housing development in their settlement. Table 1 indicates the values used for further analysis.

Table 1. Level of Satisfaction

Value	Remarks
0.00 – 1.00	Very Unsatisfied
1.01 – 2.00	Unsatisfied
2.01 – 3.00	Moderate
3.01 – 4.00	Satisfied
4.01 – 5.00	Very Satisfied

(Source: Adapted from [5])

2.3.4 The population of Orang Asli. The population of Orang Asli. Kg.Sejagong comprises of 249 Orang Asli from 52 families. Based on the age structure, the highest number of population is the age between 15-24 years old (74 people) followed by those in the age group 5-14 years old, (59 people). The two age structures that have the lowest number of people, (age group of fewer than five years old and age group of more than 55 years old). From the interview with the Tok Batin, most are from the third and fourth generation. Kg. Sejagong has a few second generations left.

Table 2. Education Level of Population

Age	**Education Level**						
	Primary School	**Secondary School**	**Uni/ College**	**Never Attended School**	**Still Schooling**	**Total**	**%**
< 5 years	-	-	-	6	4	10	4.02
5-14 years	-	-	2	-	57	59	23.69
15- 24 years	1	13	29	-	31	74	29.32
25-34 years	-	13	3	-	-	16	5.62
35-44 years	-	23	1		-	24	8.43
45-54 years	7	28	-	11	-	46	10.44
> 55 years	3	7	-	10	-	20	18.07
Total	11	72	35	27	102	249	
(%)	4.42	28.92	14.05	10.84	40.96		100

Table 2 shows that the majority, i.e., 40.96% of the population are still schooling. The second highest is the population that finished their secondary education (28.92%), followed by those who obtained university or college education (14.86%). Only 27 people (10.84%) never attended school. Based on the interview with the respondents, they stated that in their day, education was not a necessity. However, some from the age group of 15-24 years old had scholarships from JAKOA to study in the university or college.

2.3.5 Level of Satisfaction – Home Space for Family Activities. Based on Table 3, the total heads of household who were very unsatisfied about home space for family activities in 2015 was 5.8% and this figure decreased to nil. In addition, 9.6% of household heads reported that in 2015 they were unsatisfied about home space. However, in 2017, this figure decreased to 3.8% of the respondents who reported that they were unsatisfied with the level of home space. The reason may be due to the renovation of the majority of the houses for more home space for family activities with better conditions, and some family members had moved out thus creating more space for the remaining family members.

Table 3: Satisfaction Level of Home Space for Family Activities

Frequency of Satisfaction	Home Space for Family Activities			
	2015		2017	
	Total	Percentage (%)	Total	Percentage (%)
Very Unsatisfied	3	5.8	-	-
Unsatisfied	5	9.6	2	3.8
Moderate	21	40.4	16	30.8
Satisfied	10	19.2	14	26.9
Very Satisfied	13	25.0	20	38.5
Total	52	100.0	52	100.0

Table 4: Satisfaction Level of Safe Home Environment and Housing Layout

Frequency of Satisfaction	Safe Home Environment			
	2015		2017	
	Total	Percentage (%)	Total	Percentage (%)
Very Unsatisfied	-	-	-	-
Unsatisfied	7	13.5	1	1.9
Moderate	21	40.4	14	26.9
Satisfied	13	25.0	19	36.5
Very Satisfied	11	21.2	18	34.6
Total	52	100.0	52	100.0

The table 4 above shows that currently, most of the Orang Asli in the study feels safe living in their settlement. The current layout of the houses within Kg Sejagong is linear. In 2015 however, 53.9% felt moderately satisfied and unsatisfied with this aspect but currently (2017) only 26.9 % feels unsatisfied. According to families, they feel safer because they are closer to each other. Although there are some reported cases of theft, the number of cases is decreasing.

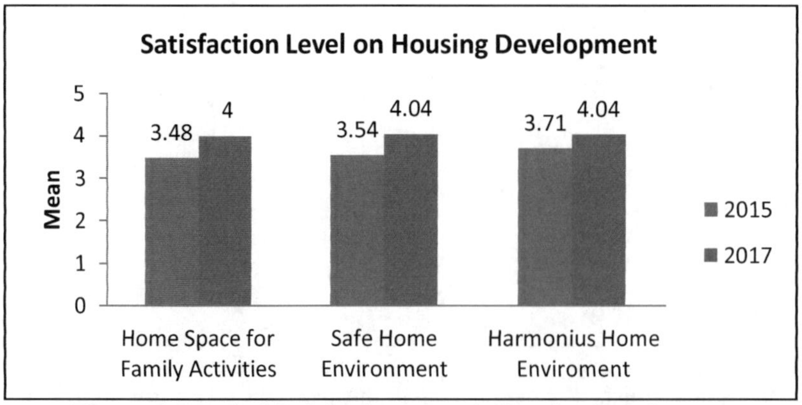

Figure 5. Level of Satisfaction towards Housing Development

Figure 5 highlights the average value of respondents' opinion of the three aspects related to satisfaction level of housing development within their settlement. Using descriptive analysis, the 52 Respondents were asked to rate the relevant question using the value on the scale. The data were analysed into frequencies and converted to a useful mean value by analysing the responses using the given score to summarise the perceived satisfaction level. This gives indication if respondents were generally satisfied or not, by how high or low the mean is according to the year identified. In general, the mean value of each aspect of satisfaction indicates a value higher than 3.00 interpreted as respondents feel satisfied regarding aspects related to development. The bar chart shows that their level of satisfaction regarding housing development had improved from two years ago. The overall mean value of satisfaction level of housing development improved from a mean value of 3.57 to the average value of 4.03, i.e., 12.8% increase in satisfaction level.

2.4 Conclusion
The overall analysis suggests that the Orang Asli Community's sense of satisfaction towards housing development has improved. However, their needs to adhere to their social traditions and norms such as living adjacent to relatives, surrounded by nature and woodlands within a low-density environment must be respected. It is also important to maintain the "informal" or traditional kampung layout of the housing development while ensuring this community has access to basic amenities and infrastructure such as water, telecommunication, and electrical supply. Further research on these aspects is required to incorporate the projected population and sustainable socio-economic opportunities.

References
[1] Jaura R 2010 Of Indigenous Peoples, Climate Change and Rural Poverty. Retrieved from Development Watch An Alliance for Monitoring International Cooperation:Ministry of Rural and Regional Development
[2] Harun and Idris 2012 Akademika **82** 65
[3] Suki Mee Malaysia 2009 Betascript Publishers Dez
[4] Gifford R, Steg L and Reser JP 2011 Environmental Psychology, in IAAP Handbook of Applied Psychology (eds P. R. Martin, F. M. Cheung, M. C. Knowles, M. Kyrios, J. B. Overmier and J. M. Prieto), Wiley-Blackwell, Oxford, UK.
[5] Diener E, Emmons RA, Larsen RJ and Griffin S 1985 *Journal of Personality Assessment. The Satisfaction with Life Scale* **49** 71

Contractor's mistakes during tendering

A I M Kamil[1], N Othman[1], H Adnan[1], ZA M Zaki[2]

[1]Department of Quantity Surveying, UniversitiTeknologi MARA, 40450 Shah Alam, Malaysia
[2]Faculty of Built Environment, UniversitiTeknologi Malaysia, Malaysia.

*abdulizzkamil@gmail.com

Abstract. The occurrence of overrun in cost and time in a project that is usually caused by contractors, will generally make peopleput the blameonthem. This happensas a result of the lack of knowledge possessed by the contractors in determining the solution to be undertakendue totheir mistakes, especially when it comes to tendering process as what had happened in the cases that were reported in Supreme Court of Canada. This paper aims to identify the types of mistakes, factors that contribute to the mistakes and the responsiveness to the mistakes so that the contractor will be able to manage and control the mistakes, as well as the solution, along with preventing them from happening, and in making decision whether or notthey may withdraw from the bidding when the fault has been discovered or whether they should request for leniency from the client for a second chance in re-submitting the bid after the discovered inaccuracy. This research has been done based on cases that were related to the contractor's mistakes during tendering which were collected from the Supreme Courts of Canada. The data have been analysed and interpreted by legal methodology, in which the result shows that the ability level of the contractors to respond to the mistake is unsatisfactory,and the fault is continuouslybeingmade. The findings are expected to educate and provide awareness to the contractors relating to the matters discussed.

1. Introduction

It is a global phenomenon in the construction industry when delay occurs in a project and it has become a real issue that continuously increases by years. Delay usually occurs when contractors fail to provide a comprehensive project management [7]. Therefore, problems such as disputes over cost and time overruns are mostly influenced by the contractors' negligence. About 17.3% of public sector's projects in Malaysia are considered sick as stated in the statistic in 2005 [7]. Occurrence of overrun in cost and time in a project that is caused by the contractorsis the result produced due to the lack of knowledge by the contractors in determining the response to be undertakenowing to their mistakes especially when it comes to tendering process.

Wrong interpretation and lack of understanding in agreement and contract will commit ignorance of implied terms within the contracts [3]. Generally, mistakes maybe associated with negligence, misrepresentation and fraud as the mistakes may be in a good faith or bad faith as it may causeconfusion to suffered parties. Thethree frequent mistakes recognised by the English law contract are, common mistake where both parties commit the same mistake, unilateral mistake where only one party commit the mistake and finally, the mutual mistake where the parties have intentionally crossed the mistake.

The tendering process has been allocated with cost and time consequences, whereby the tendering process is used as it is not to burden the tenderer with the requirement of excessive information and to ensure that the tenderer can focus on necessary works required [6]. The process of tendering is to guarantee that the real competition is achieved in terms of financial strengthsand technical experience related to assessment in the final submission of tendering. The tendering process is a process that involves two parties; the employer and the contractor, as a manner to achieve an offer and accepting to the agreement of works.

1.1. Problem statement

The mistakes made by contractorsin tenderingareunforgivable as it is clearly mentioned in the contract, that the contractors should burden the risk that is transferred to him and they should acknowledge the disclaimer provided in the contract. When the contractorsmake mistake, most oftenthey will charge a higher pricethanpreviously established to cover for their mistakes. In addition, when they are at fault, they normally want to represent themselves as the innocent party who unintentionally made the mistakesand claim those blunders as misrepresentation.

Estimating the error in tendering can result in cost overrun while the project is being executed as the contractorswould only realise that they are making mistakes after encounteringfinancial difficulties during the execution of work [4]. Most often, the lowest tenderer is the person who is likely to commit mistakes and resulting in delay of the project,only then the contractor would realise that they have miscalculated, as they have submitted unreasonable low bids that is insufficient to carry out the works to the project completion [9]. The tendering method used is to provide competition and to reduce the cost of the project, therefore,many inexperienced contractors or 'low bidder' who participate in the tendering and bump into financial difficulties, would automatically force the client to pay additional cost to recover the default.

Most contractors nowadays are too confident and greedy in pursuing projects, although, as contractors they should verify and make sure that all the information or document that they receive are correct. If the contractor did not clarify the discrepancy before the submission of the tender, any mistake in the contract will be at his own risk. Therefore, all the discrepancy shall be checked and confirmed before the submission of the tender as the profit or loss of the contractor depends on his meticulousness during tendering. Mistakes in tendering are a nightmare to the contractor as the probability of being stuck in financial quandary is very high because of the mistakes. Most of the contractors who are facing problems due to their mistakes, do not know how to address those faults when they are finally being disclosed. In their point of view, the most common cause of dispute in construction is cost overrun, therefore the easiest solutions they can come up with to conceal their error, are to withdraw the bids, terminate the contractor, flee from the project and try to manipulate claim and variation order. In this situation communication plays the biggest role as both client and contractor need to contribute the best outcome for the finest solution.

1.2. Aim and objective of study

The research's aim is to identify the mistakes made by contractor as well as the contributing factors during the tendering process so that the mistakes can be managed, controlled and the parties involved will be able to use the knowledge to work on the solution.

1.3. Scope of the research

The scopes for this study will focus on the aspects where the research will be a depth study on the mistakes made by contractors during tendering but post contract will not be touched as mistakes can be found in many stages of construction. Next, this study will also cover law cases that are related to the mistakes in tendering and contractors' awareness of the importance of cost estimation during tendering, along with their responsibility in addressing the issue. The research will most probably collect data from legal cases in Canada and the researcher will focus on faults in tendering, as well as specifically delving into the bidding during tendering process. Canada Courts is selected because the author finds that there are no case recorded involving bidding mistakes and tender error in Malaysia courts through LexisNexis (also known as Malaysian Law Journal). However, related cases was found recorded in Canada Court and they are practicing English Common Law.

2. Literature review

Tendering is a common process for a client to select the best contractor to execute the works. In selecting a contractor, a negotiation can be used even though the tendering is fused with competition, astraditional open tender method is popular and widely used as the procurement in most construction procedure [1]. The tendering process is to make sure that the concurs is fair and square as the client, most often than not, will award the contract to the lowest bidder considering that they always seek for cheaper price offered,in order to save more money. In asimpler word, the tendering can be understood as a process or activity that involves two parties; the employer and the contractor in amanner toachieve an offer and to agree on an arrangement ofthe works.

However, the implementation of open tender can create confusion to the contractors as many unfamiliar organisations willtake part and it will generate risksfor the client in finding a genuine contractor to run the project [8]. This situation is bad news for the industryas people will perceive the industry as money making industry thus will downgrade the quality of the construction. Recently the implementation of open tender is seen to decline as the process doesn't guarantee a 100% success rate in the construction industry. If we take a look in the aspect of competition, the contractor may submit a dishonest bid since sometimes the bid submitting is not a bona-fide submission and this process has been criticized since 1940 and the selective tender has taken over the open tender process [2].

The case of "M.J.B. Enterprises Ltd. V. Defence Construction (1951) Ltd., [1999] 1 S.C.R. 619" has proven that the use of this clause in this case, initiated an action for breach of contract, claiming that its bid should have been accepted as the lowest compliant bid. In defending its acceptance of the qualified bid, the owner relied on a privilege clause in the tender documents which stated that "the lowest of any tender would not necessarily be accepted." At the Supreme Court of Canada, the central issue was whether the privilege clause allowed the owner to disregard the lowest bid in favour of another tender, including a non-compliant bid.

Generally, contractors can be categorised in various background. They may be professional contractors who may have come from academic background and specific skills related to construction, politician contractors that become contractor after being awarded a project by political activities, a business man contractor who enters the construction for the sake of business diversification or the nouveau riche contractor. Today, construction industry is seen as one of the fastest money-making industries, therefore the contractor business has grown rapidly and the major challenge has to be faced by the contractors is tough competition. Most of the contractors who encounter problems that are related to the mistakes do not know how to counteract to the issue after it is being discovered. As the competition is tight, it will influence the contractor to submit unrealistic cost, if we look from the competition's point of view, the contractor may submit a dishonest bid as sometimes the bid submitting is not a bona-fide submission and this process has been criticised since 1940 and the selective tender has taken over the open tender process [2].

3. Methodology

3.1. Preparation of research proposal

Problem statement was identified by observing the current trend and issue. Based on the problem statement, formation of aim, objective and scope of research to be determined.

3.2. Literature review

In this chapter the researcher uses secondary data that is provided from journals that have mostly been retrieved from Sciencedirect as the main source. Other than that, researcher has developed the topic through readings of journals, books and trusted webs. The researcher has organized the literature review into parts to present the content as the researcher tries to synthesis the data in which he will reshuffle and re-organize to recap the important data according to the guiding concept of research.

3.3. Method of data collection

Document analysis is identified as the most appropriate method to be used in collecting data as it is in qualitative form which allows the data to be interpreted to coordinate to the research objective. The method is suitable for this research due to the necessity of the researcher to assess the law journal in meeting the research objective. The purpose of document analysis is to study the analysed data using text analysis to find themes and descriptions which is implied into the scope of the study. The Primary data was collected from Lexum which is the tool for the main data collection. Lexum is a site that provides the main public source of Supreme Court of Canada in the original published format, content of decision and judgement for the courts hearing. The data collected from Lexum is through legal cases and legislation where the legal cases referred to the transcripts that contain published written legal opinions. The written legal opinion is an issue of law in which the case is being litigated in judge's courtroom as the data is needed to fulfil the significance and the scope of the study. Lexum is a database that is designed to deliver services of hosting data from many Canadian legal institutions, including the Courts Administration Services, Canada's Department of Justice and The Federation of law Societies of Canada. For this research the researcher uses the keyword of "Contractor's Mistake", "Bid Mistake", Error in Tendering" and "Unilateral Mistake in Tendering" to access the law case in order to collect data based on the theme and scope of the research. The Cases were gathered from the Courts of Canada, including Supreme Courts of Canada, Ontario Courts, British Columbia Courts Manitoba Courts and Saskatchewan Courts as seen in Table 1 and Table 2.

Table 1. Table shows the result for keywords 'Bid Mistakes' in Lexum

NO	CASES	REFERENCING	Court of Judgments
1	The Queen (Ont.) v. Ron Engineering	[1981] 1 SCR 111	Supreme Court of Canada
2	Corpex (1977) Inc. v. The Queen in right of Canada	[1982] 2 S.C.R. 643	Supreme Court of Canada
3	Calgary v. Northern Construction co.	[1987] 2 SCR 757	Supreme Court of Canada
4.	Toronto Transit Commission v. Gottardo Construction Ltd	[2005] O.J. No. 3689 (C.A.),	Court Of Appeal For Ontario
5.	Derby Holdings Ltd. v. Wright Construction Western Inc	[2003] SKCA 87	Court of Queen's Bench of Saskatchewan

Table 2. Table shows the result for keywords 'Tender error' in Lexum

NO	CASES	REFERENCING	Court of Judgments
1	M.J.B. Enterprises Ltd. V. Defence Construction (1951)	[1999] 1 SCR 619	Supreme Court of Canada
2	BG Checo International Ltd. V. British Columbia Hydro and Power Authority	[1993] 1 S.C.R. 12	Supreme Court of Canada
3	Bank of Montreal v. Bail Ltee	[1992] 2 SCR 554	Supreme Court of Canada
4.	Nova Scotia Construction Co. v. The Quebec Streams	[1933] SCR 220	Supreme Court of Canada

	Comission		
5.	Doe et al. v. Canadian Surety Co.	[2003] SCR 1	Supreme Court of Canada

4. Results and Discussion

Ten (10) legal cases in Canada Courts were distributed, only five (5) were relevant to discussion which are the cases that use "Tender error" as the keyword in table 1 meanwhile another five (5) cases that are not relevant to be discussed are the cases under "Bid Mistakes" keyword in table 2.

Table 3. Table shows the summary of the data

Case Referred	Water Resources Commission. V. Ron Eng. [1981]	Corpex V. The Queen Canada, [1982]	Calgary V. Norther n Const [1987]	Tt Commissi on V. Gottardo Con [2005]	Derby Hold V. Wright Const [2003]	Total
Contract	A	N/A	B	B	A	N/A
Type of Mistake						
Bid Calculation	√	√	√	√	√	**5**
Assumption		√				**1**
Clerical Error			√			**1**
Provide Document				√		**1**
Factor of Mistake						
Carelessness	√		√			**2**
Competitiveness	√		√	√	√	**4**
Rely to Client	√	√				**2**
Communication					√	**1**
External Factor		√				**1**
Less Information		√			√	**2**
Decision						
Won	√	√			√	**3**
Lost			√	√		**2**

Table 3 describes the type of mistakes, contributing factors and the response to the mistakes. There are four mistakes to be identified that can be classified as the most frequent mistakes to happen during tendering. The four mistakes are mistake in bid calculating, which contributed about 62% in the percentile and this mistake is followed by wrong assumption in information, 13%, clerical error, 13% and errors that are related to documents, the final 12%. Next, there are six factors to be identified as the main factor of the mistakes. The main factorsthat contributed to the mistakes are competitiveness in winning the bid which is 33%, followed by carelessness of the contractor in submitting the bid and making assumption, 17%, too relying on the information and budget provided by the client which is 17%, providing incomplete information in tendering, 17%, lack of communication, 8% and external factor contributed the less,which is 8%. Most of the time, the contractors are not tootrustful to own up to their mistakes as eventually two out of five cases lost as the contractors do not know whether or not they can withdraw the bids when they made some mistakes and the fact that they should beresponsiblein taking riskswhile preparing tender. In this case, the contractor should acknowledge the kind of situation where they can withdraw their participation and when can they be declared as non-compliant bidder. The tenderparticipant could not withdraw the tender for a particular period of time as provided in the instructions. If the tender was not accepted within that period of time, the partaker could not recover any deposit or tender bond that had accompanied the tender.

5. Conclusion

Based on the study conducted, it shows that the top four most common mistakes made by contractorsas regards to the sample of cases in Canada were bid calculation mistake, clerical error, wrong assumption and documents error. It can be clearly distinguished that, the top four mistakes can usually occur in normal practices of tendering process in building construction. As an informed contractor, one should have the knowledge in the interpretation of form of tender as they should be able to fulfil the terms provided, the specification requested and also in pricing the items, soft or hard cost. Also, contractors should be aware about the importance of proposing a low bid as to give a higher chance of winning the bid in a competition. Even so, based on the data collected, none of the contractors managed to adapt to real costing technique as the contractors seem to make mistake by omitting the important cost that should be maintained in the cost estimation. The presumptuous contractorswho are too confident in winning the tender competition finally resulting in disappointment as they mistakenly omitted the real cost that could influence the total cost of the contract and it led to the failure in performing the execution of the works. The desperationin getting the project and facing such big competitors, the contractors may submit unrealistic cost. It shows that the contractors take things a little too easily by allowing the cost to go over the previously estimated cost and charge the client without hesitation for a cover up rather than doing a detailed calculation. The contractorsare not aware when they discover their mistakes, therefore it is usually a little too late for the contractors to withdraw their submission. Based on the sample case, three out of five contractors won the case regarding the matters discussed. It is not such a positive result as it can be seen that the contractors do not know whether they are bound to the bid when no contract has been signed or whether they can withdraw their submission.

6. References

[1] Sambasivan M and Soon YW 2007 *International Journal of Project Management. Causes and effects of delays in Malaysian Construction industry* **25** 517
[2] Jose Rivera 2016 *Unilateral Mistakes in a Contract LegalMatch Law Library* Retrieved August 8,2016, from http://www.legalmatch.com/law-library/article/unilateral-mistakes-in-a contract.html
[3] Roland 2011 *Tendering for construction projects*, Retrieved from https://www.thenbs.com/knowledge/tendering-for-construction-projects, retrieved on 11th April 2016.
[4] Michael Stone 2011 *Cost Plus Contracts, Time and Material* Contracts, Retrieved from http://www.markupandprofit.com/articles/cost-plus-contracts-time-and-material-contracts, retrieved on 11th April 2016.
[5] Ugochukwu SC and Okolie KC 2013 *Investigation Into The Factors Influencing Tendering Duration Stipulations By Public Building Clients In Nigeria* **3** 15
[6] Chinyio E 2011Working Paper Proceedings.
[7] Smith NJ, Merna T and Jobling P 2006 *Managing risk in construction projects*. Oxford: Blackwell
[8] Griffith A, Knight A and King A 2003 *Best practice tendering for Design and Build projects* (London: Thomas Telford)

IOP Conf. Series: Earth and Environmental Science **117** (2018) 012020 doi:10.1088/1755-1315/117/1/012020

A comparative study on the Environmental Impact Assessment of industrial projects in Malaysia

E A Marmaya[1] and R Mahbub[1]

[1]Faculty of Architecture Planning and Surveying, Universiti Teknologi MARA, 40450 Shah Alam, Selangor

*e_anyzah@yahoo.com

Abstract. In the past decade, mankind has been manipulating the natural environment to better suit its needs for providing buildings and infrastructure for residential, commercial, business and industrial purposes. The rapid industrialization that has taken place has generated several issues regarding the environment. Therefore, managing environmental risks in construction projects has been recognized as an important process to achieve the project objectives in terms of time, cost, quality, safety and environmental sustainability. The aim of this research is to assess the environmental impact of industrial projects to the surrounding areas. The impact to the environment can be categorized into several aspects such as ecosystem impact, natural resources impact and public impact. This research employs the quantitative approach, that is, a questionnaire survey targeted at the occupants living in the surrounding areas of the case study location, namely the industrial sites in Sabah Ammonia Urea (SAMUR), Sipitang, Sabah and Lynas Advanced Materials Plant (LAMP), Gebeng Pahang. The findings of the research show that the two projects are perceived to have negative environmental impact especially for land pollution and green-house gas emissions.

1. Introduction

The construction industry is rapidly growing because of an increase in standard of living, demands for infrastructure projects, changes in consumption habits, as well as natural increase in population (Nagapan et al., 2010). A range of issues have arisen through construction activities, affecting the environment, such as harmful gases, noise, dust, solid and liquid waste (Chen et al., 2000). According to Altoryman (2014), many construction projects suffer from mismanagement despite continuous improvement in the field of project risk management. Generally, the lack of implementation of standard risk management methods in the construction industry has led to construction projects that suffer from poor performance and consequently give negative impact to the environment. The purpose of this study is to review the issues and impact of construction activities to the environment, especially for industrial projects where the likelihood of environmental deterioration occurrence may be highest. This paper begins with a review on the background of the case studies and the categories for the environmental impact assessment.

2. Case Studies

2.1 Sabah Ammonia Urea (SAMUR)

The SAMUR project is located at Sipitang Oil and Gas Industrial Park, Sabah. The project aims at strengthening PETRONAS Chemicals Group's (PCG) position as a key regional fertilizer player, while supporting Malaysia's Economic Transformation Programme that focuses on the agriculture sector as part of the National Key Economic Areas. The Sabah Ammonia Urea (SAMUR) project consists of an ammonia plant, urea and granulation plant, as well as utilities and jetty facilities. The urea plant produces 1.2 million metric tons per annum (mtpa) of granulated urea and the ammonia plant produces 740,000 mtpa of liquid ammonia (Petronas.com.my, 2017)

2.2 Lynas Advanced Materials Plant (LAMP)

The Lynas Advanced Materials Plant is one of the largest and most modern rare earths separation plants in the world. The LAMP is designed to treat Mt Weld concentrate and produce separated Rare Earths Oxide (REO) products for sale to customers in locations including Japan, Europe, China and North America. The LAMP is located at the Gebeng Industrial Estate near Kuantan, Malaysia, close to the Kuantan deep-water port. The plant is built on a 100 hectares site and has been designed and built in two phases. The most valuable product produced at the LAMP is praseodymium/neodymium (Lynascorp.com, 2017)

3. Literature Review

3.1 Impact of Construction Activities towards the Environment

The environment is threatened severely by many problems, some of which are caused by the activities of construction projects (Ijigah et al, 2013). Rapid industrialization and the population explosion in Malaysia has led to the migration of people from villages to cities and thus increases human settlement in the growing cities and towns. This generates several issues related to the environment. Any development project plan to improve the quality of life has some built-in positive and negative impacts; and as such, the development project should be planned in a manner that has maximum positive impacts and minimum negative impacts to the environment (Kaur and Arora, 2012). Environmental impacts are categorized into three areas which are public impact, natural resources impact and ecosystem impact. (Li et al., 2010; Chang et al., 2011 and Zolfagharian et al., 2012)

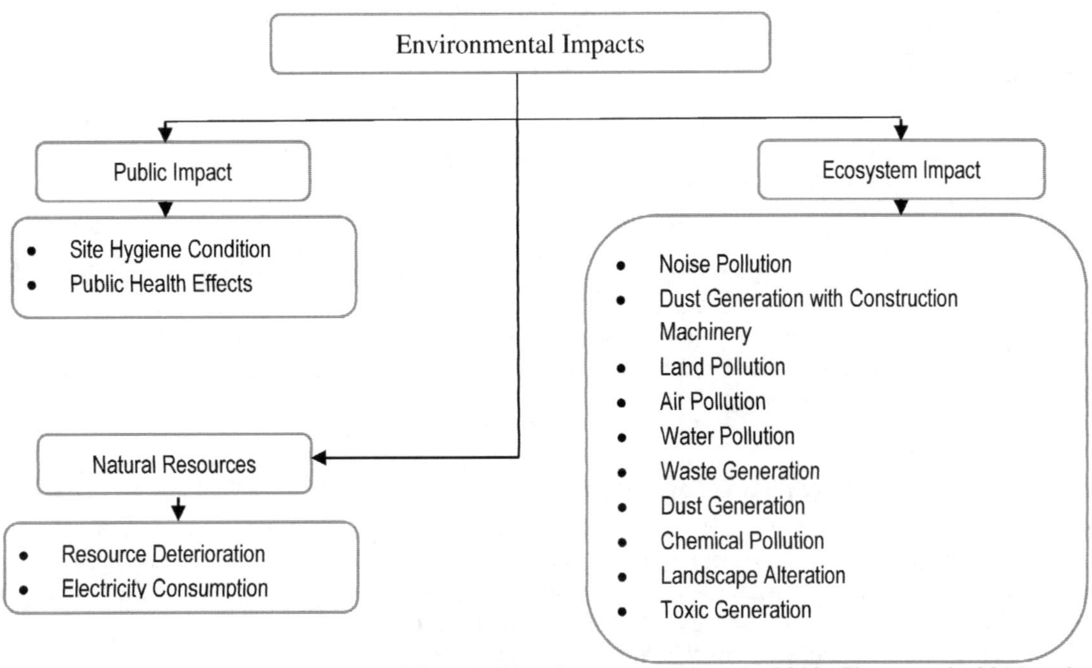

Figure 1. Environmental Impacts of Construction Processes (Li et al., 2010; Chang et al., 2011 and Zolfagharian et al., 2012)

4. Research Methodology
A quantitative method research approach is employed in this study to provide the basis for primary data in investigating the level of environmental impact within the surrounding areas of the case study projects.

4.1 Data Collection
The data collection method was accomplished through an on-site questionnaire survey, divided into 5 sections.

4.1.1. Research Population. Population is defined as 'units (people, employee or members) that have the chance to be included in the survey sample' (Groves et al., 2009). The population for both projects is the community around the site area. This community is from various villages close enough to the site to be affected by the construction activities that has taken place (SAMUR and LAMP projects).

4.1.2. Sampling Frame. Sampling may be defined as the selection of some parts of the population on the basis of which a judgment or inference about the entire population is made (Kumar et al., 2013). The community within the areas around the SAMUR and LAMP projects were considered in this research to investigate the impact of construction activities to the environment and thus, determine strategies for identifying risk factors assessment for sustainable building.

4.1.3 Instrument and Measurement
The research instrument used in this study was a questionnaire containing open – ended and closed – ended type questions. A 5-point Likert scale was used in gathering some of the data, using a measurement scale of 1-5, with 1 being "having the least impact" and 5 being "having the greatest impact". The questionnaire uses both English and Bahasa Malaysia to ensure validity and reliability of responses.

4.2 Method of Data Analysis

4.2.1 Comparative Study. According to Goodrick (2014), comparative case studies cover two or more cases in a way to produce more generalizable knowledge and undertaken over time emphasizing comparisons within and across contexts. Comparative research can be traced to a long history that has gained much attention in current research due to globalization and technological advances on cross-national platforms (Azarian, 2011).

4.2.2 Quantitative Method. Quantitative approaches can help to provide large representative samples of communities reliably assert cause and effect relationships among constructs as well as confirm or disconfirm theoretical hypotheses and summarize numerical data (Fassinger and Morrow, 2013). Quantitative statistical analysis for the questionnaire was performed using Statistical Package for Social Sciences (SPSS) version 20.

4.2.3 Mann-Whitney. Mann-Whitney U test is a non-parametric statistical technique. It is used to analyze differences between the medians of two data sets. It can be used in place of a t-test for independent samples in cases where the values within the sample do not follow the normal or t-distribution but also when the distribution of values is unknown (Milenovic, 2011). The data for this research was analysed using Mann-Whitney statistical test to enable comparisons to be made between the two case studies of SAMUR and LAMP. In addition, the normality tests performed, that is, Kolmogorov-Smirnov and Shapiro-Wilk tests of normality, have indicated that the distribution is not normal and therefore a non-parametric test such as Mann-Whitney should be used.

5. Results and Findings

Four-hundred questionnaires were distributed and collected, with a response rate of 100%. The results for the Mann-Whitney U test for both case studies are as shown in Table 1 and 2 below.

5.1 General Environmental Concerns

Table 1. Mann-Whitney U test with Mean Rank

Ranks				
	Activities	N	Mean Rank	Sum of Ranks
Site Hygiene Condition	SAMUR	200	283.07	56613.00
	LAMPS	200	117.94	23587.00
	Total	400		
Public Health Effects	SAMUR	200	257.58	51516.50
	LAMPS	200	143.42	28683.50
	Total	400		
Resource Deterioration	SAMUR	200	290.30	58059.50
	LAMPS	200	110.70	22140.50
	Total	400		
Electricity Consumption	SAMUR	200	296.38	59275.00
	LAMPS	200	104.63	20925.00
	Total	400		
Noise Pollution	SAMUR	200	233.42	46683.00
	LAMPS	200	167.59	33517.00
	Total	400		
Dust Generation (Machinery)	SAMUR	200	245.08	49016.00
	LAMPS	200	155.92	31184.00
	Total	400		
Land pollution	**SAMUR**	**200**	**297.39**	**59477.50**
	LAMPS	200	103.61	20722.50
	Total	400		
Air Pollution	SAMUR	200	193.82	38764.00
	LAMPS	200	207.18	41436.00
	Total	400		
Soil Erosion	SAMUR	200	271.24	54247.00
	LAMPS	200	129.77	25953.00
	Total	400		
Water Pollution	SAMUR	200	293.78	58756.00
	LAMPS	200	107.22	21444.00
	Total	400		
Waste Generation	SAMUR	200	290.94	58187.00
	LAMPS	200	110.07	22013.00
	Total	400		
Dust Generation	SAMUR	200	256.90	51379.00
	LAMPS	200	144.11	28821.00
	Total	400		
Chemical pollution	SAMUR	200	293.75	58750.50
	LAMPS	200	107.25	21449.50
	Total	400		
Landscape Alteration	SAMUR	200	287.85	57569.00
	LAMPS	200	113.16	22631.00
	Total	400		
Toxic Generation	SAMUR	200	291.84	58367.50
	LAMPS	200	109.16	21832.50
	Total	400		
Green House Gas Emissions	SAMUR	200	165.74	33147.50
	LAMPS	**200**	**235.26**	**47052.50**
	Total	400		
Flooding	SAMUR	200	294.80	58959.50
	LAMPS	200	106.20	21240.50
	Total	400		

IOP Conf. Series: Earth and Environmental Science **117** (2018) 012020 doi:10.1088/1755-1315/117/1/012020

Table 2. Mann-Whitney U test Results

	Mann-Whitney U	Wilcoxon W	z	Asymptotic Significance (2-tailed)
Site Hygiene Condition	3487.000	23587.000	-15.039	.000
Public Health Effects	8583.500	28683.500	-10.284	.000
Resource Deterioration	2040.500	22140.500	-16.074	.000
Electricity Consumption	825.000	20925.000	-17.614	.000
Noise Pollution	13417.000	33517.000	-7.869	.000
Dust Generation (Machinery)	11084.000	31184.000	-9.346	.000
Land pollution	622.500	20722.500	18.013	.000
Air Pollution	18664.000	38764.000	-1.432	.152
Soil Erosion	5853.000	25953.000	-13.410	.000
Water Pollution	1344.000	21444.000	-16.792	.000
Waste Generation	1913.000	22013.000	-16.353	.000
Dust Generation	8721.000	28821.000	-10.406	.000
Chemical pollution	1349.500	21449.500	-16.893	.000
Landscape Alteration	2531.000	22631.000	-15.730	.000
Toxic Generation	1732.500	21832.500	-16.557	.000
Green House Gas Emissions	13047.500	33147.500	-6.672	.000
Flooding	1140.500	21240.500	-17.782	.000

From Table 1, it can be concluded that the SAMUR project reported the highest median on land pollution, indicating that the pollution may be caused by the construction activities that has taken place. The cross-tabulation results of these variables, as shown in Table 3, shows that land pollution has affected the farmers living in the surrounding area. For the LAMP project, the highest median is on greenhouse gas emissions, which can cause global warming.

5.2 Cross Tab for Occupation and Resource Deterioration

Table 3. SAMUR cross tabulation table

		Resource Deterioration					Total
		very low effect	low effect	neutral effect	strong effect	very strong effect	
Occupation		5	12	39	16	58	130
	Business	0	2	3	4	7	16
	Carpenter	0	1	0	0	0	1
	Farmer	**0**	**2**	**7**	**0**	**22**	31
	Fisherman	0	2	7	3	10	22
Total		5	19	56	23	97	200

Table 3 shows that among the occupations of respondents, the strongest impact from resource deterioration is towards farmers. Resources deterioration affected the farmers' earning because the SAMUR project produces ammonia, a chemical that, according to the respondents, has affected their fruit trees which has not been producing as well as before.

Table 4. LAMP cross tabulation table

		Resource Deterioration			Total
		very low effect	low effect	neutral effect	
Occupation		64	88	2	154
	Business	15	13	0	28
	Fisherman	2	15	1	18
Total		81	116	3	200

Table 4 shows results for the LAMP project between occupation and resource deterioration. The two main occupations of the people in the surrounding area were not affected by the construction activities.

5.3 Discussion on findings

From the results, it can be shown that the environmental impact to the area surrounding the SAMUR project is more critical compared to the LAMP project, due to land and water pollution caused by chemical by-products. Based on the findings, the highest median in the SAMUR project is land pollution, followed by flooding. This is cross supported by interview data, in which the community stated that floodings frequently occur due to landscape changes and disposal of waste into the small river.

The highest median for LAMPS project is greenhouse gas emissions, followed by air pollution. Green-house gas emissions and air pollution are mostly due to the heavy smoke from the factory and also dust from the road caused by the movement of lorries transporting products in and out of the factory. Based on the interviews with the community, chemicals used by the LAMPS project also emit a foul smell which caused discomfort amongst the community.

6. Conclusion

This study investigated the environmental impact due to construction activities for two industrial sites, SAMUR in Sabah and LAMP in Pahang. The results demonstrate that land pollution and greenhouse gas emissions are the main concerns for the respondents living in the surrounding areas. These results can be a useful basis for developing assessment tools to assist construction practitioners to improve their on-site environmental performance. Furthermore, understanding the risk level of environmental impacts caused by construction processes and activities can be valuable when considering measures for mitigating such impacts, which may lead to successful sustainable performance and management. The outcomes of this study can help organizations and managers prepare proper sustainability plans and increase knowledge on levels of environmental impacts caused by industrial projects.

Acknowledgements

The authors would like to extend their utmost appreciation to all participants for sharing their valuable time and support. The authors also acknowledge the Universiti Teknologi MARA GIP grant 600-IRMI/GIP 5/3 (0070/2016) for funding the research.

References

[1] Altoryman A 2014 The school of Mechanical, Aerospace and Civil Engineering: *Assessment Identification in Gulf Region.*

[2] Azarian R 2011 *International Journal of Humanities and Social Science. Potentials and limitations of comparative method in social science* **1** 113

[3] Backstrom C H, and Hursh-Cesar G 1981 Survey research, Macmillan publishing Co.

[4] Chang Y, Ries RJ and Wang Y 2011 *Energy Policy. The quantification of the embodied impacts of construction projects on energy, environment, and society based on I- O LCA* **39** 6321

[5] Chen Z, Li H, Wong CTC 2000 *Journal of Construction Engineering and Management. Environmental Management of Urban Construction Projects in China* **126** 320

[6] Fassinger R and L Morrow S 2013 Toward Best Practices in Quantitative, Qualitative and Mixed Method Research: A Social Justice Perspective. Vol. 5 (no.2).
[7] Goodrick D 2014 Comparative Case Studies, Methodological Briefs: Impact Evaluation 9, UNICEF Office of Research, Florence
[8] Ijigah E, Jimoh RO, Aruleba B, and Ade A 2013 *Civil and Environmental Research. An Assessment of Environmental Impacts of Building Construction Projects* **3** 1
[9] Kaur M, and Arora S 2012 *Journal of Mechanical and Civil Engineering. Environment impact assessment and environment management studies for an upcoming multiplex- a Case Study. IOSR* **1** 22
[10] Kumar M, Talib S, Ramayah T 2013 *Business Research Methods.* Selangor: Oxford
[11] Li X, Zhu Y and Zhang Z 2010 *Building and Environment. An LCA-based environmental impact assessment model for construction processes* **45** 766
[12] Lynascorp.com 2017 *Lynas Advanced Materials Plant (Lamp) | Lynascorp.* [online] Available at: https://www.lynascorp.com/Pages/Kuantan-Lynas-Advanced-Materials-Plant.aspx [Accessed 27 Agust. 2017]
[13] Groves MR, Jr FP, Couper M, Lepkowski J, Singer E and Tourangeau R 2009 *Survey Methodology.* 2nd ed. (WILEY)
[14] Milenovic Z 2011 *Application Of Mann-Whitney U Test in Research of Professional Training of Primary Scholl Teachers* **6**
[15] Nagapan S, Rahman A 2012 *International Conference on Civil and Environmental Engineering for Sustainability. Construction Waste Management: Malaysian Perspective* (Johor Bahru, Malaysia)
[16] Petronas 2017 *Sabah Ammonia Urea (SAMUR) | PETRONAS.* [online] Available at: http://www/petronas.com.my/our-business/downstrea,/OKP/Pages/SAMUR.aspx [Accessed 27 Agust. 2017]
[17] Zolfagharian S, Nourbakhsh M, Irizarry J, Ressang A and Gheisari M *2012 Construction Research Congress. Environmental impacts assessment on construction sites* pp.1750-1759.

A study on integration of *Waqf* Real Estate and *Zakat*: A qualitative investigation for *Asnaf Muallafs*' welfare

S N M Ali[1], S Mokhtar[1], A H Mohd Noor[2], N Johari[1], N S Fauzi[1] and N A Salleh[1]

[1]Faculty of Architecture Planning and Surveying, Universiti Teknologi MARA, Seri Iskandar Campus, Seri Iskandar, 32610, Perak, Malaysia
[2]Faculty of Business and Management, Universiti Teknologi MARA, Campus Alor Gajah, Km 26 Jalan Lendu, 78000 Alor Gajah, Melaka, Malaysia

*sitin190@perak.uitm.edu.my

Abstract. This paper aims to explore the integration of *waqf* real estate and *zakat* in the social development of *asnaf muallaf.* Due to the significant integration of *waqf* and *zakat* in Malaysia, Al-Saadah Complex, Negeri Sembilan serves as a good case for this present research. The study relies on qualitative research, where primary data was collected from semi structured interviews with Al-Saadah Complex officers. The qualitative research was chosen because it represents public views and perspectives. Data was recorded, transcribed and analysed using thematic analyses for multiple interviews.From the result, it was revealed, from the Al-Saadah Complex officers' perspective, there is a need for an integration between *waqf* and *zakat* to oversee the welfare of the *asnaf muallaf.* The officers are more concerned with the quality of services they give to the *asnaf muallaf.* Finally, we find the integration is a positive collaboration within the SIRC concerned but there is still limited development of this concept in Malaysia.

1. Introduction

Asnaf muallaf is one of eight beneficiaries of zakat. Islam takes care of *asnaf muallaf* as one of the cornerstones for spreading *da'wah* in Islam. *Asnaf muallaf* are ambassadors who should be given special protection, especially from the financial point of view. The rapid development of Islam in Malaysia shows the importance of zakat for *asnaf muallaf.* Upon their Islamic conversion, *asnaf muallaf* may lose their jobs and may be ostracized by their families. Hence, lies the significance of zakat and waqf in providing adequate food and shelter to these *asnaf muallaf.* The role of zakat and waqf in this paper is very synergistic in defending the fate of *asnaf muallaf.*

Different from *zakat*, *waqf* is defined as an endowment made by Muslims to seek Allah's favour without compulsion. *Waqf* plays an important role in helping poverty reduction in many ways. It could be one of the tools to support the needy. Any inadequate in *zakat* will be accommodated by *waqf*, which shows that *waqf* and *zakat* complement each other to help the needy. The integration of *zakat* and *waqf* has been adopted in different policies and strategies in different countries. It may not be seen as the best strategy yet in Malaysia, since the integration of *waqf* and *zakat* has not been fully explored in literature reviews and empirical research. Through this study, the researcher wants to explore the extent to which the integration of *zakat* and *waqf* impacts the welfare of the *asnaf muallaf.*

2. Integration of *Waqf* Real Estate and *Zakat* Approach in State Islamic Religious Councils

Integration is defined as an action or process of integrating a system or organization which depends on the objective. In Islamic social economics context, it is defined as the process of harmonizing both *zakat* and *waqf* concepts to contribute towards poverty reduction. In this paper, integration is defined based on the concept adopted by CSR integration which is considered as a value-driven process, where values, culture and ideology are the driving forces[1]. The integration of *waqf* real estate and *zakat* is an approach that also synergizes with other instruments such as *sadaqat* and public funding [2]. The main limitation of this integration is awareness among the practitioners especially State Islamic Religious Councils (SIRCs) in its implementation. It is believed the qualitative research is the best method to investigate the implementation of this integration in SIRCs.

Realizing the importance of integration between *waqf* real estate and *zakat* as well as considering the proposition of many scholars, the present study recommends the following conceptual integration framework for the *asnaf muallafs*' welfare:

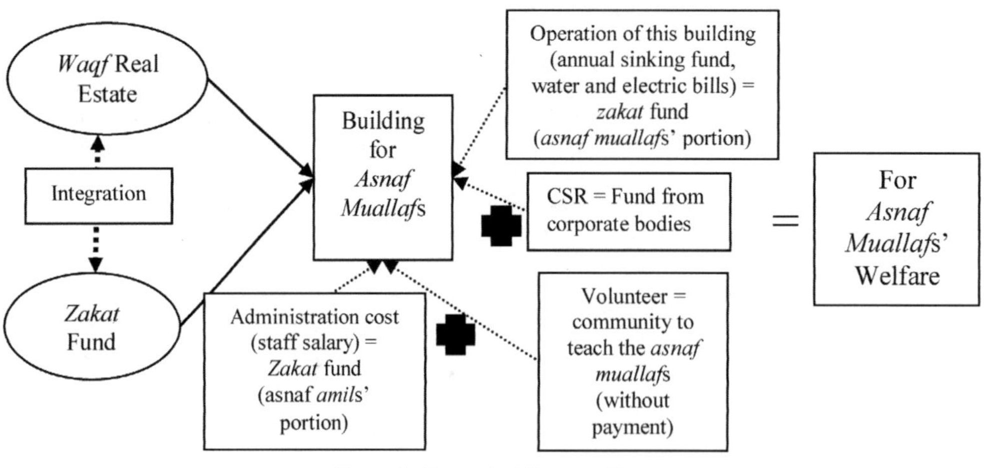

Figure 1. Conceptual Framework

3. Research Methods

The consideration behind this research is to explore the integration of *waqf* real estate and *zakat* to serve *asnaf muallaf* welfare. A single case study is the best choice to be explored in this research, ones because when a single case study is used, the researcher can get deeper understanding of the subject[3]. Thus, the study enhanced the qualitative methods by conducting semi structured interviews with 4 officers in Al-Saadah Complex. The uniqueness of qualitative research is focused on a specific group of people[4]. For the purpose of this research, the officers at Al-Saadah Complex, Negeri Sembilan were chosen to be the participants. In qualitative research, the determination of sample size is dependent on research needs under which the researcher wants to study[5] [6]. Qualitative

methodologies, details and description of particular issues, will produce a wealth of comprehensive data on a small number of participants [7].
This case study clearly demonstrates research on integration of *waqf* real estate and *zakat* is accurate to use qualitative methodologies. Qualitative method allows the researcher to ask questions while watching the body language of the participants. Here are the procedures implemented by the researcher to dissect this case study based on some famous writers especially Robert K.Yin:

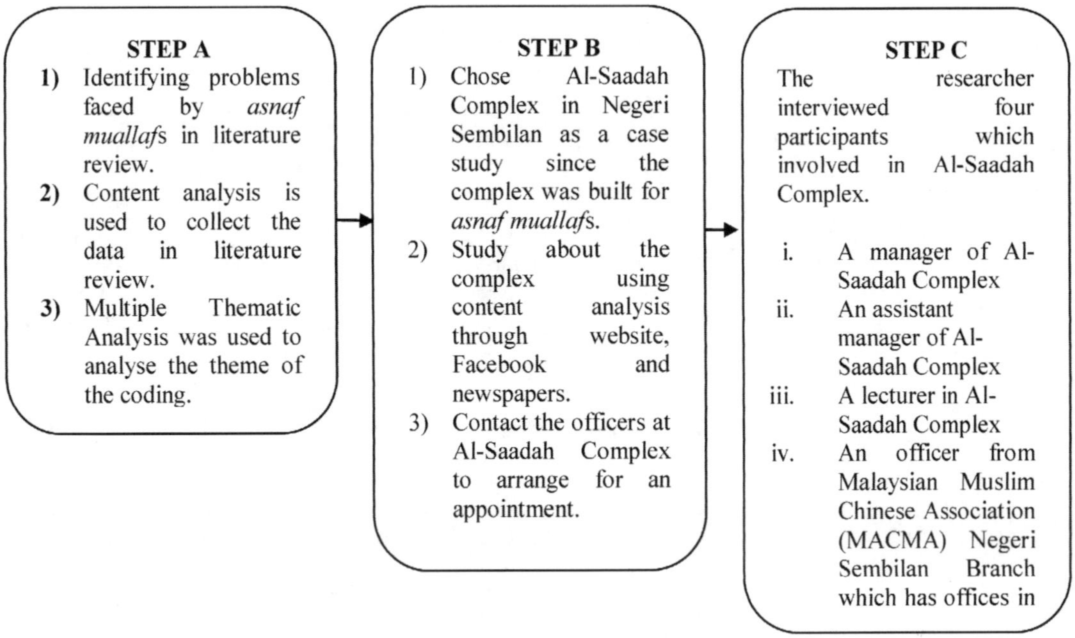

Figure 2. Process to select participants

Qualitative research is very unique because the procedure differs significantly from subject to subject [8]. Figure 2 shows the process to operate and select participants in this research. Conducting this type of research is a strategy to enhance answer quality by selective process. A major setback in this type of qualitative research is it requires a great deal of patience on the part of the interviewer to set up and complete the face-to-face interview. Usually the researcher will approach the participants by calling them and setting up appointments. This may take up to three sessions to ensure the appointment was successfully set up. The requests to interview were also sent via email and followed up through WhatsApp Messenger to get quick response. WhatsApp has created a way to feel sense of belonging, distance and closeness with participants of the interview [9]. Selection of participants in this research is based on their designation in the management of the Al-Saadah Complex.

Figure 3 illustrates in detail the dates and the time when the interviews were organized. The researcher initiated the interview with participant D in June 29, 2017 but the researcher failed to interview other participants because they were unavailable at that time. However, on August 15, 2017 and August 17, 2017, the researcher succeeded in interviewing participants A, B and C. The semi-structured interview method was chosen because the focus is on the interview design, covering a series of broad themes to be used during the interview to guide and direct the question towards the topics and issues about which the interviewer wants to know. Normally interview guides vary from highly scripted to relatively loose [10]. After the interviews and data recording are completed, the data will be transcribed into a paper script [8]. It is quite challenging to transcribe 4 participants within a short time. The researcher spent almost three days for each participant to transfer all recording into a script, totaling approximately 12 days to settle all transcribing process. This is inclusive of breaks for every 5 minutes ranges in 15 minutes of transcription.

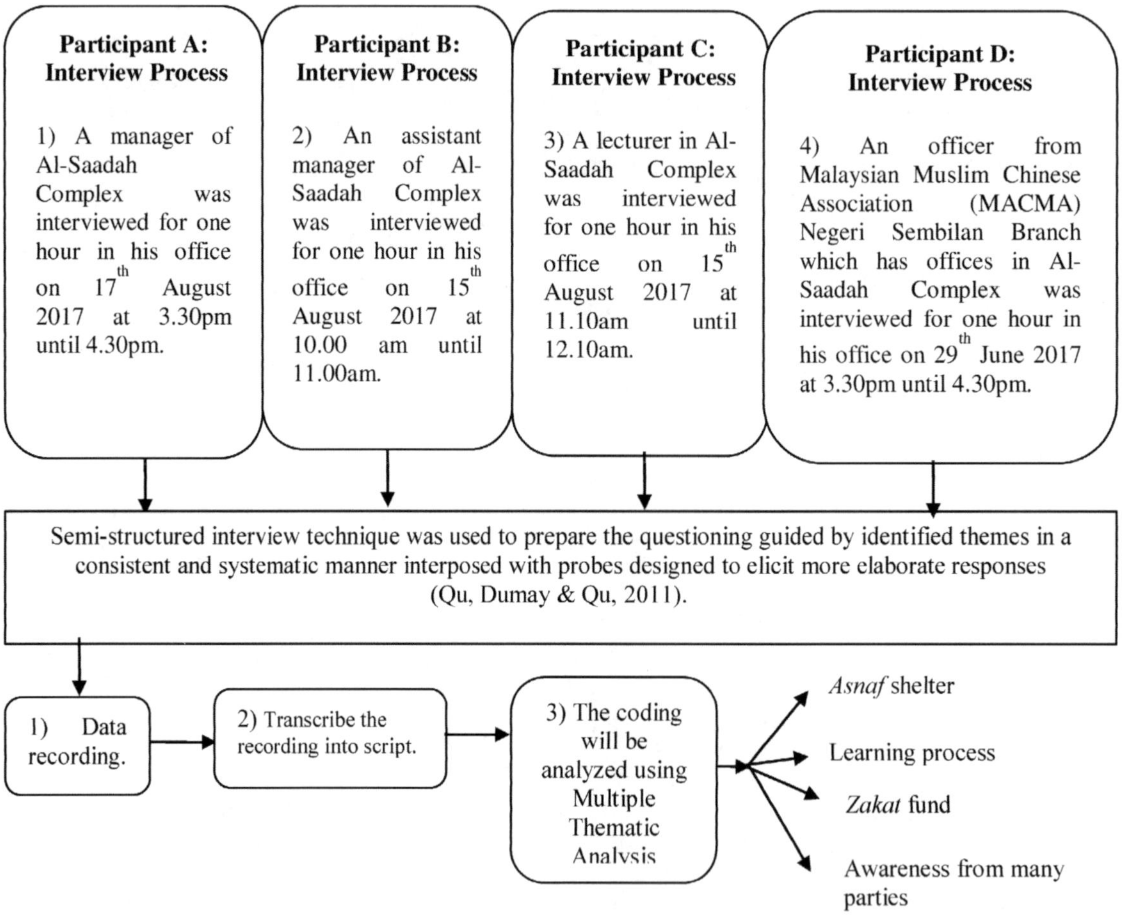

Figure 3. Data collection, data recording, data transcribing process and multiple thematic analysis process

4. Findings

On a holistic level, this study's findings of integration of *waqf* real estate and *zakat* are indispensable in providing protection to *asnaf.* Thus, the findings support the integration of *waqf* real estate and *zakat* to provide the best for *asnaf muallaf* well-being. The data themes show that the integration of *waqf* real estate and *zakat* in this case study is also influenced by the attitude of *asnaf muallaf* that are really in need for this kind of protection.

Table 1 indicates that the integration of *waqf* real estate and *zaka*t is successfully implemented in Al-Saadah Complex. For example, the development of Al-Saadah Complex is not only for the *asnaf muallaf* to gain shelter, but they are also able to improve their Islamic knowledge through attending religious classes. The interesting fact about it is that the complex has also benefited *asnaf* from other states to study here. Through this religious certification program, the *muallaf* are taught using the university system. They are required to complete assignments and there is a final examination that will be examined by Al-Saadah lecturers. For the time being, they have five lecturers and twenty students.

Based on this study too, *zakat* payers need not be hesitant about the rumors that stated the whereabouts of your *zakat* money. The complex expenses reach RM500,000 including building operations, *asnaf* expenses and staffing. The success of *zakat* and *waqf* gives a good picture of the development of Islamic *da'wah*, especially in Malaysia. *Waqf* real estate that has been developed using the turnover from *zakat* funds shows how beautiful Islam is in enhancing the quality of life among all levels of society.

Table 1. Summary of Transcription of Four (4) Participants According to Theme

Participants	Theme 1: *Asnaf* Shelter	Theme 2: Learning process	Theme 3: *Zakat* fund	Theme 4: Awareness from many parties
A	"This place is a house for them"	"They learn Islam without any payment and we give them certificate to prove they have gone through all the syllabi"	"This complex was built on waqf land and without zakat fund, the operation cannot be done"	"That's why I said, *zakat* funds and *waqf* development should become top management agenda, we can build many things not only this complex "
B	"They come here voluntarily; we separate the male and female *asnaf*s. We provide 3 times a day for food.	"All lecturers have degrees in Islamic Education, we have experienced lecturers working here as *jihad.* We don't pay high salary, but they accept it"	"We receive *zakat* fund from *Baitulmal* to operate, but we are managed under *Perbadanan* Waqf "	"Cooperation from other SIRCs is also very good, they send *asnaf muallafs* and spend on them to learn Islam in a systematic way ".
C	"Someone left the family solely to come here, so we give them protection."	"I give full commitment to teach the *muallafs*, because I know, the performances of the students depend on my lecturing"	"*Zakat* fund from each state will sponsor their students to study here, we estimate expenses for each student are about RM10K each"	"I have a degree in Arabic from Jordan. When I came back to Malaysia, I saw this advertisement for lecturers for this complex. Even though the salary is not high but I'm happy to work here. I give attention to students, and they really love to learn Islam ".
D	"MACMA has a special office in this complex. We can gather all Chinese Muslims all over Seremban here.	"This place is not only for *muallaf*s or Muslims but we also invite all, who like to learn Islam, because we organize many talks here "	"All activities here use *zakat* fund, that is why Islam is very beautiful, the rich people should pay *zakat*, that is not the rule in my previous religion ".	"I hope other places also can build like this, we have one in Ipoh, so that many *asnaf*s will get the benefits. We Chinese Muslims need support from government to make sure Islam can grow among us".

5. Conclusion
This study makes an important contribution to the integration of *waqf* real estate and *zakat* literature. Significant contributions have been made by exploring the potential development on *waqf* real estate

by having financial engagement from *zakat* fund. The ability to solve the financial problem is also valuable interactions between this two noble concepts. Specifically, the majority of literature is focusing on microfinances aspect to integrate with *waqf* and *zakat*. This paper supports the wider view of development in integrating *waqf* and *zakat* to address the social needs of *asnaf muallaf* in Malaysia to impact *da'wah* expansion. Finally, the people who work in *waqf* and *zakat* line, are not ordinary people who sacrifice their lives for Allah SWT.

References

[1] Marques-Mendes A and Santos MJ 2016 *Social Responsibility Journal. Strategic CSR: an integrative model for analysis* **12** 363
[2] Amuda YJ 2013 *International Journal of Trade, Economics and Finance. Empowerment of Nigerian Muslim Households through Waqf, Zakat, Sadaqat and Public Funding.* **4** 419
[3] Yin RK 2011*Qualitative-Research-From-Start-To-Finish.* New York London: The Guilford Press.
[4] Edwards R and Holland J 2013 *Research Methods Series. What is Qualitative Interviewing? "What is?"* **7**
[5] Boddy CR 2016 *Qualitative Market Research: An International Journal. Sample size for qualitative research Clive* **19** 426
[6] Hyde KF 2000 *Qualitative Market Research: An International Journal. Recognising deductive processes in qualitative research* **3** 82
[7] Creswell JW 2013 Research Design: Qualitative, Quantitative, and Mixed Methods Approaches. Research design Qualitative quantitative and mixed methods approaches
[8] Kumar N and Sharma S 2016 Survey Analysis on the usage and Impact of Whatsapp Messenger
[9] Qu SQ, Dumay J and Qu SQ 2011 *The qualitative research interview.*

IOP Conf. Series: Earth and Environmental Science **117** (2018) 012022 doi:10.1088/1755-1315/117/1/012022

The effects of landscaping on the residential cooling energy

A Misni[1]

[1] Faculty of Architecture, Planning and Surveying, Universiti Teknologi MARA, Puncak Alam, Selangor, Malaysia

*alamahmisni@gmail.com

Abstract. This paper examines the effectiveness of landscaping on the air-conditioning energy saving of houses in a tropical environment. This case study involved looking at the construction and landscaping of three single-family houses in three sections of Shah Alam, Selangor, Malaysia. The houses ranged in age from 5 to 30 years old, which provided different examples of construction and maturity levels of the surrounding landscaping. Landscaping affects the thermal performance as well as on the air-conditioning energy of houses, in how it provides shade, channels wind, and evapotranspiration. While the construction of the three houses was similar, they were different in size and design, including their landscape design. These houses were chosen because they are representative of single-family tropical houses and landscaping styles in Malaysia since 30 years ago. Three houses were chosen; the 30-year-old house, the 10-year-old house, and the 5-year-old house. In a tropical country, landscaping is used to reduce the effects of the hot and humid climate. The houses spent 15-45% of the electricity cost on cooling. These results were influenced by human factors and the surrounding landscaping. Every type of vegetation, such as trees, grass, shrubs, groundcover, and turf, contributes to reducing air temperatures near the house and providing evaporative cooling.

1. Introduction

Strategically placed vegetation around a building has long been recognized as a means of cooling. However, the potential for air-conditioning energy saving has not been widely recognized and quantified. Few researchers have measured the energy reductions caused by different types of vegetation in a variety of locations and climate. Energy saving up to 55% has been measured in relation to air-conditioning, although 25–50% is more common [1-3]. Large saving was found in both dry and humid climate regions. The design of external spaces needs to be prioritized because conditions there will influence the building interior. Proper placement of trees, shrubs, vines, grass, and turf can greatly reduce the temperature around a building, and create a comfortable environment. The effects on energy use depend on general climate, building type, tree species, and tree location [4]. This study will investigate the effects of different amounts of tropical landscaping around houses on their ability to provide a cool and comfortable environment, furthermore it closely related to cooling energy use.

1.1. Air-conditioning system

Air conditioning may provide and maintain a desirable indoor atmospheric environment irrespective of outdoor conditions [5]. There are seven main processes and functions required to achieve full air-conditioning: cooling, heating, ventilating, humidifying, dehumidifying, air movement and cleaning, [6]. The requirements and importance of these seven processes of air-conditioning vary with the climate. The desired indoor atmospheric condition for human comfort applications usually involves a temperature range of 18–22°C in winter and 21–24°C in summer and a relative humidity level of about 40–60% and a high degree of air purity [5]. The application of the particular type of air-conditioning system depends upon a few of factors such as condition and size of the area to be cooled and the total heat generated inside the enclosed area. The building designer would consider all the related factors and suggest the cooling system most suitable. Window air conditioners are one of the cheapest and the most popular and of all air conditioning systems. The window air conditioner can be used for almost all types of small spaces and commonly used for small rooms in a house. The component of this air conditioner contains the condenser, compressor, expansion valve/coil, fans, cooling coil and evaporator. All of these components were enclosed in a single box. This unit is a standard of the window frame and is fitted in a slot made in the wall of the room.

1.2. Energy use

One of the essential energy sources are used in many countries in the world is electric energy. Currently, the cost of electric energy has been drastically increased. In Malaysia, the average rate demand of electric energy is 11% per annum demand and will be increased each year, parallel to the development and standard of living in this country [7]. The consumption of electricity includes the application of electrical appliance, machinery, lighting and air conditioning system. With the development of services system, air conditioning is now widely used especially in a hot and humid climate in Malaysia. In the tropical region, the active part of air conditioning systems is the energy required to drive the cooling. There is no heating system for buildings in tropical country of Malaysia as for the climate is high in temperature and humidity throughout the year. Recently, central cities in Malaysia experienced urban heat islands effects, parallel to the increase of urban population, as well as improved living standard was contributed to the dramatic increase of air conditioner used especially for residential buildings in urban areas [7]. Ratings for air conditioning run by the number of British Thermal Units (BTU), (1-ton refrigeration = 12,000 Btu/h). The Energy Efficiency Ratio (EER) is a measure of how efficiently of air conditioning; a higher number means the system is more efficient [8]. A regular air conditioner for the home usually has a 5,500 BTU to 14,000 BTU per hour range.

1.3. The roles of vegetation

The roles of vegetation in urban microclimates are varied; directly by providing shade and channelling wind flow, and indirectly by evapotranspiration of water content, and the modification of exchange and storage of heat between urban surfaces [1,4 & 9]. Vegetation is all the plants of a place, including trees, shrubs, vines, groundcovers, grasses, and turf/lawn. Shade trees are one of the versatile components of our surrounding landscapes. The Large canopy of trees can enough to shade building envelopes include roofs and walls and to increase indoor comfort. During natural photosynthesis, solar radiation is absorbed by the tree canopy and next generating evapotranspiration that cools the leaves and comforts the surrounding air. Air movement then disperses this cool and comfort air resulting in an overall surrounding cooling effect. Trees and shrubs can be planted to funnel/deflect wind away from specific/required areas; both horizontal and vertical concentrations of foliage can modify air movement/flow patterns. In hot and humid tropical regions, the wind should be channelled for cooling and to provide relief from high temperature and humidity. A few large/shade trees with spreading branches in a strategic location can allow breezes and keep the area cool and comfortable. Groundcovers, grasses and turf/lawn also generate evapotranspiration and cool the earth surfaces and the surrounding air. Careful and strategic landscape planning around buildings can reduce the amount of solar radiation heating building surfaces and can prevent reflected sunlight from entering the indoor spaces. Hence, trees, shrubs and grass will also reduce air temperatures and humidity, while channelling wind flow near the building and provide evaporative cooling.

2. Methodology

2.1. House study

The bungalow houses sampled were randomly chosen from Shah Alam city housing development. In general, bungalow house represents around 10-15% of the mixed residential development in Malaysia. The area of study is located in Klang Valley; latitude 3°North and longitude101°east, with an elevation range of 24 to 48 metres, about 25km from Kuala Lumpur capital city. Shah Alam is the capital of Selangor State and is also one of the well-planned cities in the tropical country of Malaysia. Three locations at three sections of low-density housing development were considered to be case study areas. The specific locations used were in Sections 6, 9 and 11. Three bungalow houses are labelled as old, ordinary and new houses respectively as can be seen in Figure 1.

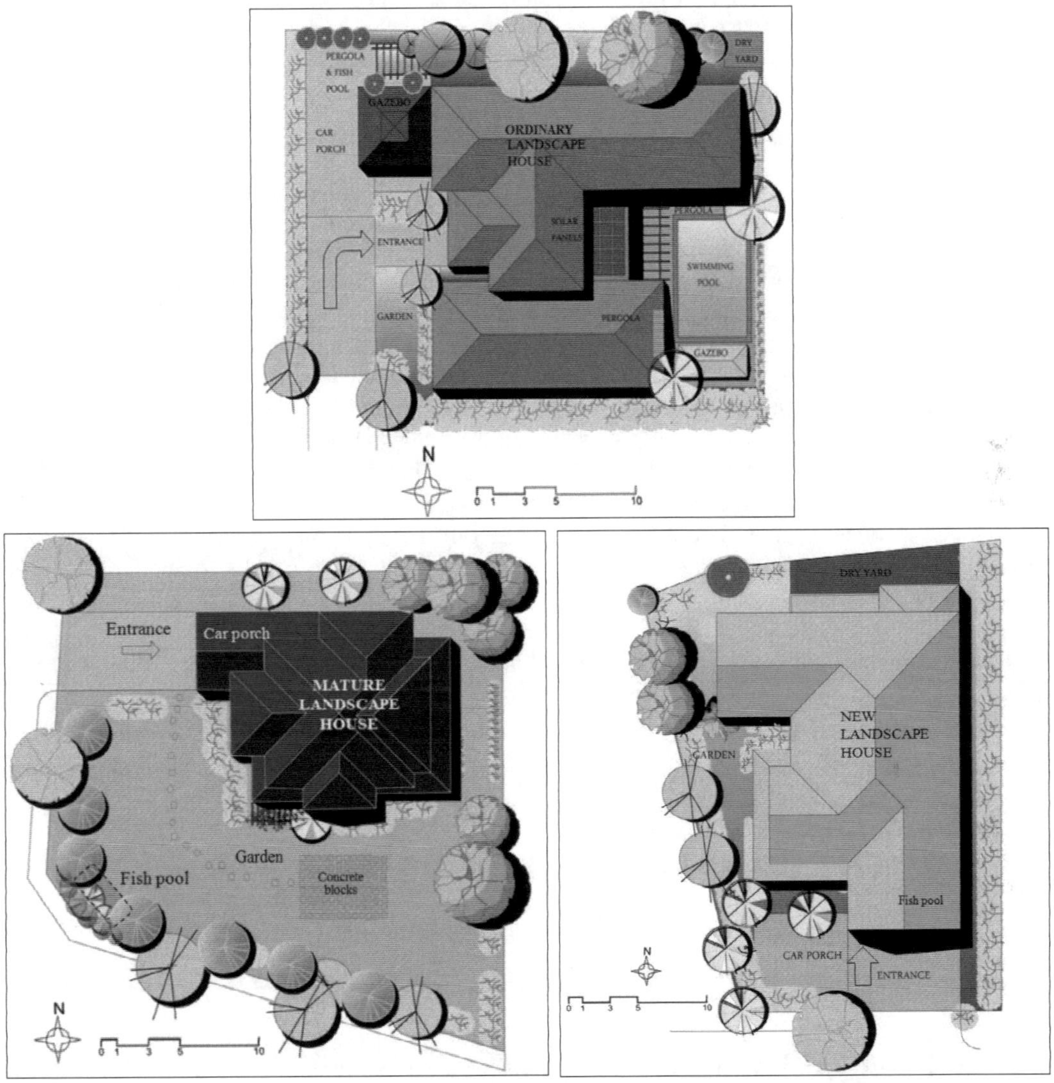

Figure 1. The chosen medium size of bungalow houses surrounded by open space

2.2. *Electric energy*

In Malaysia domestic electricity tariff was monitored and managed by National Energy Limited–TNB (Malaysian Electric Utility Company). Data of electricity used by the residential buildings were collected for the monthly amount as the cost of electrical energy used. This total of energy costs were converted into energy used in kWh units. Electricity energy used for air-conditioning also was calculated based on current local tariff in kWh. Annual domestic electricity used in the study area, a city of Shah Alam was also analysed to validate the amount of energy used for cooling along the year.

2.3. *Landscaping and energy use*

The micro scale of fieldwork and site measurements was conducted at an individual residential area. The quantitative methodology to evaluate air-conditioning energy saving was done via interviews and

IOP Conf. Series: Earth and Environmental Science **117** (2018) 012022 doi:10.1088/1755-1315/117/1/012022

building and landscaping data collections in three bungalow houses. The site measurements included recording building characteristics (building orientation/location and construction, including building envelope material and surface, etc.), plant type and biomass (species, size, canopy, leave density, location, etc.), human factors/behaviour (living styles and habits), and specific/local indoor and outdoor weather conditions (temperature, humidity and wind speed/flow). All collected data were combined and analysed to estimate the effect of vegetation on thermal performance and cooling energy use in the bungalow house located in the tropical country of Malaysia.

3. Results and Discussion

3.1 Building orientation

The three medium size of bungalow houses were of similar construction but had different main façade/building orientations, wall/glass areas, and age of landscape. The main façade/building orientations for the old house were facing north/south. The majority of the glass openings were located facing the main open space/garden area. This orientation can minimize the amount of solar radiation penetrating the internal spaces. In contrast, the main façades for the ordinary landscape house and the new landscape house were facing the east and west respectively, with the majority of their glass surfaces facing the main garden area at the south side. Immature vegetation around the main garden area (west) of the new bungalow house was not sufficient to provide shade cover to walls and glass surfaces, however the amount of shade trees and at west side help to cover low wall and earth/floor surfaces; while sparse medium maturity plants around the main garden area at east side of the ordinary house had similar effects on the surrounding ambient air. Therefore, in these house studies, the best building/façade orientation to face the open space/main garden area was facing north/south where the majority of glass opening should be located to receive diffuses solar radiation in a similar way to the old house. This can provide cooler temperatures as low as 0.3–1.2°C different compared to other two houses as shown in Table 1.

Table 1. Building envelope, surrounding vegetation, weather data, human factors and cooling energy

House	Building			Vegetation	Average daily microclimate data				Human factors	
	Façade	Glass surfaces (m^2)	Albedo (roof)	condition	Temp. outdoor (°C)	Temp. indoor (°C)	Wind speed (m/s)	RH (%)	Cooling energy (%)	Occup. number
OL	North & South	51.35	0.12	Moderate/ Mature	31.76	28.93	0.32	72.40	35.00	6.00
OD	East & West	87.30	0.30	Moderate/ Medium	32.08	29.62	0.40	77.22	45.00	6.00
NE	East & West	94.50	0.33	Moderate/ Immature	32.98	29.90	0.51	71.10	15.00	6.00

3.2. Albedo

The colour of walls at all study houses was light such as white, crème, peach and grey can provide high albedo value more than 0.3 and can reflecting well solar radiation. While roof covering is vary; in the old landscape house was brown and generated the lowest albedo value 0.12. This may affect the heat gain to the building and may increase the surrounding temperatures and directly to the indoor air temperatures. A light colour of roof covering such as light grey and light orange which were used at the new and ordinary landscape house with albedo value 0.30 and 0.33 respectively were more effective at reflecting solar radiation.

3.3. Surrounding landscaping

The three different ages of landscaping/vegetation located around the three bungalow houses create moderating effects by providing shade to the building envelopes and earth surface, channelling wind to the garden and evapotranspiration cooling to the surrounding outdoor area. In this study, the building envelope components such as roofs, walls and glass surfaces are the most exposed parts to direct solar radiation. This is due to non-strategic planning and placement of plants. Medium sized of shade trees and around/spreading form of the canopy can potentially provide shade to walls surfaces nearby. The non-strategic location of shade trees and shrubs was for the old landscape house. Most of the trees in the old landscape house were located between 15 and 20m from the house and therefore cannot provide shade to the building envelope/surface. This suggested the need for two/three trees of medium size with heights 6–9m be planted on the west and east sides' 3 to 5m distance from the house. This shade trees would cover part of the roofs and walls in early morning and afternoon, and shade most of the earth surfaces close to the house over a day. Furthermore can reduce heat gain to the surrounding garden and building, as well as decrease the surrounding temperature during the peak hour of the day. The result of this study clearly shows that small and low shrubs, groundcovers and turf located close to walls, only can provide shade to the nearest underground foundations and floors, while the walls and windows of the house are still too exposed to the direct sunlight during morning and afternoon. The hotter outdoor environment resulting from sparse and non-strategic landscaping can directly promote warmer interior spaces. The temperatures in the three bungalow houses gradually increased with the peak indoor temperature delayed by approximately three hours.

3.4. Human Factor

The human/behaviour factors for the three bungalow houses were similar in terms of the number of occupants but they had different total cooling energy consumption. In the evening, the average interior temperature for the three houses was up to as much as 29°C showing that the occupants need an active cooling system to achieve their comfort level. For example, the two bungalow houses with similar living habits; the ordinary and old houses used 100% air-conditioning for their master bedrooms, while the new house used the air-conditioning system for only 3 hours during early mid-night time, although the set temperature was low when it was used. The rest of the times the new house used a ceiling fan and the passive ventilation system. This reduced the total electrical energy used. The advantage of the new house is the house is located facing a large natural open space where the prevailing winds can refresh the surrounding air all the time. This cool and comfort ambient air directly flowed to the indoor spaces. The strategic position of immature vegetation around the new house provides cool ambient air by channelling wind to the surrounding garden and its indoor spaces.

Overall a strategic placement of vegetation at the east and west sides where direct solar radiation receives, close to the building is very important to provide shade/cover to the building envelope especially roofs, walls and windows. Evapotranspiration cooling during plant photosynthesis was produced by a combination of plants and water elements provide cool ambient air. The arrangement and positioning of plants in the garden and around houses should be compatible with prevailing wind directions and to channel and promote wind to the appropriate places around and through the interior of the house. The combination of shading by all types of plants, compatible moderate wind flow and evapotranspiration cooling can produce fresh, cool and comfortable environment.

3.5. Energy saving

Human factors are one of the main factors that influence the energy consumption for cooling in a tropical city. The living habits recorded for the three bungalow houses included the number of occupants and their ages, the total of time they used the active cooling systems during the day and night and the occupants' preferred temperature. The house will release the heat is stored in its building envelope throughout the day when the outdoor temperature around a house decreases during the night. This heat in the interior spaces contributes to the discomfort of the occupants, so using an active air-conditioning system is the only way to ease this. All of the occupants of the three houses in this study used air-conditioning at slightly different times, but mainly when most of the family members were at home. In all houses the occupants closed external openings at night for security reasons, to prevent noise, insects and to protect themselves and their house interiors from any extreme climate phenomena such as heavy rain or thunderstorms during their rest time at night. As a result, they depend entirely on active cooling systems to achieve their preferred thermal comfort levels at night.

The three bungalow houses used similar window air-conditioning systems, but they had different capacities and models. The ordinary house had eight air conditioning units, with six occupants/users; five adults and a teenager. The old house had four air-conditioning units for six occupants, all adults.

Finally, the new house had three air-conditioning units for six occupants; three adults and three children. All houses used their cooling systems extensively at night, especially in the bedrooms while the occupants were sleeping. The ordinary house used the highest amount of cooling energy, followed by the old house, then the new house. The houses used the cooling system with different temperature settings and for different amounts of time. As a result, the ordinary house spent 45% of the electrical energy on air-conditioning; the old house used approximately 35%, and the new house only used 15%. This shows that the time's air-conditioners were used and different temperatures influenced the amount of cooling energy consumption for each house. Overall, the longer cooling system was used, and lower temperature settings will affect the amount of cooling energy consumption.

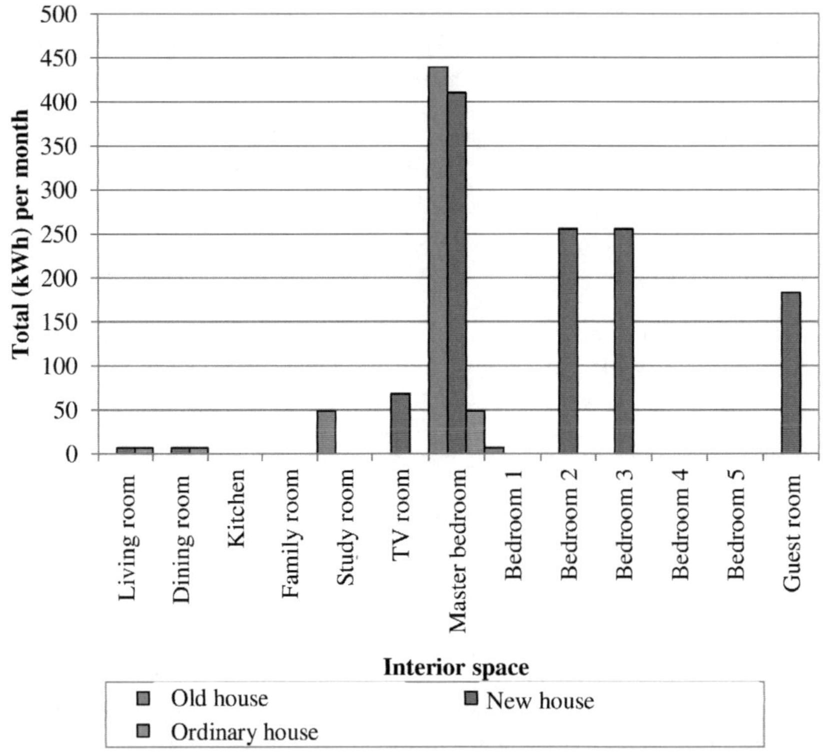

Figure 2. Cooling energy use for the three bungalow houses

In a tropical city, during the night, air conditioning is a basic need for the people being able to live/rest safely and comfortably. However, the new house is designed to use as much natural ventilation as possible to minimize the need for active cooling. The building configuration/orientation was perpendicular to the prevailing southerly wind during the daytime, while the prevailing wind at night was from the east. The new house was strategically located, with large, open green spaces to the south and west. The house was also surrounded by a private garden. These factors promoted convective cooling of the building. The design of the window openings meant the interior was relatively safe from rain or strong wind at night while providing ventilation. For example, the master bedroom was located on the first floor, and the windows were protected by shading devices and a veranda roof. All of these factors mean that the use of natural ventilation probably contributed to less cooling energy being used in the houses.

4. Conclusion

The construction methods materials used and of all houses were similar. They used typical styles of tropical architecture and design. The medium size of each house was built in different eras. The building orientation of the old and ordinary houses was east/west, while the new house was orientated north/south. The main areas to protect from direct solar radiation during morning and afternoon is the east and west sides of the houses, while the other sides; north and south areas needed protection from diffuse solar radiation. The size and amount of all types of plants, as well as their lushness and natural condition, were also important factors. These can affect the ability of the plants especially trees to provide shade and promote the evapotranspiration process. The air-conditioning used in their indoor spaces was in different temperature settings and for different amounts of time. The ordinary house spent 45% of the electricity cost on cooling; the old house used approximately 35%, and the new house only used 15%. This reveals that human factors and behaviour of occupants influence the cooling energy. The lower temperature settings and longer time especially during the peak time of the day of air conditioning used will affect the amount of cooling energy used.

References

[1] Misni A 2013 *Procedia - Social and Behavioral Sciences. Modifying the Outdoor Temperature around Single-Family Residences: The Influence of Landscaping* **105** 664
[2] DOE 1995 *Landscaping for Energy Efficiency.* Energy Efficiency and Renewable Energy, (046) p. 1-8.
[3] Parker JH 1987 *Landscape Journal. The use of shrubs in energy conservation plantings* **6** 132
[4] Misni A and Allan P 2010 *Sustainable Building Conference (SB10). Sustainable Residential Building Issues in Urban Heat Islands - The Potential of Albedo and Vegetation* (New Zealand: Wellington)
[5] Oughton DR and Hodkinson S 2002 *Faber and Kell's Heating and Air-conditioning of Buildings*. Ninth ed., Oxford Butterworth Heinemann.
[6] McDowall R 2007 *Fundamental of HVAC Systems*. First ed., Oxford (UK: Elsevier)
[7] Nasution H and Hassan MNW 2009 *Energy Efficiency at Usage of Air Conditioning Systems In the Effort of Electrics Energy Saving Consumption.* 27 September 2009; Available from: http://www.he4si.com/Artikel/Abstract_ROVISP05.PDF.
[8] Bailey O and R Alcock 2009 *Home Energy Brief.* 23 September 2009; Available from: http://www.p2pays.org/ref/32/31147.pdf.
[9] Akbari H et al 1992 Cooling our Communities, A Guidebook on Tree Planting and Light-Colored Surfacing, (Washington: Lawrence Berkeley Laboratory)

IOP Conf. Series: Earth and Environmental Science **117** (2018) 012023 doi:10.1088/1755-1315/117/1/012023

Remote sensing entropy to assess the sustainability of rainfall in tropical catchment

M R Mahmud[1,2], M N M Reba[1,2], J S Wei[1,2], N H Abdul Razak[2]

[1]Geoscience & Digital Earth Centre, Research Institute for Sustainable Environment, Faculty of GeoinformatIon and Real Estate, Universiti Teknologi Malaysia,81310, Skudai, Johor Bharu, Malaysia
[2]Department of Geoinformation, Faculty of GeoinformatIon and Real Estate, Universiti Teknologi Malaysia, 81310, Skudai, Johor Bharu, Malaysia

*rizaludin@utm.my

Abstract. This study demonstrated the utility of entropy computation using the satellite precipitation remote sensing data to assess the sustainability of rainfall in tropical catchments. There were two major issues need to be anticipated in monitoring the tropical catchments; first is the frequent monitoring of the rainfall and second is the appropriate indicator that sensitive to rainfall pattern changes or disorder. For the first issue, the use of satellite remote sensing precipitation data is suggested. Meanwhile for the second issue, the utilization of entropy concept in interpreting the disorder of temporal rainfall can be used to assess the sustain ability had been successfully adopted in some studies. Therefore, we hypothesized that the use of satellite precipitation as main data to compute entropy can be a novel tool in anticipating the above-mentioned conflict earlier. The remote sensing entropy results and in-situ river level showed good agreement indicating its reliability. 72% of the catchment has moderate to good rainfall supply during normal or non-drought condition. However, our result showed that the catchments were highly sensitive to drought especially in the west coast and southern part of the Peninsular Malaysia. High resiliency was identified in the east coast. We summarized that the proposed entropy-quantity scheme was a useful tool for cost-effective, quick, and operational sustainability assessment This study demonstrated the utility of entropy computation using the satellite precipitation remote sensing data to assess the sustainability of rainfall in tropical catchments.

1. Introduction

Tropical catchments are facing serious threat from changing climate especially drought and spatio-temporal rainfall pattern disorder (Van Loon & Laaha, 2015; Yuhaslin, 2012). There were two major issues need to be anticipated; first is the frequent monitoring of the rainfall and second is the appropriate indicator that sensitive to rainfall pattern changes or disorder. Depending on the rain gauge to perform frequent and operational monitoring is sufficing, however, that was not the option for region that experienced data-conflict situation (DCS). Data-conflict situation can be defined as region that facing either one or more of this problems which includes sparse ground measurement, undefined quality of ground data, limited data sharing, and limited data inter-operability. DCS were commonly found in developing countries or rural areas or remote areas (Musiake, 2003). To resolve this conflict, an alternative measurement is necessary. We suggest the use of satellite precipitation as substitute.

Unlike precipitation radar, satellite remote sensing that carried precipitation radar provides cost-effective, consistent, and fine resolution measurement. Satellite precipitation has evolved significantly where the latest Global Precipitation Mission (GPM) satellite provide half an hour temporal resolution and 0.1 degrees resolution measurement (Hou et al. 2014). The most operational satellite precipitation, Tropical Rainfall Measuring Mission (TRMM) has been widely used to support from global to microscale hydrology studies (e.g., Mahmud et al., 2015, Collischonn et al., 2008). Moreover, their status as public domain data which enabled user to retrieve and utilized it for non-commercial purpose meet the status quo of our first issue of DCS.

Regarding our second issue, most of the drought based indices or techniques to monitor rainfall surplus or deficit, were centralized only to quantity. Another drawbacks for the established index such as Standard Precipitation Index (Gutman, 1999) or Palmer Index (Gutman, 1991) is their reliance to the historical data to obtain good prediction results. This is another disadvantage condition for areas with DCS. To resolve this, a decent approach introduced by Kawachi et al. (2001) can be helpful. They had successfully used the entropy concept in interpreting the disorder of temporal rainfall and later integrate it with the actual rainfall quantity to create sustainability matrix. We hypothesized that if this approach can be repeated using the precipitation data acquired from remote sensing satellite, it can anticipate the two main issues that we are facing earlier.

Therefore, we conducted a study to test the hypothesis. The main objective is to investigate the utility of entropy computation using the satellite precipitation remote sensing data to assess the sustainability of rainfall in tropical catchments. This study consists of three major phases, first is the computation of both rainfall and entropy. After that, the sustainability matrix will be constructed. The third phase consist of verification with the in-situ data. We will consider two case condition in this study, one is normal (non-drought) and second is drought. This is to ensure that our technique is sensitive to represent the extreme variation of rainfall disorder. Peninsular Malaysia is selected as our experimental site. This study would be useful as support tool to measure the productivity of tropical catchments.

2. Material and method

2.1 Data

2.1.1 Satellite precipitation data. We used satellite precipitation from Tropical Rainfall Measuring Mission as primary data. This data was selected due to its high operational condition; fine spatial resolution and high temporal coverage. The monthly rain rate product namely TRMM 3B43 will be used. It will then have converted into total monthly rainfall by multiplying with the corresponding number of days in each day. Two datasets were created, one is during non-drought condition (year 2001) and another is during intense drought (year 1998). This year was selected based on the global anomalies given by National Observation Ocean Administration (NOAA).

2.1.2 Geo-location of reservoir and catchments. The location of the reservoirs and catchments were obtained from the hydro-spatial database of Geoscience and Digital Earth Centre, Universiti Teknologi Malaysia. Those data were in geospatial layers of ArcGIS shapefile format. The reservoir layer consists of the shape of reservoir and its attributes including area, perimeter and size. Meanwhile, the boundary lines of each reservoir's catchment areas were obtained from separate layer.

2.2 Water catchment and reservoir in Peninsular Malaysia

Peninsular Malaysia is situated at the center of Southeast Asia, neighboring Thailand at north, Singapore at the southern and facing the Sumatera Island of Indonesia on its west. It has humid tropical climate with annual rainfall average is 2200mm and temperature ranging from 26-33 degree Celcius. Peninsular Malaysia humid tropical climate was heavily influenced by Asian monsoon.

Relatively, the period from February to April is the driest (~150mm/month) where convective rainfall dominates. It is followed then by the wet period starting roughly in June until September (~220mm/month) resulted by the southwest monsoon. November to January is the wettest months (~320mm/month) due to the moist air masses from the northeast monsoon. The freshwater resources were entirely depending on surface reservoir (95%), primarily lakes (Fig.1). There are about 91 reservoirs in Peninsular Malaysia. The largest reservoir is Kenyir (36844 hectares) and their catchment reached 500 square kilometer. The growing demand and frequent drought has led to critical water supply conflict. An effective monitoring of the catchment productivity is crucial to support sustainable development.

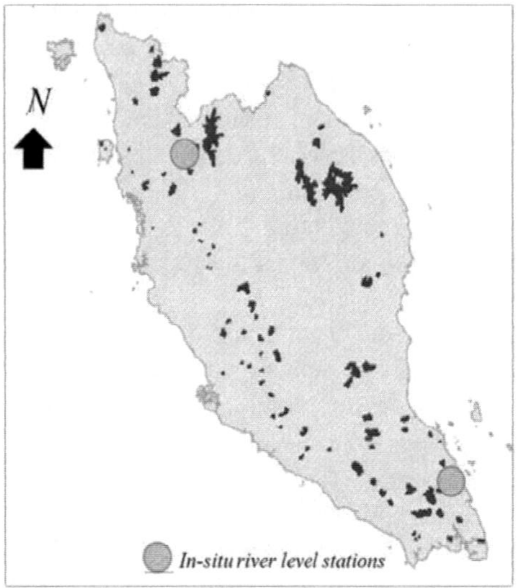

Figure 1. Study area and in-situ river level stations

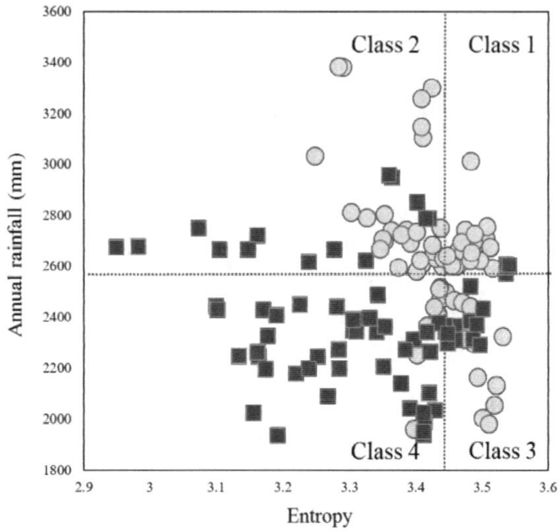

Figure 2. Entropy-quantity matrix. Class 1: Abundant rainfall and perennially available, Class 2: Abundant rainfall but concentrated at specific season, Class 3: Moderate rainfall but perennially available, and Class 4: Low rainfall and not perennially available.

2.3 Calculating the entropy value from the satellite precipitation

The monthly rainfall estimates were used as the primary input to compute the entropy value. Because the tapping behavior of the reservoir was highly characterized by monsoon season that occupied specific months, therefore it is significant that the monthly rainfall was used. The entropy computation to rainfall developed by Kawachi et al. (2001) was adopted. The following equation is described in equation below.

$$E = \sum_{i=1}^{n} \left[\frac{r}{R} * \log_2 \frac{r}{R}\right] \qquad \text{Equation 1}$$

where R is the annual rainfall from satellite estimates, r is the monthly rainfall from satellite.

In interpreting the entropy, the higher value would indicated consistent order of the temporal variation of heavier rainfall. The low entropy value then suggest that the temporal variation is highly disorder with depleting quantity.

2.4 Validation of the remote sensing entropy and rainfall

To verify the remote sensing derived entropy value, we observed the in-situ river discharge. Two stations were selected where each represent different monsoon climate region in Malaysia (Fig. 1).

Low satellite derived rainfall and entropy would indicated a low discharge and vice versa. We evaluated the scenario during the drought and non-drought.

2.5 Quantity-entropy matrix to assess the sustainability of the catchment

We used the quantity-entropy matrix to graphically assess the sustainability of the catchments. There were four regions in this 2D matrix, Class 1 to 4. Each region in this matrix represents different scenarios for the catchment. Table 1 provides the detail. Because we plot two set of samples in different scenario (drought and non-drought), this entropy hypothetically could indicates the condition of the catchments more intensively besides detecting the rainfall deficit. The changing disorder of the temporal rainfall especially the skew distribution can influenced the infiltration of the rainfall which affect the quantity of the reservoir's volume.

Table 1. Sustainability matrix class based on entropy and rainfall quantity

Class	Description
1	The rainfall is considered as abundant and perennially available
2	The rainfall quantity is assumed to be large, however characterized by concentrated heavy rainfall on certain period. Plausible to experience flood.
3	The rainfall is assumed to be moderate but perennial
4	The water resources is considered low and not perennially available

3. Results

Table 2 informed us that drought had affected the sustainability of the rainfall in majority catchments in Peninsular Malaysia. During the drought, 63% of the catchment were classified into class 4, compared to only 14% during the non-drought condition. Figure 2 had graphically support this claim. Only 4% of the class 1 catchments remained including Bernam, Batu, Tinggi and Selangor catchments (Figure 3). Maps of the catchments sustainability class indicated that the 1998 drought had affected most of the areas (Figure 3). The intense areas that were severely affected was the southern where most of the catchments score dropped from class 1 and 2 to class 4. The less affected area is the east coast.

Both in-situ river level in the southern and west Peninsular Malaysia agreed with the entropy-quantity matrix scores (Fig. 4). In Sg. Lenggor, the river level for the first six months was relatively low compared to the non-drought season. This finding was in common with Sg. Rui in the northwest that experienced decreasing entropy-quantity score. This evidence had indicated that the remote sensing based computation of entropy and rainfall quantity was capable to represent the actual condition on the site although the observation was having non- direct contact with the subject.

Table 2. Summary of the sustainable matrix

Condition	Class 1	Class 2	Class 3	Class 4
Non drought	36	30	13	13
Drought	4	15	15	58

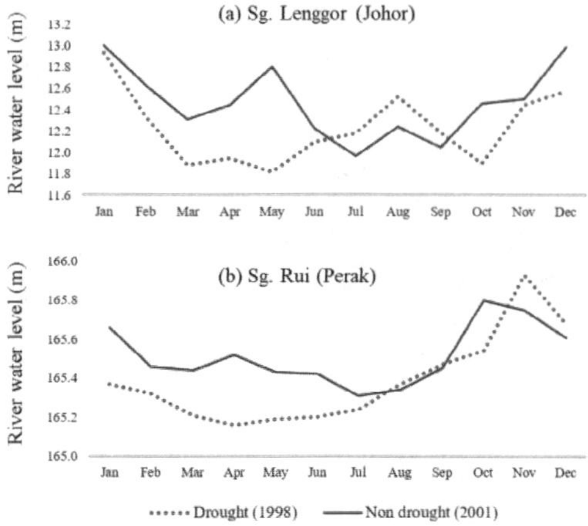

Figure 3. Sustainability class based on entropy-quantity matrix

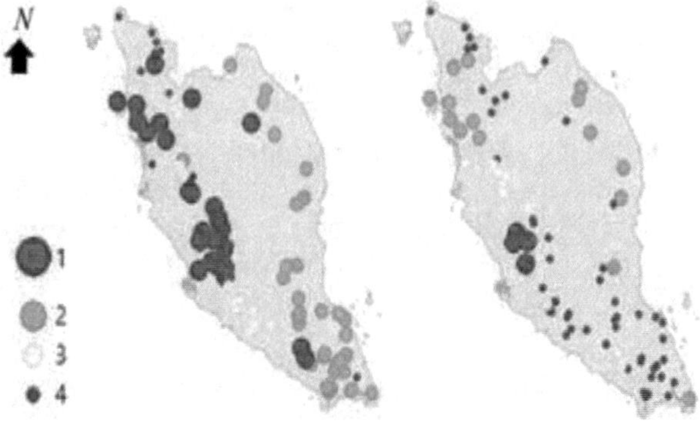

Figure 4. In-situ river level

4. Discussion

Our results demonstrated that the remote sensing based entropy and quantity was able to assess the sustainability of the catchments and comparable with actual on-site condition. This indicates that by only relying on the satellite precipitation data, a robust evaluation of the catchments productivity can be achieved. In addition, the assessment can evaluate the spatial impact of drought. Critical information regarding its intensity, geolocation and spread can be obtained. Our method can be cost-effective, operational and reliable assists to the respective users (scientist, hydrologist, hydro-engineers, managers, etc.) in supporting the primary observation method of rain gauge or precipitation radar.

Nonetheless, the improvisation was required. Our results needs to be confirmed with larger temporal samples with appropriate partition according to the monsoon season. It is suggested that the weekly and daily rainfall can be taken into account. This might accurately revealed the behavior of the catchments prior to any hydro-meteorological changes. Our main concern is that this early experiment showed promising result and subjected to further studies. Therefore, we recommend that this work to be conducted in the future.

Acknowledgements

The authors would like to thanks all the stakeholders and respective agencies particularly the Department of Irrigation and Drainage for their support regarding the in-situ data.Our utmost gratitudes goes to Universiti Teknologi Malaysia for supporting this study through the potential academic staff grant (QJI30000.2727.02K83) and Ministry of Higher Education awarded national grant (QJI30000.7827.4F773).

References

[1] Van Loon AF, Laaha G 2015 Hydrological drought severity explained by climate and catchment characteristics. J. Hydrol. 526, 3-14.

[2] Yuhaslin Y, 2012 Final report: Working group on hydrology regional association 5. World Meterological Organization, 41

[3] Musiake K 2003 Hydrology and water resources in monsoon Asia: A consideration of the necessity of establishing a standing research community of hydrology and water resources in the Asia Pacific region. Hydrol. Process, 17, 2701-2709.

[4] Hou AY, Kakar RK, Neeck S, Azarbazin AA, Kummerow CD, Kojima M, Oki R, Nakamura K, and Iguchi T 2014 The global precipitation measurement mission. Bull. Amer. Meteor. Soc., 95, 701-722, doi:10.1175/BAMS-D-13-00164.1.

[5] Mahmud MR, Numata S, Hosaka T, Matsuyama H, Hashim M 2015 Assessment of effective seasonal downscaling of TRMM precipitation data in Peninsular Malaysia. Remote Sens., 7, 4092-4111, doi:10.3390/ rs70404092.

[6] Collischonn B, W Collischonn, and Tucci C EM 2008 Daily hydrological modeling in the Amazon basin using TRMM rainfall estimates. J. Hydrol., 360, 207-216.

[7] Gutman NB 1999 Accepting the standardized precipitation index: a calculation algorithm, J. American Water Resources Association, Vol. 35, 2.

[8] Guttman NB 1991 Sensitivity of the Palmer Hydrologic Drought Index. Water Resources Bulletin 27:797-807.

[9] Kawachi T, Maruyama T, Singh VP 2001 Rainfall entropy for delineation of water resources zones in Japan. J. Hydrol., 246, 36-44.

IOP Conf. Series: Earth and Environmental Science **117** (2018) 012024 doi:10.1088/1755-1315/117/1/012024

GBI assessment checklist: Level of awareness of the contractors in the Malaysian construction industry

M A Marhani [1] and M A S Muksain [1]

[1]Faculty of Architecture, Planning and Surveying, Universiti Teknologi MARA, Shah Alam, Malaysia

Abstract. Malaysia has developed Green Building Index (GBI) as the green rating tool for buildings and townships to encourage sustainability in the construction industry. Understanding the roles and tasks of a contractor is essential in accomplishing the GBI assessment on a green building project. However, there are challenges in adapting the GBI. The objective of this paper is to identify the knowledge of contractor for each area of assessments in GBI's checklist. Data were collected through the questionnaire surveys to the selected contractors, which were involved in GBI projects. The findings revealed that most of the respondents agreed and acknowledged the impact of each criterion in the assessment. Thus, it is highly recommended to educate and change the public perception on the awareness of the green building. Aside, more research and development (R&D) process should be encouraged to accelerate the GBI certificate implementation in the construction industry. In conclusion, proper education training, more legislative and financial framework are essential for the future of green building projects.

1. Introduction

The ideology of environmental sustainability and the new paradigm of 'sustainable human development' are already spread in Malaysian construction industry in last few years. In accelerating sustainability awareness to the industry, the government of Malaysia introduced the Eleventh Malaysia Plan. In 2016, the Construction Industry Development Board Malaysia (CIDB) has established the new construction plans, which is Construction Industry Transformation Programme (CITP). One of the primary objectives of CITP is to integrate an environmental sustainability to the construction processes [1].

There are lots of fundamentals to encourage the implementation of GBI. For example, any of green building projects requires a systematic management and proper operation activities. Cooperation with many specialists and experts are also needed to upsurge the success of green building projects. Thus in 2009, the Malaysian Institute of Architects (MIA) and the Association of Consulting Engineers Malaysia (ACEM) introduced a corporation: The Green Building Index Sdn. Bhd. The main objective of this corporation is to enhance the implementation of GBI in Malaysia while considering the fundamentals in implementing GBI [2].

This paper is intended to identify the knowledge of a contractor for each area of assessment in GBI's checklist in the Malaysian construction industry. This study is also anticipating to accelerate the GBI certificate implementation in construction sector. It is significant to educate and change the public perception on the awareness of the green building and encourage the R&D process. Furthermore, it is also essential to provide a proper education training, more legislative and financial framework for the future green building projects.

2. Green Building Index (GBI) certification

GBI is used as the ordinary tool in exercising any green building development in Malaysia. Towards the end of development, the green building project will receive the GBI certification. Thus, this shows that GBI plays a major role in enhancing the sustainable development in Malaysia as the certification of GBI is recognised internationally. This practice does not only provide the advantages to the environments and eco-system, , but to the economics of the country too [3].

GBI is created to develop the green construction practical assessment for the tropical weather, environmental and development context, cultural and social needs. The GBI assessment is intended to describe green buildings by creating a standard measurement; encourage cohesive whole-building design; recognise and reward environmental leadership; transform the already built environment to decrease its environmental impact; and ensure new buildings keep on pertinent in the future, and existing buildings are refurbished and upgraded properly to remain relevant [2].

3. Methodology

This paper concentrates on the knowledge of contractor for each area of assessment in GBI's checklist. A questionnaire survey is conducted to collect the quantitative data. The respondents of this survey are the construction companies that register with the CIDB under class of G7 categories and, which are involved with GBI projects. The survey uses a 5-point Likert scale, which range from 1 = strongly disagree to 5 = strongly agree. Thirty out of one hundred questionnaires were returned, making the complete response rate of thirty percent (30%).

4. Data Analysis and Findings

This section discusses findings from the questionnaire surveys regarding the level of awareness of GBI checklist among the contractors in the industry

Table 1. Analysis of Level of Awareness of GBI checklist among the Contractors

Item	Level of Awareness	Average index	Rating
	Energy Efficiency (EE)		
1	Reducing the energy consumption in building	3.67	1
2	Implementing the building integrated photovoltaic as the building envelope	3.33	3
3	Reducing the usage of the air-conditioning in building	3.67	1
	Indoor Air Quality (IAQ)		
4	Increasing of living quality and wellbeing of occupants	3.67	1
5	Achieving the usage of low volatile organic compound material, operation of quality air filtration, appropriate control temperature, movement, and humidity of air	3.33	2
	Sustainable site planning and Management (SM)		
5	Considering the site selection and planning, access to the area of neighbourhoods, and the landscape in the perimeter area	4.00	2
7	Reducing the materials and wastage through the appropriate site management	4.00	2
8	Providing the storm water management and avoiding the environmentally sensitive areas	4.33	1
	Material and Resources (MR)		
9	Focusing on the selection of the construction materials and waste management in site	3.33	3
10	Utilising products and wastage of recycling content of materials as the primary element	4	1
11	Implementing appropriate waste management through the concept of reduce, reuse, and recycle (3R)	4	1
	Water Efficiency (WE)		

13	Recovering the water which diverted to the waste system and cleaning it to be reused in portable water consumption	4.33	1
	Innovation (I)		
14	Developing creative ideas in the element of the construction	4.33	2
15	Involving the innovation elements	5.00	1

Source: Adopted and modified from Muksain (2017) [4]

Based on Table 1, the perspective and understanding of contractor for each of the criterion stated in GBI assessment was evaluated. All the respondents have good knowledge about the checklist. For instance, the EE's criterion, the respondents agreed that the impact of EE was reducing the energy consumption in building and lessen the usage of air-conditioning, and the criteria received 3.67 as the average index. Meanwhile, the implementation of the building photovoltaic as the building envelope in the green building received the moderate understanding among the respondents.

The respondents agreed on to improvise the living quality and wellbeing of the occupants. The impact of this IAQ's criterion received 3.67 as the average index. Meanwhile, achieving the usage of low volatile organic compound material, an operation of quality air filtration, appropriate control temperature, movement, and humidity of air received only 3.33 as the average index from the respondents. Several of respondents acknowledged that SM's criterion must provide the storm water management and avoid environmentally sensitive areas in any of green building project. Furthermore, considering the site selection and planning, access to the neighbourhood's area, and the landscape in the perimeter area received an excellent understanding among the respondents. This criterion of reducing the materials and wastage through the appropriate site management had drawn 4.00 as the average index. Utilising products and wastage of recycling content of materials and implementing appropriate construction waste management through the concept 3R drawn a good understanding and received 4.00 as the average index among the respondents for the MR's criteria. Furthermore, focusing on the selection of the construction materials and waste management in site received 3.33 as an average index.

For WE's criterion, with an average index 4.33, most of the respondents believed in recovering the water which diverted to the waste system and cleaning it to be reused in portable water consumption. Furthermore, adapting specific technologies and methods to supply the water in better service with less water usage had drawn 3.67 as the average index from the respondents.

Majority of the respondents agreed that GBI involves innovation elements in any green building project. This criterion received 5.00 as the average index. Furthermore, development of innovative ideas in the construction's element received 4.33 as the average index from the respondents.

5. Discussion

5.1 Energy Efficiency (EE)

It showed that the green building project would save the thirty-six percent of energy consumption rather than the conventional project. In addition, any of green building project would reduce the electrical productivity by using direct sunlight as the medium of energy consumption. The implementation of the system could reduce the electricity consumption and layout of the building also one of the achievement of EE. Thus, the implementation of EE in the Malaysian construction industry is beneficial to the contractors and give a constructive impression to the national economic growth in enhancing the future of green building project.

5.2 Indoor Air Quality (IAQ)

Indoor Air Quality would reduce the electrical lighting consumption by installing operable window in the buildings. This is not only to increase the level of comfort of the buildings' occupants; but also utilised the usage of the natural resources and increased the ventilation and lighting control for the buildings. Furthermore, building design can be improvised with the adaptation of the usage of low volatile organic compound material, utilisation of quality air filtration, proper control temperature, movement, and humidity of air and improved the achievement of IAQ in any of green building projects. In reducing bio aerosol growth and improving the thermal performance, any of green building projects should improve its moisture management as supported by Persily [5]. This improvement can be accomplished through the envelope design and construction of the projects themselves.

5.3 Sustainable Site Planning and Management (SM)
A proper site planning should be considered in any of green building projects to sustain the positive environmental impact to the surrounding areas. Each project development should submit the structure plan for the proposed area, and it is compulsory to comply with the guideline given by the local authorities as per stated in Nizarudin, Mohd Hussain, and Tukiman [6]. Furthermore, any of green building projects should have a proper site management to reduce the construction materials and wastage on site. If any of the projects excels in dealing with this problem, it would benefit the contractors in achieving the project's goals [7].

5.4 Material and Resources (MR)
In the main process of MR's criterion, selection of materials that consists green features elements without affecting the quality of the project is crucial. Thus, the positive result could be seen in betterment on the environment, low maintenance of the building and low cost of the project as per stated in Rahardjati, Khamidi, and Idrus [8]. Besides that, the adaptation of 3R concept in enhancing the systematic construction waste management plays an important role in reducing the negative effect to the environment. There are sub-section in GBI checklist for the contractors to segregate the construction waste (e.g., plastic, concrete, timber, and glass), submit to the registered waste management company and acknowledge the consultant as proof of the implementation of the 3R concept as stated by Muksain [4].

5.5 Water Efficiency (WE)
The GBI guideline has set up the amount of water in sanitary fitting such the amount of water flow in water tap per second, hand bidet taps in the water closet, and water closet. In this WE's criterion, these processes can be visualised rather than the other criteria. Several of contractors agreed that there are different methods of constructing the rainwater harvesting system to enhance the water recycling processes at the construction site.

5.6 Innovation
The respondents stated that the innovation criterion was different between each of the green construction projects. The criterion indicated that the process of developing the creative ideas from designing stage as the trigger phrase in enhancing the adaptation of GBI. The requirements should emphasise on the objectives of the sustainable design to be more efficient in the future. Furthermore, the understanding of each of the required criteria in the checklist involved will provide a better understanding in achieving the objectives of the implementation of GBI assessments.

6. Conclusion
All the respondents, which directly involved in the green construction acknowledged the requirement of each criterion, the methods to fulfil the requirement and the impact of the GBI assessments towards the projects. The significance of each criterion had also been agreed by most of the respondents.

However, the authors believe that several improvements to the GBI assessment checklist should be made for the future green building projects as agreed by the contractors on the checklist. There is need to increase the awareness among the citizens and end-users through the education and media report. Furthermore, Lop, Che Ahmad, and Nik Zulkipli [9] stated that education and training should be included in any of green building development to be well known and acknowledged by the people. Improvement in MR's criterion, POE's criterion and the checklist should also considered the environmental interior design's system and materials. Aside, the government should increase the financial fund for the GBI project as well.

References

[1] Ministry of Works Malaysia, "Construction Industry Transformation Programme 2016-2020 2015 [Online]. Available: http://www.citp.my/.

[2] Green Building Index Sdn Bhd, "GBI Assessment criteria contents," no. March, pp. 0–57, 2011.

[3] Olubunmi OA, Xia PB and Skitmore M 2016 *Renew. Sustain. Energy Rev. Green building incentives: A review* **59** 1611

[4] Muksain MAS 2017 *GBI assessment checklist: contractor's perspective* Faculty of Architecture, Planning & Surveying, Universiti Teknologi MARA, Shah Alam

[5] Persily AK 2012 *Indoor air quality in sustainable , energy efficient buildings Indoor air quality in sustainable , energy efficient buildings HVAC&R Res.* **18** 37

[6] Nizarudin ND, Mohd Hussain MR, and Tukiman I 2010 The application of the Green Building Index (GBI) on sustainable site planning and management for residential new construction: prospects and future benefits the application of the GBI," in *SUSTED*

[7] Marhani MA, Jaapar A, Bari NAA and Zawawi M 2013 *Procedia - Soc. Behav. Sci. Sustainability through lean construction approach: a literature review* **101** 90

[8] Rahardjati MF, Khamidi and Idrus A 2011 *International Conference on Environmental Science and Technology (ICEST). Green building rating system : the need of material resources criteria in green building assessment* pp148–151.

[9] Lop NS, Che Ahmad A and Nik Zulkipli NAD 2016 *Malaysian J. Sustain. Environ. The implementation of green building in Malaysian construction industry: determination of key success factors* **1** 64

An assessment of maintainability of elevator system to improve facilities management knowledge-base

N.A. Siti[1,2], A.S. Asmone[2], and M.Y.L. Chew[2]
[1]Faculty of Architecture, Planning & Surveying, Universiti Teknologi MARA, Malaysia
[2]School of Design and Environment, National University of Singapore, Singapore

*e0000438@u.nus.edu

Abstract. Elevator system is a highly specialized machinery that requires technicians that have a wider array of knowledge in maintaining the system to be safe and reliable. While attaining reliable data of elevator malfunction become challenges, this study has filled the gap by gathering the management-maintenance issues and operational defects of elevator system. Forty-three types of operation defects were found and the consequence defects and their possible causes of occurrences were discussed. To respond to the prime challenges of maintaining elevator system provided by the industry players' perspective, a theoretical framework is established as a recommendation to improve knowledge base of defects in elevator system which comprises good practices, and solutions to rectify each defects found. Hence, this research paper has theoretically improved the knowledge base of maintainability of elevator system and provide meaningful guidelines in practical senses to the industry professionals.

1. Introduction

Building maintainability has significantly affect building's cost, risk, and performance. A review of building performance of construction industry was undertaken and it expressed building maintainability is substantially affecting building performance [1]. Mechanical and Electrical (M&E) services which include elevator system is one of the elements which contribute to a major percentage of the total construction cost of a building from initial to its operation and maintenance costs. If M&E services are facing with malfunction or breakdown, it will contribute to the entire building performance substantially influence the building's life cycle cost [2]. This system must operate in an efficient and effective manner as its operation not only affected the system's life cycle cost but also ensure safety consideration to the users.

Inappropriate maintenance of elevator system can cause rusted and cracked steel wire rope and it is one of the examples of defective in an elevator system, which can trigger accidents to happen. The most common practice of elevator inspection is depending on human visual inspection, and such practice fails to detect the faults occurred in the system [3]. Over the past few years, there have been some reported cases of injuries caused by elevator defects or malfunction. Hence, it is very important for the building owners, particularly for the facility management team to perceive it as a dominant part of building maintenance element, while not merely putting all the duty of care to the mechanical and electrical team. Meanwhile, one of the strategies to minimize the failures is by recognizing the defects and its causes. By doing so, a proper maintenance of building systems can be achieved which may delay the process of degradation of the systems [4].

This study aimed to investigate the maintainability issue of elevator system by recognizing the type of operational defects or malfunction that could be found in the elevator system and the possible causes of each defect detected. By understanding the common defects found and the possible causes, it helps the maintenance management team to make a prediction of possible more damages resulted from independent defects of each elevator's components. In addition, the maintenance-management issues were discussed to identify the prime challenges of maintaining elevator system. Finally, a recommendation framework was then tabulated with the objective to improve the knowledge base of which includes the testing and diagnostic, good practices and solutions to rectify the elevator defects.

2. Literature Review

2.1 Regulatory controls of elevator system in Singapore

As according to Singapore Standard for Lift, Escalator and Building Maintenance (SS550:2009), the maintenance for elevators system in Singapore is done in periodic regime which include daily, monthly, biannually, annually, and every five (5) years' examinations. Meanwhile, The Building Maintenance & Strata Management (BMSMA) [5], has clearly stated the duty of care for both elevators owner and elevator service in relation to the health and safety of passengers or property which may be affected by their actions, or failure to act. This include the requirement for the elevator owners need to keep all the maintenance records for a period of at least five (5) years from the date the documents are recorded or issued. Likewise, the elevators service contractors are required to submit full report involving elevator system's malfunction, and notify the Commissioner if the elevators are found to be unsafe to be operated. There is a legal requirement to report any incidents associated with malfunctions of elevator system, which include:

i. Incidents that involved injury or death where the cause of such injury or death is associated with equipment or machinery of the elevator;
ii. Failure of main drive system, except in the case of failure of the main power system;
iii. Breakage of suspension ropes;
iv. Failure of brakes, overload device, safety component and safety equipment;
v. Failure of an interlocking device for any door of the elevator way or door of the elevator car, except for the reason of safety contact does not make electrical contact.

2.2 *Maintainability issues of elevator system*

Elevator malfunctions create inconvenience to the public especially for those with disabilities and diminishes public confidence to utilize this facility [6]. Some of residents' concerns on elevator maintenance practices for the residential apartments include [7]:

i. Alternative for building residents to report where there are repeated elevator malfunctions
ii. Delayed rectifications due to shortage of elevator parts
iii. Insufficient qualified manpower to conduct repair works
iv. Servicing works done by penalized contractor and still fails to do an effective service
v. Organize an authority lift expert team in or working with the industry to improve insufficient qualified manpower to service the elevator system

To acquire a reliable accident statistics become challenges as the users generally would not make a report when they encountered minor accidents, reports involving major accidents being decentralized, and most of the elevator companies do not always consent to involve in reporting any breakdown or malfunctions. Meanwhile, it is vital to have an accurate and reliable statistics and records of accidents as it can help to prevent the accidents from occurring in future by assisting the relevant parties to identify the causes of the accidents. In conjunction, proposals to improve guidelines and regulations shall be improved, by identifying common malfunctions that caused accidents in order to improve safety features [8]. The recent years of media-released accident reports involving elevator malfunction in Singapore varies from causing injuries to the users and even fatal accident and make it even more unfortunate, some cases still under investigations and the causes remained unknown [9]. Figure 1 is some examples of elevator accidents in Singapore.

IOP Conf. Series: Earth and Environmental Science **117** (2018) 012025 doi:10.1088/1755-1315/117/1/012025

	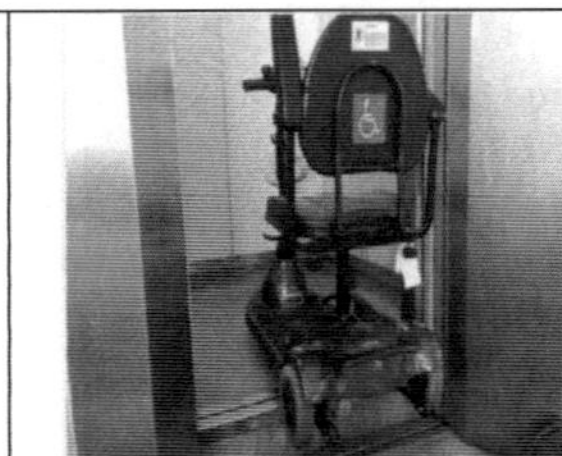	
User's hand was severely injured as the elevator sensor unable to detect dog leash attached to the victim's hand [10]	Poor levelling of elevator and landing doors causing fatal accident to the impaired user [11]	Elevator door remained open while travelling while outer and inner panels were closed [12]

Figure 1. Examples of elevator-accident cases in Singapore

3. Research Methodology

3.1 Research design

This research uses a sequential explanatory mixed-method design where qualitative method was used in the initial phase followed by quantitative method on the later phase which presented in a flowchart below.

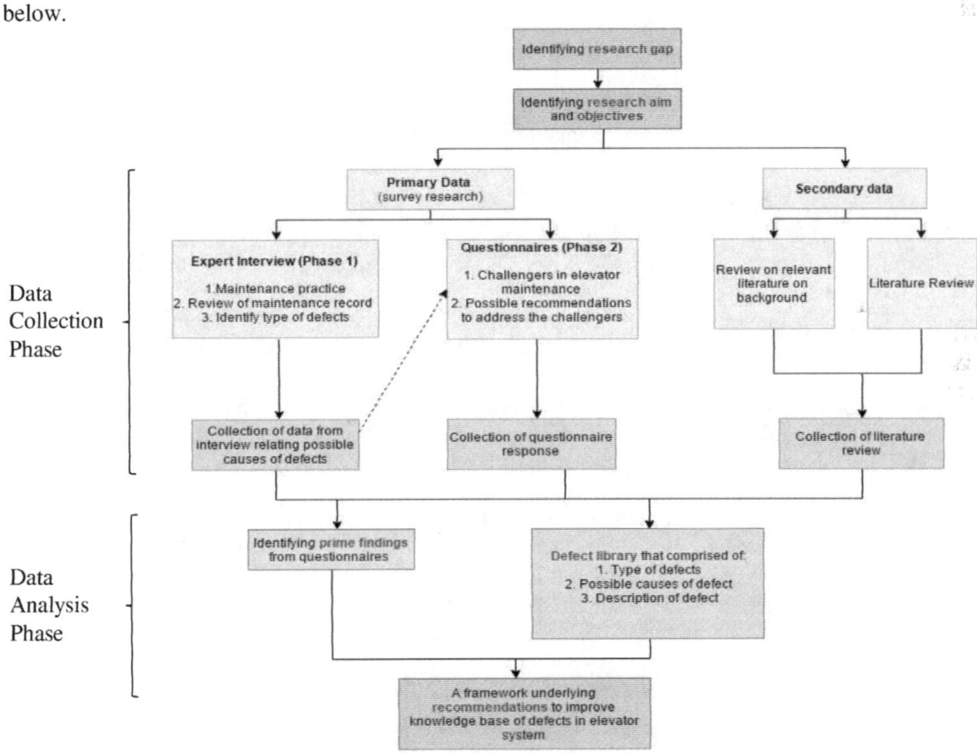

Figure 2. Flowchart of Research Methodology

The data collection phase starts by conducting secondary research. A preliminary review of the background of this topic was conducted to identify research gap, aim, and objectives. A more comprehensive literature review was then done to identify essential knowledge of elevator system and maintenance issues with regards to elevator system in Singapore.

Secondly, the primary data of this research was retrieved in two phases where the first phase was done by conducting interviews and the second phase was by distributing questionnaires survey with industry personnel. In order to achieve valuable insights from the industry professionals, the interviewees were carefully selected. Five interviews were conducted, where three of the interviewees are Senior Facility Managers, a representative from Singapore's Town Council, and a lecturer form Building Department National University of Singapore (NUS). During the interview, maintenance-management issues were raised, and common type of defects in elevator system was collected by reviewing maintenance checklist records. Defect library was then tabulated comprised from both primary and secondary data, stating type of defects, possible causes of defects and description of defects. Thereafter, the findings from the interviews conducted were used to structure the content of the questionnaires in order to validate the finding.

The questionnaires were structured in two sections where the first section was to identify the prime challenges in maintaining elevator system, the second section was to identify prime recommendations to address the challenges. Survey questionnaires were sent out electronically to the industry players in Singapore and out of sixty-six (66) responses, only thirty (30) surveys were usable as the remaining were partially completed hence the data were not used to evaluate the survey. From a total number of respondents who able to complete the survey, sixteen (16) of them are the Facility Managers, six (6) from developer companies, five (5) from elevator contractors' companies and the remaining three (3) are building owners. The data was then compared, ranked and discussed from both primary and secondary research sources to enable to determine the severity of each challenges in maintaining elevator system and the effectiveness of suggested solutions to address the challenges. Finally, survey findings are discussed and a recommendation framework is tabulated to improve knowledge base of elevator systems. The framework consists of testing and diagnostic method, good practices, and correction for each defect type found in the earlier section.

4. Data analysis

4.1 Discussion of industry professional interviews and survey questionnaires

Based on the interview conducted, there are similarities responses given by the industry expert interviewees pertaining management challenges in maintaining elevator system. According to the experts' judgment, the knowledge gap among the technicians sent to do servicing work restricting them to be able to detect premature wearing signs of elevator components. In addition, there is a limitation on the availability of skilled technicians and engineers in Singapore, due to low interest of people to participate in the elevator industry and finally, maintenance record of faults and defects of elevator system provided by the elevator contractors is insufficient and unreliable to be referred. From the interview, maintenance records were reviewed, and some common operational defects were identified. Literature study was used to find the further description of the defects and possible causes of each operational defect found, partially shown in Figure 4 and a summary of common operational defects of elevator system are tabulated in Figure 3.

The questionnaire survey was then sent out to the industry players in Singapore to validate the findings from the interview with the objective of identifying the prime challenges in elevator system maintenance. Three prime challenges in maintaining elevator system were found, which include:

a) Insufficient maintenance records by the technicians sent to do servicing works
   In the usual practices of periodic maintenance records, a detail description of system's faults will not be recorded, nor photos will be attached. In contrast, reporting in detail the condition of components faults is very important as a record to assist both building managers and contractors to achieve agreement in the case of a period of replacement of components.

b) Limited availability of skilled manpower

As elevator system is being a very specialized system, the qualified technicians and engineers are limited hence the price competitiveness are remaining low. Consequently, this scenario promotes the industry players to keep the consultation charge to be high.

c) Lack of knowledge among the technicians sent to do servicing works
The technicians sent to do servicing works are lack of knowledge to be able to detect the early signs of wearing of components in elevator system. As the elevator system is to be modernized nowadays, the software used to control the performance of the system is even more complex and the technicians sent to do servicing works may not be fully quipped with knowledge to manage these advanced technologies implemented in the system.

In conjunction, the respondents agreed that the three most perceived to be effective solution to address the challenges are by providing maintainability grading system, raising awareness to the public on how to use the elevator appropriately, and providing the elevator system faults library. In order to respond to challenges mentioned above which include deficiency on the records of operational defects of elevator system, a framework which comprised of diagnostic and testing methods, good practices, and method of rectifications are provided with the objective to improve the knowledge base of the industry players, specifically among the Facility Managers. An example of tabulated framework is shown in Figure 5. A comprehensive explanation on each of the elevator defect which include possible causes of defects and the framework can be retrieved in the Maintainability of Buildings website (http://www.hpbc.bdg.nus.edu.sg/).

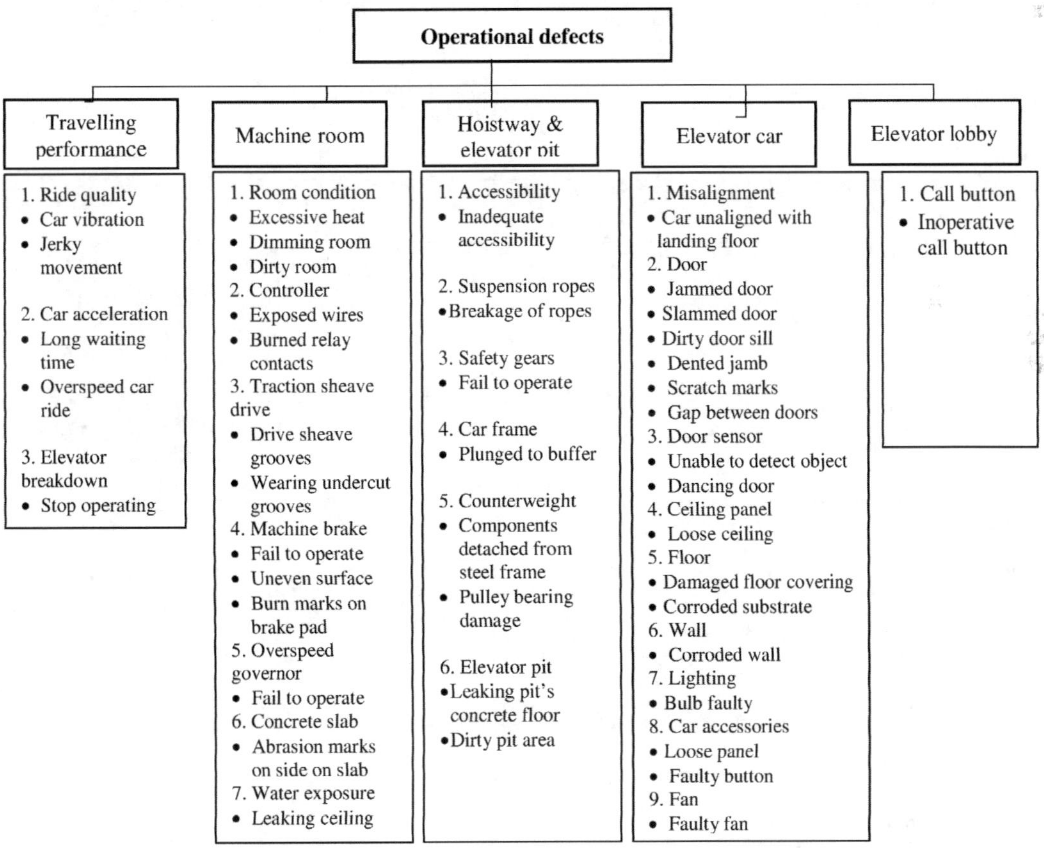

Figure 3. Common operational defects found in the elevator system

IOP Conf. Series: Earth and Environmental Science **117** (2018) 012025 doi:10.1088/1755-1315/117/1/012025

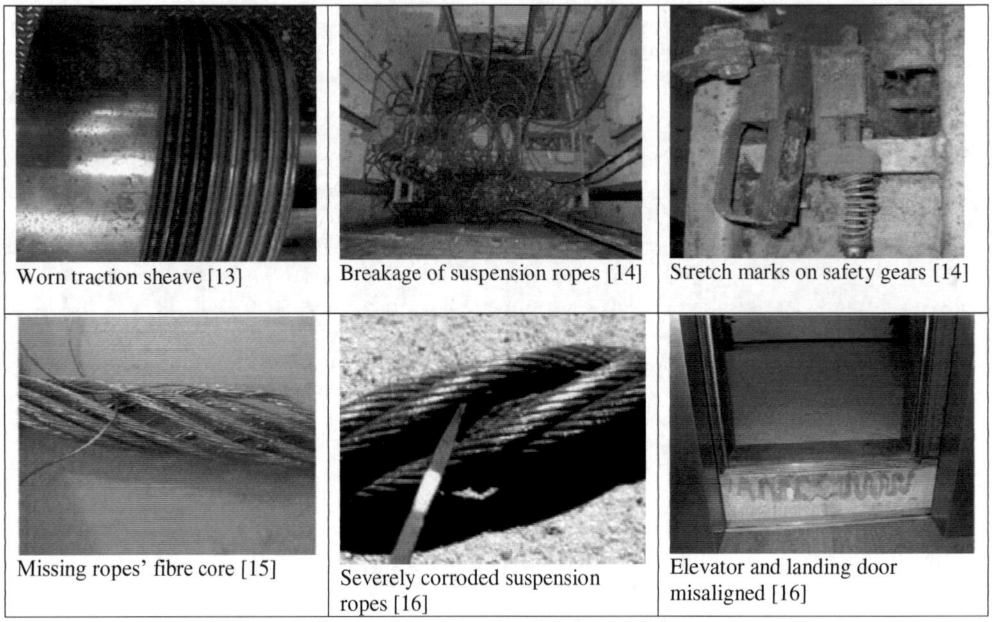

Worn traction sheave [13]
Breakage of suspension ropes [14]
Stretch marks on safety gears [14]
Missing ropes' fibre core [15]
Severely corroded suspension ropes [16]
Elevator and landing door misaligned [16]

Figure 4. Some of the operational defects of elevator system

Element	Diagnostic/testing method and good practices	Good practices	Corrections	Diagram
1. General elevator traveling performance				
Ride Quality: 1.Car vibrations 2.Jerky movement	Elevator vibration analysis is a method to measure ride quality for elevator. The hardware tool system combined with a software will allow the technicians to measure vibration and sound and the measured value shall be referred to the ride quality standards, such as ISO 18738:2012 (PMT, 2016)	The unit for vibration is mili-g (mg), and one mg equals ca. 0.01 m/s². For lateral vibrations, the standards required: i. ≤ 15 mg (ISO MPtP) ii. ≤ 10 ±3 mg (ISO A95) For vertical vibrations, the standards required: i. ≤ 25 mg (ISO MPtP) ii. ≤ 15 ±5 mg (ISO A95) The following standards are applicable to be referred for vibrations ride quality (Schindler, n.d.): i. ISO 2631-1-1997 – Mechanical vibration and shock –Evaluation of human exposure to whole-body vibration – Part 1: general requirements ii. ISO 18738:2003 – Lifts (elevators)- Measurements of lift ride quality iii. ISO8041:1990 and Amd.1:1999 Human response to vibration – Measuring instrumentation	To avoid discomfort motion of rides, in the case where displacement of air pressure in the hoistway created lateral vibrations, the hoistway shall have a clear area at least 15% of the car platform area for elevator speeds up to 5 m/s and this figure shall be increased by 5% for each additional 0.5 m/s increase in car speed. To reduce lateral vibration produced from the roller guide, it is recommended to use Electric Active Roller Guide that reduces vibration produced by 50%. It comprised with an accelerometer that detects car vibration during operation, and the actuators will cancel the vibration through a controlled electromagnetic force (CastleLift, 2016)	 Physical Measurement Technologies EVA-625 Elevator Vibration Analysis system (PMT, 2016)

Figure 5. An example of tabulated framework to improve knowledge base of elevator defects available in the website

5. Conclusion

In brief, this research has provided the information from the perspectives of industry professionals with regards to maintainability issues of elevator system both management and operational defects. Aided with the secondary data, the possible causes and consequence defects were discussed to enable to understand the impact of defects on each component on the overall performance of elevator system. Responding to the issues of insufficient information on elevator faults from the service records, these gathered information could be used by the industry players to be referred. In addition, the possible solutions to address the challenges in maintaining elevator system were also evaluated to recognize effectiveness of each solution from the industry players point of views. Finally, with this study, maintainability of elevator system can be better understood by gathering the information pertaining elevator's good practices as according to standard, diagnostic and correction methods and hence improving the knowledge base of elevator maintenance, to be used both academic and practical use and value to this research.

References

[1] N. de Silva, M. F. Dulaimi, F. Y. Y. Ling, and G. Ofori, "Improving the maintainability of buildings in Singapore," *Build. Environ.*, vol. 39, no. 10, pp. 1243–1251, Oct. 2004.
[2] N. . Khiyon and S. . Mohamed, "Life cycle costing of mechanical and electrical services for private finance initiatives projects in Malaysia," *Int. J. Tomogr. Simul.*, vol. 28, no. 3, pp. 96–103, 2015.
[3] P. . Tse and J. . Chen, "Effective guided wave technique for performing non-destructive inspection on steel wire ropes that hoist elevators," vol. 19, pp. 309–320, 2015.
[4] L. S. Pheng and D. Wee, "Improving maintenance and reducing building defects through ISO 9000," *JQME J. Qual. Maint. Eng.*, vol. 7, no. 1, pp. 6–24, 2001.
[5] BMSMA, "Building Maintenance and Strata Management Act, (Lift, Escalator and Building Maintenance) Regulations 2016," 2016.
[6] G. Ong, "Lift accidents: What to know and how to react - The Middle Ground," 2016. [Online]. Available: http://themiddleground.sg/2016/05/19/lift-accidents-what-to-know-and-how-to-react/.
[7] S. O. Municipal, "Oral Answer by Ministry of National Development on lift breakdowns in HDB housing estates," 2016. [Online]. Available: http://www.mnd.gov.sg/mso/parliamentary-11-jul-2016-lift-breakdowns.htm. [Accessed: 26-Aug-2017].
[8] ElevatorWorld, "The Importance of Accident Statistics," *ElevatorWorld*, 2013.
[9] V. Lim, "One month after fatal lift accident, son of deceased says: 'I just want some answers', Latest Singapore News - The New Paper," 2016. [Online]. Available: http://www.tnp.sg/news/singapore/one-month-after-fatal-lift-accident-son-deceased-says-i-just-want-some-answers. [Accessed: 29-Aug-2017].
[10] AsiaOne, "Dog hanged to death from leash after owner accidentally left it outside lift," pp. 1–8, 2015.
[11] L. Chia and J. One, "Elderly man dies following lift incident in Pasir Ris," no. May 16, 2016.
[12] ChannelNewsAsia, "Punggol HDB block's lift operation suspended due to faulty door panel," pp. 1–7, 24-Jan-2016.
[13] Regrooving, "Sheave Regrooving - Renown Electric," 2016. [Online]. Available: http://www.renown-electric.com/field-services/sheave-regrooving/.

[14] EMSD, "Technical Report on the Lift Incident on 25 October 2008 at Shin Nga House , Fu Shin Estate , Tai Po," no. December, 2008.
[15] EMSD, "Technical Investigation Report on Lift Incident at 478-480 King's Road, North Point, Hong Kong," 2013.
[16] Elevatorbob, "elevatorbob's Elevator Pictures - Elevator Sills - Page 1," 2010. [Online]. Available: http://www.elevatorbobs-elevator-pics.com/elev_sills_p1.html.

An empirical analysis of strategy implementation process and performance of construction companies

F I Zaidi[1], E M A Zawawi[2], R M Nordin[2] and E M Ahnuar[2]
[1] Centre of Research and Postgraduate Studies, UiTM
[2] Centre of Studies for Construction, Faculty of Architecture, Planning and Surveying, University of Technology Mara (UITM), Selangor, Malaysia

*fazaihsanzaidi@gmail.com

Abstract. Strategy implementation is known as action stage where it is to be considered as the most difficult stage in strategic planning. Strategy implementation can influence the whole texture of a company including its performance. The aim of this research is to provide the empirical relationship between strategy implementation process and performance of construction companies. This research establishes the strategy implementation process and how it influences the performance of construction companies. This research used quantitative method approached via questionnaire survey. Respondents were G7 construction companies in Klang Valley, Selangor. Pearson correlation analysis indicate a strong positive relationship between strategy implementation process and construction companies' performance. The most importance part of strategy implementation process is to provide sufficient training for employees which directly influence the construction companies' profit growth and employees' growth. This research results will benefit top management in the construction companies to conduct strategy implementation in their companies. This research may not reflect the whole construction industry in Malaysia. Future research may be resumed to small and medium grades contractors and perhaps in other areas in Malaysia.

1. Introduction

Strategy implementation is the most critical part of strategy planning. It is defined as the conversion of strategies into actual action plans [1]. It is admittedly tedious and more complicated than formulating the strategy itself [2]. Successful strategy implemented enables for a company to improve its overall performance [3]. Strategy implementation also helps a company to gain its competitive advantage [3]. However, a good strategy formulated but not being implemented successfully will not serve its purpose [4].

Construction industry in Malaysia is growing throughout years. There are roughly about 64,000 construction companies were registered with CIDB as of August 2017. However, studies have shown that the failure rate of Malaysian construction companies is critical due to improper strategy implementation by the construction firms [5]. Previous research showed that 90% companies failed to implement strategies successfully due to the mistakes made in the strategy implementation process [6]. Major obstacles in strategy implementation process are to transfer the strategy from managerial level to lower organisational levels employees [6]. Unawareness of lower organisational levels employees towards company strategy may be the major issue which contribute the strategy implementation inefficiency [6].

Strategy implementation can influence the whole texture of a company including its performance [6]. However, many construction companies in Malaysia have yet to formalize the strategic process [7]. Therefore, a study on relationship between strategy implementation process and performance of construction firms needed to be carried out. This research will be able to provide the empirical relationship between strategy implementation process and performance of construction companies.

2. Literature review

2.1. Strategy implementation process

Strategy implementation has become an essential part of business strategies for companies to stay internationally competitive [8]. A formulated strategy can only benefit the company if it is successfully implemented [9]. To implement strategy successfully there is several processes needed to be considered [10].

A good strategy implementation naturally starts with a good strategic input [11]. The management team has to ensure the employees are aware of the company's vision and mission and understands it well [11]. Acceptance and strong understanding of company's strategy by employees will helps the strategy implementation runs smoothly and effectively [12].

Strategy implementation only works when there is a clear strategic action plan [1]. Strategic action plan consist of who does what, when, at what cost [11]. Therefore, building a strategic action plans helps to segregate task, provide datelines and estimate the cost.

Allocation of resources is another critical activity in strategy implementation [1]. Without resources, strategy cannot be implemented smoothly [1]. Resources allocation for strategy implementation must align with the strategic action plan [4]. Resources allocation in strategy planning includes human resources, physical resources and technological resources [13].

Employees are a human being and humans do not like to move out from the comfort zone [9]. During initial strategy implementation, usually employees will resist to make any changes in their working routine [1]. Training helps to eliminate employees' resistance toward changes and helps to adapt employees into implementing the strategic action plan [11].

Successful strategy implementation can also be influence by the involvement of employees in decision making [1]. An organic organisation structure helps to increase employees' involvement in decision making by providing low formality in reporting relationship and shared in authority with employees in decision making process [14].

Reward is closely related to the successful of strategy implementation [1]. Rewards helps to give positive stimulation and helps to influence employees' behaviour [15]. Either monetary or non-monetary incentives, rewards help to motivate the employees to implement the company's strategy [15].

Corporate culture is another human factor which influences company performance. Corporate culture as "a set of norms and values that are widely shared and strongly held throughout the organization" [16]. Corporate culture as the way group learning, solving its problems and adapt to the external environment [17].

2.2. Strategy performance

Construction's companies need to stay competitive to survive this increasing globalize competitive construction industry. Effective strategy implementation enables for construction firm to improve the company's performance [3].

Successful strategy implementation will maximize the revenue of the company and helps to give positive profit growth [18]. Revenue is the total amount revenue within a specific time period which usually is 12 months whereas; profit is the net income of the business [19].

Employees' growth is another positive indication that a company is improving its performance [20]. When a company successfully grows its business, more employees are required to fulfil company needs for the growing capacity [19].

IOP Conf. Series: Earth and Environmental Science **117** (2018) 012026 doi:10.1088/1755-1315/117/1/012026

3. Relationship between strategy implementation process and the construction companies' performance

Strategy implemented definitely give impact to companies' performance [21]. A successful implementation of strategies will influence the company performance to a better state [21]. However, there are still a lack empirical evidence proving the relation [22]. Therefore, this research provides the empirical relationship between strategy implementation process and performance of construction companies. These hypotheses guided the direction of the study:

H1. There is strong positive relationship between strategy implementation process and construction companies' performance

H1a. There is strong positive relationship between strategy implementation process and construction companies' profit growth

H1b. There is strong positive relationship between strategy implementation process and employees' growth in construction companies

4. Methods

4.1. Sampling and data collection

The research focuses on the relationship of strategy implementation process with the performance of construction's company in the Klang Valley, Selangor which is the one of most rapid development areas in Malaysia and focus point for most of the construction companies in Malaysia [23]. The competition between construction companies in this area is definitely high and required a proper strategy implementation to enable the survival of the construction companies.

It was assumed that, not entire population of small and medium construction companies conduct strategy implementation [24]. Hence, this research targeted contractors registered as Grade G7 with CIDB Malaysia because almost all large construction companies conduct strategy implementation [7]. In this pilot study, twenty-five (25) samples of questionnaires survey were collected.

4.2. Research constructs

Conceptual framework was developed based on the findings from previous research. The framework is shown in 'figure 1'. The framework emphasizes on the relationship between strategy implementation process and construction companies' performance. Previous research showed that, profit growth and employees' growth are critically influence by strategy implementation process. Thus, construction companies that established such strategy implementation process have high potential in gaining positive company's performances.

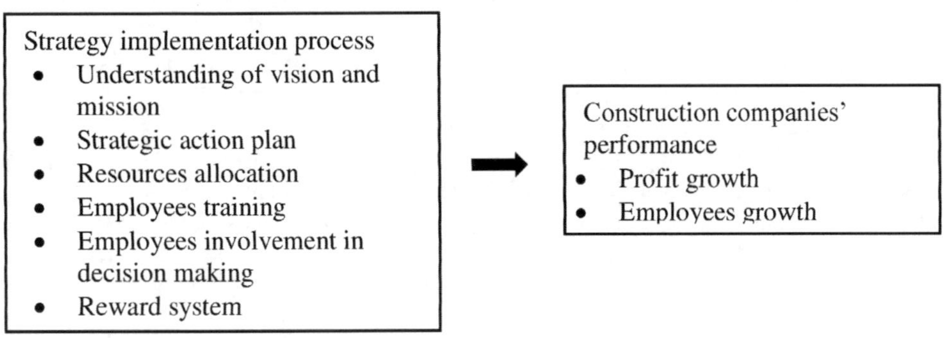

Figure 1. Conceptual framework of strategy of implementation process and the construction firms' performance.

5. Discussions of results

Respondents for the questionnaires survey consist top management such as Head of Strategic Planning, CEO of construction's company, Administration Manager and Human Resources Manager. Statistical Package for the Social Sciences (SPSS) Software was utilized as tool to analyse the data.

The Spearman's correlation analysis was carried out. The result shown in 'table 1'. It can be seen that there is a relationship between strategy implementation and the performance of construction companies.

Table 1. Pearson correlation analysis between strategy implementation process and the construction companies' performance.

		Profit Growth	Employees' Growth
Ensuring vision & mission are understood by the employees	Pearson Correlation	.471*	.533**
	Sig. (2-tailed)	.018	.006
	N	25	25
Making strategic action plan to achieved the strategic objective	Pearson Correlation	.397*	.394
	Sig. (2-tailed)	.049	.052
	N	25	25
Allocating sufficient resources for strategic action plan	Pearson Correlation	.618**	.429*
	Sig. (2-tailed)	.001	.032
	N	25	25
Providing sufficient training for employees to carry out the action plan	Pearson Correlation	.641**	.560**
	Sig. (2-tailed)	.001	.004
	N	25	25
Involving employees in decision making	Pearson Correlation	.600**	.482*
	Sig. (2-tailed)	.002	.015
	N	25	25
Rewarding employees	Pearson Correlation	.427*	.503*
	Sig. (2-tailed)	.033	.010
	N	25	25
Strong corporate culture	Pearson Correlation	.422*	.532**
	Sig. (2-tailed)	.036	.006
	N	25	25

**. Correlation is significant at the 0.01 level (2-tailed).
*. Correlation is significant at the 0.05 level (2-tailed).

The results of ensuring vision and mission are understood by the employees during strategy implementation is that they have the pure understanding on the importance their contribution to construction company. This will helps to successfully implement the strategy. The construction companies will then able to increase their the profit growth. The expansion on profit growth allow the companies to grow and expand their horizon. Thus, more employees are employs which contributes to the construction firm employee growth. In addition to that, the employees also will be motivated when they know that they can contributes towards achieving the companies vision and mission.

Strategic action plan enable for construction companies to achieve the strategic objective during strategy implementation. Proper strategic action plan helps the construction companies increase profit growth. This is as showed in 'table 1'. Strategic action plan helps to increase the profit growth because a good action plan implemented able give a construction firm sense of direction in achieving it vision and mission. However, strategic action plan does not influence the employees' growth.

'Table 1' also highlighted that allocating sufficient resources for strategic action plan is crucial in order to increase profit growth. This is because without sufficient resources allocation, the strategy implementation cannot be implemented properly. Nevertheless, allocating sufficient resources will not impact employees' growth. Although, proper resources allocation also seems to give impact towards the employee growth but not as powerful as the impact of ensuring vision and mission are well understood by the employees.

The result shows that providing sufficient training for employees to carry out the strategic action plan helps the construction firms to extremely increase in both companies profit and employees' growth.

This variable shows the most significant correlation between strategy implementation process and the construction companies' performance. This is proven in 'table 1' where the Pearson's correlation coefficient is at the highest for both profit growth (.641) and employees' growth (.560) which exceed large correlation coefficient (r = .5) [25]. It can be deduced that, employees training can increase the capabilities to execute the action plan which then, increase the profit growth as well as the employees' growth.

The involvement of employees in decision making motivates employees towards the achievements of the companies' strategic objectives. The result shows that, involvement of employees in decision making impacted profit growth as well as the employee growth. However, the correlation is higher towards profit growth.

Similarly, rewarding employees helps to motivate employees in working toward achieving the construction company objective. Thus, it resulted in an increase to the profit growth and the employees' growth in construction companies. However, the correlation is found to be not too significant as others. Therefore, rewarding employees is not that crucial in strategy implementation process.

Having strong corporate culture also helps to motivate employees in working toward the construction companies' strategic objective. The result shows that it helps to increase both the employees' growth and companies' profit growth.

In the nut shell 'table 1' has shown that, the construction companies G7 in Klang Valley, Selangor have conducted strategy implementation and it has contributed to positive construction companies' performance both in terms of profit growth and employees' growth.

5.1. Conclusions and recommendation

This research acknowledges and supports the relationship between strategy implementation process and construction company's performances. The empirical evidence shows that there is significant relationship between strategy implementation process and construction companies' performances. This research also proved that, the G7 construction companies in Klang Valley conducted strategy implementation and this has result a positive construction companies' growth. However, this research may not reflect the whole construction industry in Malaysia because it is only focuses on G7 in Klang Valley, Selangor. Therefore, a future research may be resumed to small and medium grades contractors and perhaps in other areas in Malaysia.

References

[1] Mohamed ZA, Ann HJ, Yee WF. *Strategic Management*. Shah Alam: Oxford Fajar; 2013.

[2] Kachaner N, King K, Stewart S. Four best practices for strategic planning. Strategy & Leadership. 2016;44(4):26-31.

[3] Wibowo MA, Astana INY, Rusdi HA. An Analysis of Bidding Strategy, Project Performance and Company Performance Relationship in Construction. Procedia Engineering. 2015;125:95-102.

[4] Hassan H, Ali A, Rahmat I. Organisational and Managerial Capabilities Factors in Relation to Strategic Planning within the Malaysian Quantity Surveying Firms. Built Environment Journal, UiTM. 2008.

[5] Bakar AHA, Tabassi AA, Abdul Razak A, Yusof MN. Key Factors Contributing to Growth of Construction Companies: A Malaysian Experience. World Applied Sciences Journal. 2012;19.

[6] Gębczyńska A. Strategy implementation efficiency on the process level. Business Process

Management Journal. 2016;22(6):1079-98.
[7] Bakar AH, Tufail MA, Yusof MN, Virgiyanti W. Implementation of Strategic Management Practices in the Malaysian Construction Industry. Pakistan Journal of Commerce & Social Sciences. 2011 Jan 1;5(1).
[8] Jiang N, Carpenter V. A case study of issues of strategy implementation in internationalization of higher education. International Journal of Educational Management. 2013;27(1):4-18.
[9] Raps A. Strategy implementation – an insurmountable obstacle? Handbook of Business Strategy. 2005;6(1):141-6.
[10] Stewart RA, Mohamed S, Daet R. Strategic implementation of IT/IS projects in construction: a case study. Automation in Construction. 2002;11(6):681-94.
[11] Allio MK. A short, practical guide to implementing strategy. Journal of Business Strategy. 2005;26(4):12-21.
[12] Mišanková M, Kočišová K. Strategic Implementation as a Part of Strategic Management. Procedia - Social and Behavioral Sciences. 2014;110:861-70.
[13] Hashim MK. Strategic Management: Text & Case. 2nd edition. Singapore: Thomson Learning; 2008
[14] Voberda HW, Morgen ER, Reinmoellers P, Hitt AM, Irealand RD, Hoskisson RE. *Strategic Management: Competitiveness and Globalization (Concept and Cases)*. UK: South-Western Cengage Learning; 2011.
[15] Bartol KM, Srivastava A. Encouraging Knowledge Sharing: The Role of Organizational Reward Systems. Journal of Leadership & Organizational Studies. 2002;9(1):64-76.
[16] Guiso L, Sapienza P, Zingales L. The value of corporate culture. Journal of Financial Economics. 2015 Jul 1;117(1):60-76. Available from, DOI: 10.1016/j.jfineco.2014.05.010
[17] Pearce, JA, Robinson, RB. *Formulation, Implementation and Control of Competitive Strategy*. Boston: Mc-Graw Hill; 2000.
[18] Fukawa N, Zhang Y. Understanding B2B Relationships Between an Open-Source Firm and Application Developers: Sharing Profits from Applications and In-Application Advertisements. In: Obal MW, Krey N, Bushardt C, editors. Let's Get Engaged! Crossing the Threshold of Marketing's Engagement Era: Proceedings of the 2014 Academy of Marketing Science (AMS) Annual Conference. Cham: Springer International Publishing; 2016. p. 459-.
[19] Robert AB, Mark H, Thomas W. Small business performance: business, strategy and owner-manager characteristics. Journal of Small Business and Enterprise Development. 2013;20(1):8-27.
[20] Chau VS. The relationship of strategic performance management to team strategy, company performance and organizational effectiveness. Team Performance Management: An International Journal. 2008;14(3/4):113-7.
[21] Kohtamäki M, Kraus S, Mäkelä M, Rönkkö M. The role of personnel commitment to strategy implementation and organisational learning within the relationship between strategic planning and company performance. International Journal of Entrepreneurial Behavior & Research. 2012;18(2):159-78.
[22] Özer F, Tınaztepe C. Effect of Strategic Leadership Styles on Firm Performance: A Study in a Turkish SME. Procedia - Social and Behavioral Sciences. 2014;150:778-84.
[23] Hong TT. Neighborhood preferences of house buyers: the case of Klang Valley, Malaysia. International Journal of Housing Markets and Analysis. 2011;4(1):58-69.
[24] Cheng WH, Kadir KA, Bohari AM. The Strategic Planning Of Smes In Malaysia: A View Of External Environmental Scanning. International Journal of Business & Society. 2014 Dec 1;15(3).
[25] Onwuegbuzie AJ, Daniel LG. Uses and misuses of the correlation coefficient.

Exploring the social reality of waqf community using social media: the Netnography approach

Ahmad Shazrin Mohamed Azmi[1,3], Noor Rosly Hanif[1], Siti Mashitoh Mahamood[2]

[1] Faculty of Built Environment, University of Malaya, 50603 Kuala Lumpur, Malaysia
[2] Syariah and Law Department, Academy of Islamic Studies, University of Malaya, 50603 Kuala Lumpur, Malaysia
[3] Centre of Studies for Estate Management, Faculty of Architecture, Planning and Surveying, Universiti Teknologi MARA, 40450 Shah Alam, Selangor, Malaysia

*nroslyhanif@um.edu.my

Abstract. The emergence of waqf institutions in Malaysia has attracted the attention of many. Considerable attention has been paid by various parties including the Federal government. In tandem with the revival of waqf, it has driven the emergence of a group of people who are passionate in matters pertaining to the development of Waqf (herein referred to as the waqf community). According to the Theory of Structuration, interaction between the agency and structure would induce changes. Therefore, this waqf community, being a part of the agency, is capable to influence the structure and bring about changes. Triggered by the existence of this waqf community, this study is interested in exploring their social reality. Having observed that a majority of people from this Waqf community had formed a specialized Whatsapp group dedicated to discussing matters pertaining to waqf, this study has considered the usage of netnography as an instrument to observe their social reality. Despite having limitations in narrowing down the research, this study was able to extract nine findings on their behaviour namely; their sensitivities, unity, spirit, willingness, commitment, the disclosure of information, influence, focus, gaps and frictions. Overall, the approach can be improved further by strategizing a refinement on the scope and technique in making observations.

1. Introduction

Waqf in Malaysia received greater attention during the tenure of the Prime Minister Tun Abdullah Ahmad Badawi, who had the vision to improve the waqf institutions in the country and wanted to see greater involvement of waqf in socio-economic development. Federal agencies like the Department of Awqaf, Zakat and Hajj (JAWHAR) and the Malaysian Awqaf Foundation (YWM) were all established during the tenure of Tun Abdullah Ahmad Badawi as the platform to facilitate SIRCs to develop the waqf institution in the country further. Since then, waqf has received numerous attention from various parties such as academics, law-makers and practitioners, financial and capital market regulators, financial institutions, non-governmental organisations, cooperatives, property developers, ministries, and the public at large. Also, there is a positive change in the SIRCs who are placed as the sole trustee in administrating the waqf, whereby all SIRCs have now seemed to strategize their ways to revitalize the waqf through better administration structure, regulating enactments, imparting public awareness, developing more waqf lands, introducing many waqf products, providing innovative platforms for the public to be involved in waqf, opening up for more strategic partnering and many more.

Presently, there is a burgeoning waqf community in Malaysia that consists of people as mentioned above who are passionate about waqf. These people can be said to have a great understanding on waqf and strongly believe in the greatness of this Islamic social instrument, not only for this world but also for the hereafter. Intrigued by the growing interest to know more about this waqf community, this study intends to explore the social reality of this group by trying to further understand their culture and behavior among themselves. The strategy is to use social media as the instrument under the netnography approach. Conducting the observation in the virtual world may pose some limitations, but relatively it is the most convenient and cheapest approach to be utilized.

2. Brief overview of waqf

Waqf can be described as a religious endowment in Islam. It is a type of donation, but with religious influences and highly governed by its principles and laws that are not shared by other types of charity instruments. It can be defined as "to hold and prevent the consumption of an asset to reap or extracting the derive benefits or usufructs to serve the righteousness or philanthropy objectives" [1]. The objective of waqf is mainly to assist the disadvantaged groups and to provide a general wellness in the society [2]. Waqf can be said as a more robust charity instrument because it has been characterized by exclusive principles of perpetuity, irrevocability, and inalienability [3]. It should also be noted here that waqf should not be treated the same as other trust instruments. As waqf is influenced by the belief of *Tawheed* (oneness of Allah), properties which has been donated according to the principles of Waqf is now being held in the name of Allah, whereas conventional trusts is held by the trustee [3,4].

In the context of Malaysia, it is vital to understand that the administration of Islamic religious affairs (waqf being one of them) is under the purview of the State Islamic Religious Council (SIRC) of each respective state. As Malaysia adopts the principle of Federalism, there is a separation in the administration affairs between the Federal government and State government. Since the new millennia, waqf has received some greater attention from various parties including the public. It is interesting to see more people are now interested in waqf. Some are directly involved in the management and making the waqf grow; while others are giving support in the form of various approaches. Embedded by religiously motivated spirit, these people have become part of a distinctive waqf community who are ready to be involved and introduce changes to the waqf institution. The involvement of people from various affiliations and backgrounds have made the institution of waqf in Malaysia achieve progress. They are involved in applying pressure to the regulators through ideas and actions to lead the changes and open up for more opportunities. The interactions between the waqf community and the existing regulatory structure by way of fulfilling and manipulating the governing administration and legal framework, regulations, policies, as well as the bureaucracy structures, have met the idea of the Structuration Theory. These interactions also form parts of the reality that constitutes the nature of waqf institution in Malaysia.

3. Methodological aspects

Ontologically, the reality of any concerned subject can be understood when a researcher has delved into the process of finding the reality. In common understanding, the reality can be objective or subjective, with the possibility of having multiple realities for the concerned subject [5]. Depending on the position of the researcher, the stance for the reality existence could be in two forms, one, as independently and waiting for discoveries; and on another stance, the reality is created by the perceptions and interpretation of the social actors [6,7]. Understanding the nature of the subject would help the subsequent processes in the research including the selection of research approach and instrument.

In regards to this article, the objective is set as to explore the social reality of the waqf community. In the context of Structuration Theory, the said waqf community would be the agent or agency who will practice within the existing structure of the social system, which is characterized by the constraints and enablement that would reproduce practices inserted in time and space [8]. However, the process to observe the social reality of the waqf community may entail some difficulties due to diverse backgrounds and being scattered in many locations. A more effective way needs to be explored by taking advantage of the advancement of current technology.

4. Netnography as research instrument

This article is considering the usage of social media as an instrument to observe the social reality within the waqf community. It was derived from the virtual ethnography [9] or netnography approach [10] that emerged in the mid 90's. The term 'netnography' was derived from ethnography but is mediated by the contemporary computerized communications technology, or Internet-based communications to do online fieldwork [10,11]. It is a more appropriate instrument to provide an unobtrusive observation, connecting people with diverse interest and backgrounds, enabling an extra level of anonymity in researching for sensitive information, and utilizes relatively less time and cost consumptions [12,13]. It is also suggested for the researcher to engage and have full participation in the culture being studied to perform focused observations and explore the 'thick descriptions' of virtual communities and cybercultures such as the social power, attitudes, perceptions, feelings, and their ways of life [11,13,14]. However, attention should also be given to the boundary between reality and the virtual world by observing the similarities and differences of the subject matter, which resulted from different communication modes [14].

Social media that is specific to the online social platform has rapidly grown and evolved due to the increasing number of internet users. It has become a virtual platform for people to meet and communicate about the subject of their interest. Past studies have revealed that social media has been used mostly by teenagers and young people [15]. However, these social media now have been used by almost everybody regardless of age bracket. The mobile instant messaging application such as WhatsApp, Telegram, WeChat, Line, Viber, Kakao Talk, Yahoo Messenger, Facebook Messenger and many more have become parts of cross-platform communication applications that have been appreciated by billions of users. Among the important features of these mobile instant messaging applications are the ability to minimize communication costs, communicate internationally, instant content delivery, ability to set read notification, shares a variety of media, location sharing, communication without cellular service and integration with a desktop platform [16]. Among the available mobile instant messaging applications, WhatsApp has become the largest and most popular application with 1.2 billion users worldwide (as of May 2017) that resulted in more than 1 billion Whatsapp groups being created, which delivered 42 to 64 billion messages and shared 1.6 billion photos daily (as of February 2016) [17]. Malaysian has been ranked as one of the top countries that uses WhatsApp with 60% of penetration into its population as of 4th quarter of 2016 [18].

5. Utilizing WhatsApp for research

As addressed in this paper, the exploration of the social reality of waqf community will be conducted on the WhatsApp platform. One WhatsApp group has been created by one of the waqf staff in JAWHAR on 8th January 2014 at 9.39 pm and he became the administrator of that group. He subsequently added more members from JAWHAR, YWM, SIRCs and a few related people from different organisations. Initially, the group was used as a platform to informally communicate among members about matters related to JAWHAR's affairs that are mostly related to waqf and zakat. It was a closed group at that time. However, some members had left the group for unknown reasons as the researcher was not in that group since the beginning. A light conversation with one of the formermembers of the group had revealed that among the reasons for their departure from the group were; feeling uneasy with the way the discussions had been carried out among members who like to criticize others, discussing out of context issues, discussion of sensitive issues which had offended their feelings, and the lack of interest in discussing the matter on social media.

The researcher was only added to that WhatsApp group circa August 2015 after the group administrator decided to make the group more open. Since then, more members joined with a variety of backgrounds and affiliations who were interested in the growth of waqf in the country. More members were appointed as co-administrators, including the researcher. Examining the researcher's mobile device that was used since 22nd September 2015, which was not long after becoming the member of that group, yielded some WhatsApp statistics referring to that group. Based on the retrieved statistics (as at 7th September 2017), the group currently has 180 members and contains 9,982 messages, 1,792 photos, 3 GIF media, 364 videos, 32 voice messages, and 23 documents. The daily discussions in the group are considered as moderate, fluctuating according to the issues.

6. Findings and analysis

Observing the behavior of the group members based on the daily discussions and materials that were shared in that group, this study able to identify a few pointers as the following: -

- *Disclosure of information* - Most members did not thoroughly disclose information pertaining to their organisation. They would prefer to contribute to the discussions based on their personal opinions. Nevertheless, they would participate when the discussions were related to their organisation.
- *Sensitivity* - For some members who have been involved in waqf for so long or are attached to waqf governing agencies, they were cautious in delivering their comments but would be sensitive when having to discuss matters regarding to the authority or power to manage waqf. However, for several members, they were at times too pessimistic on the issue and tended to criticize others.
- *Influence* - Some members would react aggressively on influencing others to accept or adopt their ideas, instruments, mechanisms or initiatives that they have worked out at their organisation. It is not bad at all because the main intention was to see the fast progress of waqf growth. Also, it has been noted that members with an honorific title such as Datuk or Dato' and academic titles of Professor, Associate Professor or Dr. (Ph.D.) may impart some influence on other members.
- *Focus* - The discussions often emphasised on the clarity of the issues and to find some hints for improvement. But, there were times where the discussions were jumping from one to another without clear focus due to too many things to discuss or probed by members. Although everybody realized the possibility of very small percentage of those ideas had been brought forward to the responsible agencies by the members, most of them seem to agree that the discussions were good for their knowledge enrichment.
- *Unity* - The members seemed to be very united in promoting any waqf agendas. Members appropriately supported ideas and actions from other members. Some were willing to take the lead and bring the force to the mainstream media by appearing on television programmes or write in newspaper columns which would be shared in the group.
- *Gaps and frictions* - Little gaps have been sensed between the members who are representing the Federal agencies and SIRCs. Both parties seldomly discussed topics or issues that they had in common or were both involved in. Rather, they became extra careful when to express anything related to each other, or refrain from commenting when other members raise the topic.
- *Commitment* - Some members found the opportunities to extend the potential of waqf through various great ideas. Any invites for programmes, events, discussions or brainstorming sessions outside the WhatsApp platform, although in a casual manner, would be accepted by members at their expense.
- *Willingness* - Some members are holding strategic positions in the Government Agencies or having good connections with authorities in the country. So, in certain circumstances, within their capacity or capability, they were willing to bring up any proposal for the betterment of waqf. Usually, they will pursue or push for policy changes or complementary mechanism. But require for perseverance and time, considering that government cannot simply change any policy.

- *Attachment* - Members were also personally very supportive of each other. It is done by way of congratulating or consoling each other in the event of success or grievance. Motivations and advice were also being regularly shared.

Overall, the observation of the social reality of waqf community by using social media is possible to be conducted. Applications such as WhatsApp are able to place the targeted population together and study their culture or behavior virtually. However, this study has some limitations that certainly affect the reliability of the instrument and validity of the findings, such as:

- No time frame has been set prior the study. The observation was conducted continuously ever since the researcher was in that group.
- The researcher has realized that not everybody was actively involved in every daily discussion. Despite no thorough checking on the level of members' activeness, it somehow reflected the notion of 1% heavy contributors, 9% intermittent contributors and 90% lurkers for the rules of participation inequality in social media and online communities [19].
- No pre-set condition has been made prior to the study as a control. The behavior of the members was naturally observed.

Despite being able to explore some of the social realities that exist amongst the waqf community, this study prefers not to put much weight on the findings considering that this study was exposed to many limitations. Confirmation should be made with other research instruments.

7. Conclusion

The existence of a waqf community that is willing to work for the betterment of waqf institutions in Malaysia was intrigued this study to explore their social realities as part to know their behaviors and cultures. The usage of WhatsApp under the netnography instrument helped the study to observe this waqf community. Although in a virtual platform, the textual conversations between members of that social media were expressive enough to reveal themselves. The ability to extract nine important findings from the virtual platform was considered as important. However, this study cannot claim the approach that was used as superior due to various limitations. It can be improved by implementing a better strategy with refinement on its scope and technique to do the observations or having conciliation with other research instruments to confirm the findings.

References

[1] Kahf M. Financing the Development of Waqf Property. In: Essential Readings in Contemporary Waqf Issue. Kuala Lumpur: CERT Publications Sdn. Bhd.; 2011. p. 163–202.

[2] Abdul Hamid Mar Iman, Mohammad Tahir Sabit Mohammad. Waqf Property: Concept, Management, Development, and Financing. First. Johor Bahru, Malaysia: Penerbit UTM Press; 2014. 300 p.

[3] Razali Othman. Institusi Wakaf Sejarah Dan Amalan Masa Kini. Kuala Lumpur: Dewan Bahasa dan Pustaka; 2013. 195 p.

[4] Razali Othman, Rohayu Abdul Ghani. Wakaf vs Trust: Siapa Pengaruh Siapa. In: Wakaf Pendidikan Tinggi: Prospek & Cabaran. Shah Alam, Selangor, Malaysia: Institut Kajian Zakat Malaysia (IKaZ), Akademi Pengajian Islam Kontemporari (ACIS), Universiti Teknologi MARA (UiTM); 2016. p. 133–51.

[5] Creswell JW. Qualitative Inquiry and Research Design: Choosing among Five Approaches. 2nd edition. Thousand Oaks: SAGE Publications, Inc; 2007. 416 p.

[6] Saunders MNK, Lewis P, Thornhill A. Research Methods for Business Students. 6th edition. Harlow, England ; New York: Financial Times/ Prentice Hall; 2012. 728 p.

[7] Neuman WL. Social Research Methods: Qualitative and Quantitative Approaches. Seventh. England: Pearson Education Limited; 2014. 608 p.

[8] Giddens A. The Constitution of Society: Outline of the Theory of Structuration. Cambridge, MA: Polity Press; 1984. 444 p.

[9] Hine C. Virtual Ethnography. London; Thousand Oaks; New Delhi: SAGE Publications Ltd; 2000. 192 p.
[10] Kozinets RV. "I Want to Believe": A Netnography of the X-Philes' Subculture of Consumption. Adv Consum Res. 1997;NA-24:470–5.
[11] Kozinets RV. On Netnography: Initial Reflections on Consumer Research Investigations of Cyberculture. Adv Consum Res. 1998;NA-25:366–71.
[12] Kozinets RV, Handelman J. Ensouling Consumption: a Netnographic Exploration of the Meaning of Boycotting Behavior. Adv Consum Res. 1998;NA-25:475–80.
[13] Langer R, Beckman SC. Sensitive research topics: netnography revisited. Qual Mark Res Int J. 2005 Jun 1;8(2):189–203.
[14] Wang Y-S, Lee W-L, Hsu T-H. Using netnography for the study of role-playing in female online games: Interpretation of situational context model. Internet Res. 2017 Jun 8;27(4):905–23.
[15] Noa Aharony, Tali Gazit. The importance of the Whatsapp family group: an exploratory analysis. Aslib J Inf Manag. 2016 Mar 2;68(2):174–92.
[16] Katie Elson Anderson. Getting acquainted with social networks and apps: WhatsApp-ening with mobile instant messaging? Libr Hi Tech News. 2016 Aug 1;33(6):11–5.
[17] DMR. 65 Amazing WhatsApp Statistics [Internet]. DMR. 2017 [cited 2017 Sep 6]. Available from: https://expandedramblings.com/index.php/whatsapp-statistics/
[18] Statista. WhatsApp: mobile usage penetration by country [Internet]. Statista: The Statistics Portal. 2017 [cited 2017 Sep 6]. Available from: https://www.statista.com/statistics/291540/mobile-internet-user-whatsapp/
[19] Nielsen J. Participation Inequality: The 90-9-1 Rule for Social Features [Internet]. Nielsen Norman Group. 2006 [cited 2017 Sep 6]. Available from: https://www.nngroup.com/articles/participation-inequality/

Quality of life among residents in a sub-urban area. Case study: Puncak Alam, Selangor, Malaysia.

Oliver Ling Hoon Leh[1], Norseha Mohd Mahbot[1], Nur Asma Aqmalina Hadzaman[1], Marlyana Azyyati Marzukhi[1], and Jamalunlaili Abdullah[1]
[1]Environmental and Social Health Research Group, Faculty of Architecture, Planning and Surveying, Universiti Teknologi MARA (UiTM), 40450 Shah Alam, Selangor, Malaysia.

*oliverling.my@gmail.com

Abstract. Along with the sub-urbanisation, people working in urban can stay in the sub-urban areas. Sub-urban housing areas provide cheaper and larger houses and more greenery environment. However, the residents are required to travel in longer distance. The effect on Quality of Life (QOL) due to the migration to sub-urban areas may not be positive even with the better environmental quality and lower direct housing cost. Puncak Alam, a new sub-urban area in Selangor, Malaysia had been chosen as the study area to examine the change of QOL among the residents after they moved into the study area. Through a questionnaire survey, the satisfaction of residents on the various aspects/indicators of QOL were examined. Through the statistical analysis, it is found that slightly more than half of the respondents felt that the QOL was dropped after they were moved to the study area. The sub-urban area did provide better quality for their residents. It had increased respondents' satisfaction on most of the indicators in the aspects of economic, social and physical. However, due to the decreasing of satisfaction in job opportunities, family life, social activities, safety, and transportation system, most of the respondents were felt that their QOL were dropped.

1. Introduction

The Quality of life (QOL) is a broad multidimensional concept that includes subjective evaluations of both positive and negative aspects of life [1]. It is a term used to describe many aspects of life and its enjoyment or lack thereof [2]. The measurement of QOL is related to the human perception of ability to meet daily needs, physical activities, and well-being [2]. QOL is also defined as a social welfare component that includes all the important aspects of human, *i.e.* nutrition, shelter or housing, health, education, security, social stability, recreation, physical environment, transportation, arts, and economy [3, 4]. For World Health Organization [5], the main four elements of QOL are identified as health, personal safety, social relationship and family life. Thus, QOL covers the perception/satisfaction of people on their life for the aspects of physical (basic needs & environmental quality), economic (income, affordability, jobs, etc.) and social well-being (relationship, happiness, values, etc.).

A high quality of life is very important in an area because it is one of the effective measures in assessing the progress of a community and development. QOL is also an important determinant in describing the population of a country that was enjoying a comfortable standard of living. In the context of built environment, sub-urbanisation may affect the QOL of people. Under the sub-urbanisation development, people working in urban are living in the sub-urban areas. Moreover, housing in the sub-urban areas are cheaper with larger land size and more greenery environment. However, the residents are required to travel longer in terms of distance and time to/from working places which are located in city/urban areas. Furthermore, the rapid urbanisation in Selangor including the sub-urbanisation has led to extreme changes in the state land use pattern. Its impact towards the macro environmental quality is worrying [6]. Arguably, the effect on QOL due to the migration to sub-urban areas may not be positive. For example, the residents of Puncak Alam (study area) are taking more than 45 minutes to commute to the nearest major cities, *i.e.* Shah Alam and Klang, and more than an hour to Subang Jaya, Petaling Jaya and Kuala Lumpur. The connecting roads between Puncak Alam and Shah Alam/Klang were experiencing serious traffic congestion particularly during the peak hours, *i.e.* morning and evening, when Puncak Alam's residents are travelling to/from work. Based on the questionnaire survey, 69% of respondents agreed that Puncak Alam was facing the problem of traffic congestion.

Figure 1. The view of traffic congestion at the Persiaran Mokhtar Dahari, the main connecting road to Puncak Alam from Guthrie Corridor Expressway

2. Research method

2.1. Scope of research and the analysis

The aim of this study is to examine the changes of QOL among residents in a sub-urban area (Puncak Alam). The examination of the level of QOL was carried out using indicators for economic, physical and social aspects (Table 1). Flynn, Berry and Heintz [7] stated that indicators are attitudinal, which may give a complete picture of contemporary society. The understanding of QOL indicators in measuring the well-being is necessary since its aims to evaluate society and add substantially to the regnant economic indicators that are favoured by some policymakers [8]. Its reflect people's objective circumstances in a given cultural or geographic unit [9]. For this study, data were analysed using Frequency, and Cross-tabulation tests as available in Statistical Package for Social Science (SPSS) software. The relationship analysis was carried out to examine the influence of distance to workplaces and changes of QOL.

Table 1. Aspects and indicators Used to Measure Quality of Life

Aspects	Indicators
Economic	Satisfaction of respondents on: Financial status; Cost of living; Job opportunities
Physical	Satisfaction of respondents on: Comfortable of living place; Safety of living area; Road safety / Frequency of road accidents; Provision & quality of public facilities; Provision & quality of commercial area; Transportation system; Access to workplace
Social	Satisfaction of respondents on: Health; Happiness; Social activities; Neighbourhood activities; Pressure of life; Family life / quality time with families; Leisure activities

2.2. Study area

Puncak Alam as the main township development in Kuala Selangor District, State of Selangor, Malaysia has been selected as the case study area. Kuala Selangor District is one of the less urbanised districts in Selangor. It is located approximately 20 kilometres in the Northwest direction from Shah Alam, the state capital of Selangor. Puncak Alam development was started in the late 90's under the management of Bukit Cherakah Development Sdn Bhd with approximately $57km^2$ in size. However, from 15 May 2001 the development was charged to Puncak Alam Housing Sdn Bhd as the principal developer. Currently, most of the developments in Puncak Alam are housing area with a small percentage of commercial and industrial units. As refer to the Figure 2, the satellite image clearly showed that Puncak Alam is located away from the major built-up area in Klang Valley (Klang – Shah Alam – Petaling Jaya – Subang Jaya – Puchong - Kuala Lumpur – Damansara). As a sub-urban area, Puncak Alam is surrounded by rural lands within the boundary of Kuala Selangor district.

Figure 2. Location of Puncak Alam in Klang Valley [10]

Table 2. Distribution of Respondents in the Study Area

Bandar Puncak Alam	Number of houses [11]	Population [11]	% of respondents
Phase 1	1,250	5,560	13
Phase 2	2,530	10,120	22
Phase 3	4,820	14,520	32
Shah Alam 2	1,580	7,900	18
Alam Jaya	1,650	6,600	15
Total	11,830	45,000	100

Table 3. Background of Respondents

Variables	Percentage (%)	Variables	Percentage (%)
Length of stay		House ownership	
5 years or less	35	Own property	81
6 - 10 years	40	Renting	16
11 – 15 years	16	Others	3
More than 15 years	9		
Income level		Distance to workplace	
RM1000 or less	2	5 km or less	13
RM1001-RM3000	38	6 – 12 km	20
RM3001-RM5000	41	13 – 15 km	16
RM5001-RM8000	13	16 – 20 km	18
RM8000 & above	6	More than 20 km	33

2.3. Questionnaire survey and sampling of respondents

For the purpose of this study, the satisfaction of residents on the various aspects/indicators of QOL was collected through a questionnaire survey. Data from the questionnaire survey was used to measure the QOL among the respondents in the study area. In general, the questionnaire covers the following aspects:

a) Background of respondents, *i.e.* gender, age, employment, educational level, income, type of house ownership, the length of stay in Puncak Alam and distance to the workplace.
b) Satisfaction of respondents on the indicators as shown in Table 1. Respondents were required to answer the questions to examine their quality of life before and after living in the study area.

By using the stratified random sampling technique, 100 units of houses were chosen from a total of 11,830 units of houses in the area, which covered different housing types and pricing. It covered general households (Malaysian only) with various socio-economic backgrounds. Table 2 shows the distribution of respondents in the study area. The majority of the respondents (65%) were living in the area more than 5 years. Meanwhile, the majority of the respondents (81%) were the house owners. From the 100 respondents, 33% of them commute to/from work more than 20km every day (Table 3).

3. The result and findings

3.1. Economic indicators in measuring QOL

In measuring QOL, one of the important aspects is the economic or financial condition of people. Nowadays, people cannot live without financial healthiness especially in the urban area. Based on the satisfaction of respondents on the financial status of their family, most of the respondents (65% to 67%) were on the moderate level before and after they moved into Puncak Alam. The satisfaction level on the financial status of their family was becoming better after they moved to Puncak Alam. Before they moved to Puncak Alam, there were 21% weakly (or very weakly) satisfied on their financial status as compared to only 10% of those who were weakly satisfied after they moved to Puncak Alam. The percentage of respondents who were strongly satisfied with their financial status was increased from 14% (before moved in) to 23% after moved into Puncak Alam.

For the level of satisfaction on the cost of living, the percentage of respondents who felt moderate and strongly satisfied with their ability to manage their cost of living were increased. Meanwhile, the percentage of respondents with the weak or very weakly satisfied with their ability to manage the cost of living was decreased from 70% to 25% only.

However, the satisfaction of respondents on job opportunities was still low after they moved to Puncak Alam. Percentage of those who answer “weak” or “very weak” was increased from 62% (before moved in) to 70% after moved into Puncak Alam. Those who answered moderate and strong satisfied were reduced in the number or percentage. It showed that even though the respondents were experiencing better financial status and cost of living, they were facing fewer job opportunities in their new area (Puncak Alam). Normally, job opportunities are available in urban areas especially the major cities in Selangor and Kuala Lumpur. Thus, they were required to work at places far from their new housing areas and not within the area of Puncak Alam.

3.2. Physical indicators in measuring QOL

In measuring the QOL, respondents’ satisfactions on the physical aspects of the study area, were also analysed. There were seven (7) physical aspects in total for this study. However, the physical aspects not showing clear decrease or increase in QOL after respondents moved to the study area.

For the aspect of comfort, percentages of respondents who felt weakly satisfied as well as strongly satisfied were decreased. At the same time, the percentage of the moderate level was increased after respondents were moved to the study area. It showed that the study area was at the moderate level for the comfort aspect. For the aspect of safety, Puncak Alam was less safe as compared to other areas. The percentages of respondents who felt weak and very weak for the level of satisfaction on safety were increased after they moved to the study area. The percentage for the moderate level was decreased from 55% to 21% only. However, in term of road safety, more respondents felt the roads were safer as compared to their previous living area. The study showed an increase in the percentage of respondents who strongly satisfied with the safety aspect of the roads in the study area.

Based on the survey, respondents were more satisfied with provision and quality of the commercial area in Puncak Alam as compared to the previous living areas. Percentages for weak as well as very weak satisfaction were decreased after respondents were moved into the Puncak Alam. At the same time, the percentage for those who were strongly satisfied was increased from 1% to 35%. It showed that the newly planned and developed Puncak Alam was providing good commercial facilities for the respondents. For the satisfaction of public facilities, it showed not a clear general trend of increase or decrease after respondents moved to the study area.

Being a sub-urban area, Puncak Alam is suffering from traffic congestion during the peak hours. Furthermore, the transportation system in the area was weakly satisfied by most of the respondents

(40% felt weak, and 38% felt very weak). There was only 22% of respondents felt moderate satisfaction on the transportation system. There was no respondent felt strongly satisfied on the transportation system. As compared to the satisfaction level before they moved to Puncak Alam, generally, the satisfaction level was dropped.

3.3. Social indicators in measuring QOL

Percentage of respondents felt strong satisfaction on their health was increased after they had moved into the study area. Meanwhile, the percentage of respondents who were weakly satisfied on their health was decreased (Table 4). Besides, a higher percentage of respondents were feeling happier after moved into the study area (Table 4). However, there were also a slightly increased number of respondents who felt very weakly satisfied on their happiness (from 0 to 4).

Table 4. Satisfaction on their health and happiness

Level of satisfaction	Health (% respondents)		Happiness (% respondents)	
	Before	After	Before	After
Very weak	0	0	0	4
Weak	35	13	19	17
Moderate	55	53	75	37
Strong	10	33	6	41
Very strong	0	1	0	1
Total	100	100	100	100

For the aspects of family life and social activities, there was a clear increase in the percentage of respondents who felt weakly satisfied. However, there was also a slight increment in the percentage of strongly satisfied. It showed that higher percentage of respondents who were moved into the study area (sub-urban area) were facing reducing of time with family and satisfaction in carrying out social activities. It might be due to the longer travelling time and distance to workplaces. However, for the aspects of neighbourhood activities, leisure activities, and pressure of life, the percentage of respondents who were strongly satisfied was increased after they moved to the study area.

3.4. Overall QOL level

When the respondents were asked about their overall opinion on their QOL changes after they moved to the study area, slightly more than half of the respondents (56%) felt that the QOL was dropped. However, there were 44% of respondents felt that the QOL was increased. There were 55% of respondents regretting their decision to move into the study area. It was due to the fewer job opportunities and longer travelling distance to the workplace. It has led to the reduction of time with family (family life) and social activities.

It showed that the new development area (the study area) did provide better quality for their residents. For example, it had increased respondents' satisfaction on most of the indicators in the aspects of economic, social and physical. However, due to the decreasing of satisfaction in few indicators (*i.e.* job opportunities, family life, social activities, safety, and transportation system in relation to the working trip), most of the respondents were felt that their overall QOL level were dropped and they were regretting their decision to move into the study area. Among the respondents, there were 35% answered that they might be moving out from the study area.

3.5. Distance to workplaces and change of "overall QOL"

Cross Tabulation analysis was carried out to examine the relationship between distance to workplaces and the changes of QOL by respondents. Table 5 showed that high percentage of respondents who travel in a longer distance (>20km) as well as shorter distance (<6km) were experiencing decrease of QOL. Among those who were living within 13 to 15 km from workplace, 50% experiencing decrease of QOL and another 50% experiencing increase of QOL. It showed that the moderate distance to workplace is having better impact on QOL.

Table 5. Relationship between Distance to Workplaces and Changes of QOL

Distance to Workplace	Percentage of respondents (%)		
	Increase of QOL	Decrease of QOL	Total
5 km or less	40.0	60.0	100.0
6 – 12 km	46.2	53.8	100.0
13 – 15 km	50.0	50.0	100.0
16 – 20 km	44.4	55.6	100.0
More than 20 km	42.4	57.6	100.0
Total	44.0	56.0	100.0

4. Conclusions and recommendation

In conclusion, slightly more than half of the respondents (56%) felt that the "overall QOL" was dropped after they were moved to the study area (sub-urban). However, there were 44% of respondents felt that the "overall QOL" was increased. The new development of sub-urban area did provide better quality for their residents. For example, it had increased respondents' satisfaction on most of the indicators in the aspects of economic, social and physical. However, due to the decreasing of satisfaction in few indicators (*i.e.* job opportunities, family life, social activities, safety, and transportation system), most of the respondents were felt that their "overall QOL" were dropped and they were regretting their decision to move into the study area. The situation of less job opportunity and longer travelling distance to the workplace at the sub-urban area, has led to the reduction of time with family (family life) and social activities. There were one-third of respondents who might be going to move out from the study area. This preliminary study has successfully indicated the changes of QOL among the residents in one of the sub-urban area in Malaysia. It is proposed that urban planning and development for a sub-urban area should be carried out with more concern on the aspects of QOL among residents, especially the aspects of job opportunities, travelling distance, social activities, and family life.

Acknowledgments

The authors would like to thank Universiti Teknologi MARA (UiTM), Malaysia for partly funding the study by given the REI grant (600-RMI/DANA 5/3/REI (8/2015)). The authors would also like to thank all the government departments, organizations, and respondents who had contributed to this study.

References

[1] Centers for Disease Control and Prevention (CDC) 2016 *Health-Related Quality of Life* (U.S. Department of Health & Human Services. Retrieved from www.cdc.gov.)

[2] Hlavacek M 2010 *Qualitative Study to Determine Quality of Life Factors based on Reported EFNEP Success Stories* (Master Thesis. University of Nebraska, Lincoln)

[3] Asmah A 2005 *Kualiti hidup dan pengurusan persekitaran di Malaysia* (Bangi, Malaysia: Universiti Kebangsaan Malaysia)

[4] Boyer and Savageau 1981 *Places Rated Almanac* (New York: Rand McNally)

[5] World Health Organization Quality of Life Group 1994 Development of the WHOQOL: Rationale and Current Status *International Journal of Mental Health* 23 (3): 24-56

[6] Nurul Ashkin M, Ling OHL, Dasimah O 2015 Urban air quality and human health effects in Selangor, Malaysia *Procedia Social and Behavioral Sciences* 170: 282-291.

[7] Flynn, Berry and Heintz 2002 Sustainability and quality of life indicators: toward integration of economic, social and environmental measures *The Journal of Social Health*

[8] Diener ED and Suh E 1997 Measuring quality of life: economic, social and subjective indicators *Social Indicators Research* 40 (1): 189-216

[9] Diener ED 1995 A Value Based Index for Measuring National Quality of Life *Social Indicators Research* 36 (2): 107-127

IOP Conf. Series: Earth and Environmental Science **117** (2018) 012028 doi:10.1088/1755-1315/117/1/012028

[10] Google 2017 *Google maps.* (Retrieved from https://maps.google.com/)
[11] Puncak Alam Residents Committee 2016 *Informal interview.*

Merging building maintainability and sustainability assessment: A multicriteria decision making approach

A S Asmone[1] and M Y L Chew[1]

[1]Department of Building, School of Design and Environment,
National University of Singapore, Singapore.

*ashan.a@u.nus.edu

Abstract. Accurately predicting maintainability has been a challenge due to the complex nature of buildings, yet it is an important research area with a rising necessity. This paper explores the use of multicriteria decision making approach for merging maintainability and sustainability elements into building grading systems to attain long-term sustainability in the building industry. The paper conducts a systematic literature review on multicriteria decision analysis approach and builds on the existing knowledge of maintainability to achieve this. A conceptual framework is developed to bridge the gap between building operations and maintenance with green facilities management by forecasting green maintainability at the design stage.

1. Introduction

Maintainability of a building is its ability to perform optimally throughout the lifespan within the minimum life cycle cost [1]. Building maintainability is reflected in effective facility utilization over its lifetime; which is achieved through thoroughly detailed designs and good construction workmanship followed through with a comprehensive maintenance regime [2,3]. Traditionally, maintainability was involved in selecting building technologies and materials which require minimal maintenance, viz. cleaning, repair and replacement [4]. Which could be specified as an expression of various physical and functional maintenance parameters [5]. These strategies were aimed to create buildings with an efficient and effective operation and maintenance by using maintenance feedback data from existing buildings [1,6].

However, there is an apparent discrepancy between building scoring systems and maintainability [7]. Parida et al. [8] noted that not utilizing available data on maintenance management is a cause for this knowledge gap. Corry et al. [9] corroborates this in their study on current buildings incorporated with complex green technologies. They claim that greener designed buildings fall short of their projected expectations due to gaps between design and performance. A further knowledge gap is identified here between sustainability practices and building operation [10,11]. Ajukumar and Gandhi [12] identified the lack of tools to incorporate green maintenance or maintainability considerations at the early design stages to be a cause of lack of sustainability during maintenance stages. This agrees with studies [13,14] that shows how green building operation and maintenance has become impractical, costly and less sustainable due to ignoring the operations and maintenance considerations. Especially since the industry understanding of maintainability of novel green building technologies is still limited as they are yet to undergo the effects of degradation and weathering [15].

To attain long-term sustainability in the building industry, life cycle sustainability of buildings has to be incorporated into the existing building stock operation and maintenance [16]. Therefore, the need to link maintainability principles with the principles of sustainability is identified; to evolve traditional maintainability to green maintainability [17,18].

2. Method

This study identifies the lack of green maintainability considerations in building project delivery as a decision making problem; proposing maintainability scoring as a fundamental approach for implementing green maintainability strategies. Figure 1 below illustrates the research framework for the current study. This paper presents findings from ongoing research in operationalizing this framework.

This paper reviews the literature on decision analysis and the potential use of multicriteria decision making techniques for this purpose. A comprehensive literature review on areas of building maintainability and sustainability has been conducted to identify green maintainability indicators which are used for the multicriteria decision making. These serve as the basis of the discussion on how to use decision making tools to model and score green maintainability by building professionals.

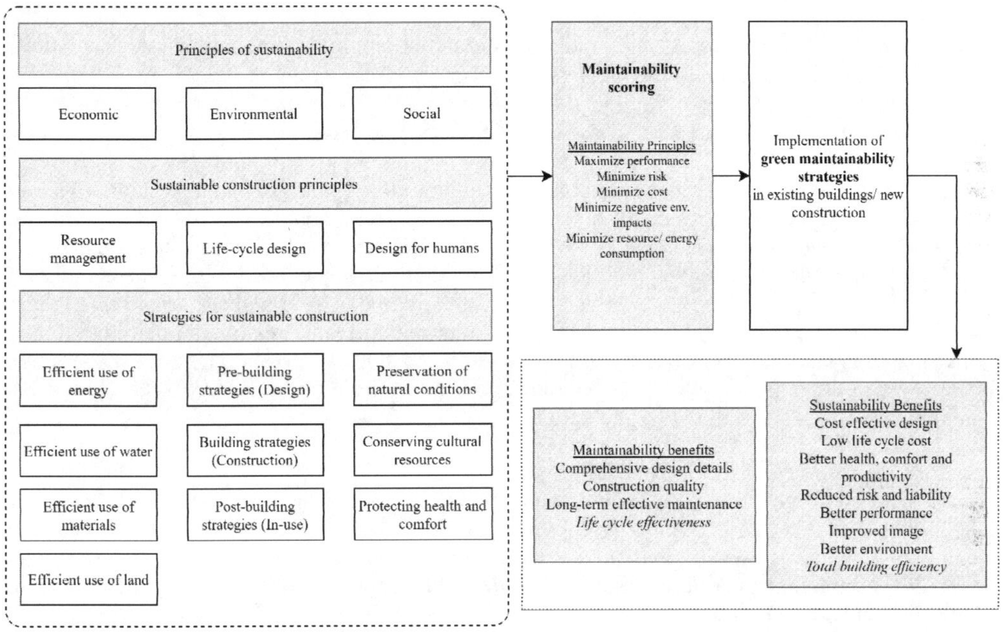

Figure 1. Research framework: Merging building maintainability and sustainability assessment

3. Overview of decision analysis

There are two kinds of decision makers; the rational and the intuitive [19]. The rational decision makers need to have sufficient knowledge and informed alternatives to operate while the intuitive decision makers tend not to reason logically when they do not hold perfect knowledge [20]. In light of these shortcomings of human decision making, computerized decision support systems can aid humans in finding solutions to problems which require systematic analysis [21].

Decision analysis can, therefore, be used as a guide to solve complex problems. Such as those often found in design and construction. Designers and builders can use decision analysis to structure these complex issues in order to foresee the consequences of their decisions (i.e. possible outcomes and likelihoods). This is made possible by identifying the uncertainty attached to a problem and presenting it in a useful way for decision makers to handle. Decision analysis also can show trade-offs of different decisions when working towards multiple objectives in a complex system. This is especially useful in instances where multiple stakeholders work together; each proponent with different values and objectives.

For an example, an architect's objective in facility design may be to design an aesthetically pleasing iconic building; while the engineers may strive to make the buildings structurally stable and safe. A facility manager involved at this stage would strive to minimize operating costs by optimizing the designs. Whereas a quantity surveyor may mainly be concerned to keep the construction costs in check. A decision analysis approach can be used to resolve the conflicting interests in such cases, and to make design decisions that are acceptable to all. This is carried out by using such tools to aid them in making informed decisions by presenting information in a structured manner [22].

4. Multicriteria decision making (MCDM)

In an effort to come up with most appropriate alternatives based on multiple criteria (i.e. maintainability considerations and sustainability considerations) multicriteria decision making (MCDM) can be used [23]. The decision support provided by multicriteria decision making approach is useful in problems with high complexity, uncertainty, and other challenges such as, varied data forms and different stakeholder with multiple interests, perspectives, and objectives which can be conflicting; such as in integrated building design domain [24]. Studies have integrated environmental sustainability considerations into decision frameworks in building related research [24–27].

4.1. Multiobjective decision making (MODM)

MODM concentrates on continuous decision spaces with several objective functions [24]. MODM tools can be used even in cases where objectives are conflicting and non-commensurable. MODM models find a unique solution to the problem by creating a vector based on the parameters pertaining to the decision, objective functions, and problem constraints. These solutions can be considered as alternatives where decision makers can choose an alternative based on how much it satisfies the multiple objectives [23].

MODM problems can be described by use of multiobjective linear programming tools which use real world factors in explaining the objective functions and constraints. These are further classified into parameters which are quantified based on expert opinions [28]. These parameters are used to approximate the Pareto frontier that indicates non-dominated solution sets (i.e. a state where a solution does not have any other alternatives which will improve one objective without negatively affecting another objective) [29]. This is achieved by calculating the vector values for the objectives for the decision parameters and by comparing each vector for its relative dominance. This is an optimization process where the objective values are deterministic if the entire problem space can be calculated in this way [29].

4.2. Multiattribute decision making (MADM)

The MADM approach centers on taking decisions based on preferences with discrete decision spaces [24]. The use of MADM is characterized by the selection of solutions from a predetermined finite number of alternatives using various traits. The number of alternatives is dependent on achievement levels of the different traits and the generation of these alternatives are constrained by the information availability and domain knowledge [28].

The MADM process usually begins by determining the evaluation criteria for the alternatives. These criteria are used to generate possible alternatives. Heuristic techniques can also be used in this regard. Afterwards, the most suitable alternative is selected based on the multiple attributes identified. Inter-attribute and intra-attribute comparisons can be made to determine these alternatives [28].

A key challenge faced by decision makers herein is when dealing with qualitative attributes as they are imprecise and problematic to quantify as human judgement is subjective [24,27]. Fuzzy set theory is therefore used to express such information, by coding the subjective parameters in linguistic terms to bring them to the decision models [24]. Fuzzy set theory fused with MADM methods are increasingly found in literature as it provides an opportunity to systematically evaluate the subjective input data [24,27].

5. Discussion

Two main approaches to multicriteria decision making are reviewed, namely MODM and MADM. In the context of the current study, a MODM approach would look at the building design with objectives of sustainability and maintainability in mind. The criterion used herein will then be used to convert the objective functions to a final utility value (e.g. life cycle cost, life cycle environmental impact). Figure 2 below presents typical indicators for green maintainability; which incorporates both sustainability and maintainability considerations in building project delivery. At this point decision makers can make the best design decisions in order to optimise the objectives, by considering trade-offs among the various objectives. Alternatively, the MADM approach uses a set of predefined design options to measure the end result of using these options in terms of sustainability and maintainability. This can be seen as a ranking approach to choose from a finite number of design alternatives.

Both approaches aid decision makers with useful information on the possible consequences of their decisions. MADM approach looks at the design issues from prior experiences of the industry, which can be useful during early design stages where project objectives can be ambiguous. However, this approach limits the creativity and innovation of the design teams as it is a more prescriptive approach to design. On the other hand, MODM is used to optimise a level of satisfaction among the objectives and constraints which allow for the design teams to innovate to find the best solution. Even so, it should be noted that various challenges still remain in using MODM in this regard. Most notably are the issues of quantification of certain parameters such as aesthetics appearance or practicality of construction methods. Methods need to be developed in order to express such criterions in a useful manner for MCDM.

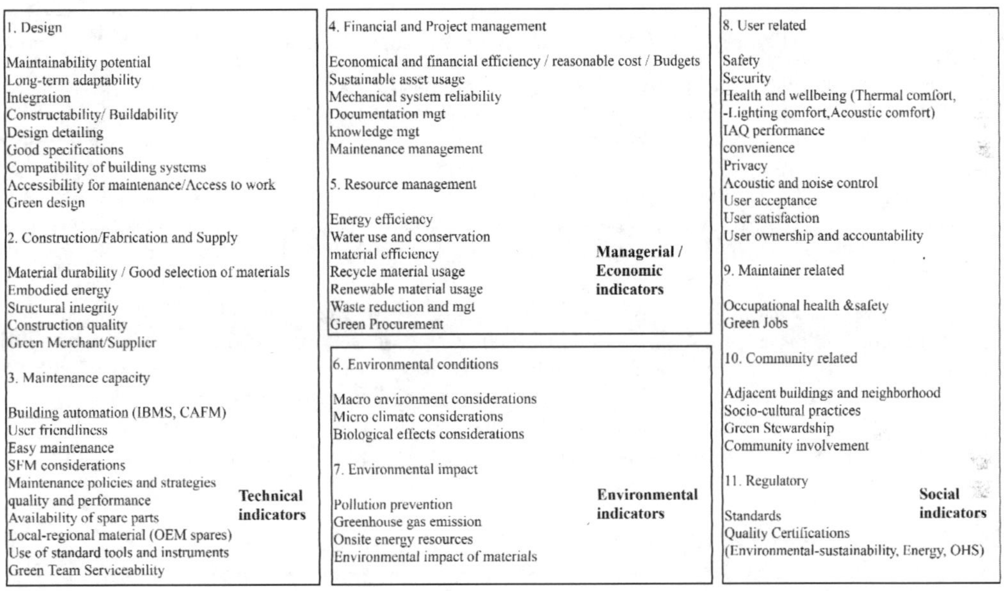

Figure 2. Green maintainability indicators for a building project

Based on these approaches, Figure 3 below illustrates the proposed MCDM workflow for green maintainability assessment. The process starts by defining project requirement and this includes inputs from all project proponents. The defined requirements are later used as objective functions for rating alternatives. Secondly, the different maintainability and sustainability criteria (e.g. technical, economic, etc.) and sub criteria are identified for different building systems and subsystems. Based on project goals, possible alternative design strategies are then evaluated. The different variables for components under building sub systems are analysed to measure cumulative weights and rank alternatives. It is proposed to view final results from the model in monetary terms to make objective comparisons by decision makers. The final result for alternatives is then used to determine the alternative which best fits project requirements.

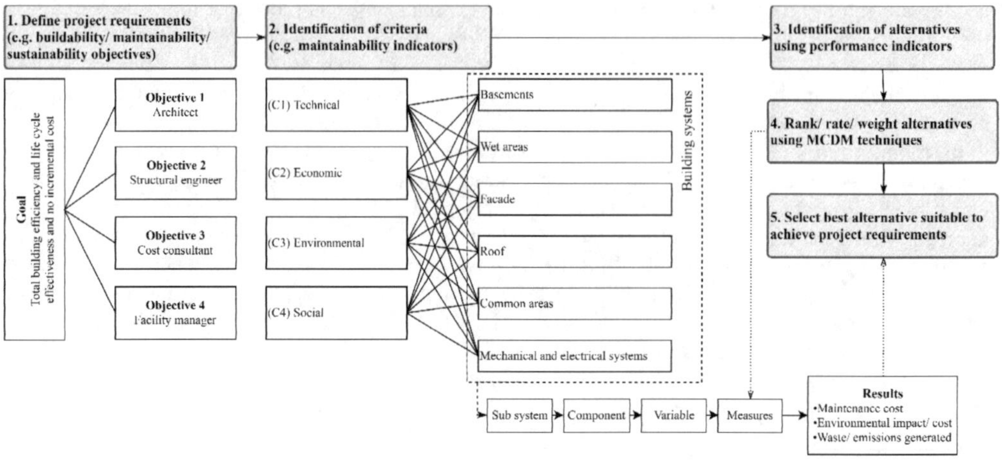

Figure 3. MCDM workflow of green maintainability analysis

Considering a scenario of designing a wet area, the maintainability depends on various factors such as water tightness, spatial, integrity, ventilation, material, and plumbing maintainability [30]. These factors are based on different variables which are affected by design decisions such as detailing, material selection, or construction workmanship. Furthermore, these decisions have impacts on environmental (e.g. embodied energy) and social (e.g. safety) aspects which also needs to be considered when planning for life cycle sustainability. In order to aid the decision makers to achieve this goal, MCDM is envisioned as the solution for modelling as it has been repeatedly to solve construction-related problems [5,31–33]. However, there are features of MODM and MADM that are both desirable and challenging for use as discussed above. Therefore, a possibility of a hybrid decision making model addressing these issues is worth taking note of in further studies. Building upon that, the future direction of the current study can be listed as below;

i. Identify green maintainability criteria for different building subsystems
ii. establish parameters to measure variables based on national/international standards
iii. Develop MCDM model to calculate equivalent cost for each alternative based on existing cost benchmarks

6. Conclusions and future work

This paper briefly reviewed multicriteria decision analysis approaches used in the construction industry which potentially be used to aid building design teams in the effort to incorporate sustainability and maintainability strategies in their designs. Each approach is useful in different kinds of decision problems, which can largely vary on available information, the criterions considered and objectives of the decision makers. As the most desirable building performance is achieved when maintainability is fused with sustainable principles early on in facility design; these decision making tools will influence the building's performance to accrue economic, environmental and social benefits over all its operation stages.

The current study is part of a study on developing a green maintainability appraisal framework which integrates maintainability and sustainability principles to elicit green maintainability. It is envisioned that this will lead to reduced failures and maintenance issues in buildings leading to reduced life cycle costs and environmental impacts to achieve a lasting sustainability in the built environment.

References

[1] Chew M Y L 2015 *Maintainability of Facilities: Green FM for Building Professionals* (Singapore: World Scientific)

[2] Lindsey T C 2011 Sustainable principles: common values for achieving sustainability *J. Clean. Prod.* **19** 561–5

[3] Wood B 2006 The role of existing buildings in the sustainability agenda *Facilities* **24** 61–7

[4] Chew M Y L and De Silva N 2003 Maintainability problems of wet areas in high-rise residential buildings *Build. Res. Inf.* **31** 60–9
[5] Das S, Chew M Y L and Poh K L 2010 Multi-criteria decision analysis in building maintainability using analytical hierarchy process *Constr. Manag. Econ.* **28** 1043–56
[6] Kumar S, Khan I A and Gandhi O P 2015 A theoretical framework for extraction and quantification of psychological attributes in design for maintainability: a team-inspired approach *Res. Eng. Des.* **26** 289–308
[7] Chew M Y L and Das S 2008 Building grading systems: A review of the state-of-the-art *Archit. Sci. Rev.* **51** 3–13
[8] Parida A, Kumar U, Galar D and Stenström C 2015 Performance measurement and management for maintenance: a literature review *J. Qual. Maint. Eng.* **21** 2–33
[9] Corry E, Pauwels P, Hu S, Keane M and O'Donnell J 2015 A performance assessment ontology for the environmental and energy management of buildings *Autom. Constr.* **57** 249–59
[10] Meng X 2014 The role of facilities managers in sustainable practice in the UK and Ireland *Smart Sustain. Built Environ.* **3** 23–34
[11] Sim Y L and Putuhena F J 2015 Green building technology initiatives to achieve construction quality and environmental sustainability in the construction industry in Malaysia *Manag. Environ. Qual. An Int. J.* **26** 233–49
[12] Ajukumar V N and Gandhi O P 2013 Evaluation of green maintenance initiatives in design and development of mechanical systems using an integrated approach *J. Clean. Prod.* **51** 34–46
[13] Lai J H K 2010 Building operation and maintenance: education needs in Hong Kong *Facilities* **28** 475–93
[14] Zainol N N, Mohammad I S, Baba M, Woon N B, Ramli N A, Nazri A Q and Lokman M A A 2014 Critical Factors that Lead to Green Building Operations and Maintenance Problems in Malaysia: A Preliminary Study *Adv. Mater. Res.* **935** 23–6
[15] Chew M Y L, Tan S S and Kang K H 2005 Contribution analysis of maintainability factors for cladding facades *Archit. Sci. Rev.* **48** 215–28
[16] Aaltonen A, Määttänen E, Kyrö R and Sarasoja A 2013 Facilities management driving green building certification: a case from Finland *Facilities* **31** 328–42
[17] Asmone A S and Chew M Y L 2016 Sustainable facilities management and the requisite for green maintainability *In proc. SMART Facilities Management Solutions Regional Focus Group Session* (Singapore)
[18] Chew M Y L, Conejos S and Asmone A S 2017 Developing a research framework for the green maintainability of buildings *Facilities* **35** 39–63
[19] Simon H A 1991 Bounded rationality and organizational learning *Organ. Sci.* **2** 125–34
[20] Wason P C 1978 Hypothesis testing and reasoning. Unit 25, Block 4 *Cogn. Psychol.*
[21] Norman D A 1978 Notes toward a theory of complex learning *Cognitive psychology and instruction* (Springer) pp 39–48
[22] Clemen R T and Reilly T 2013 *Making hard decisions with decision tools* (Cengage Learning)
[23] Pedrycz W, Ekel P and Parreiras R 2011 *Fuzzy multicriteria decision-making: models, methods and applications* (John Wiley & Sons)
[24] Wang J-J, Jing Y-Y, Zhang C-F and Zhao J-H 2009 Review on multi-criteria decision analysis aid in sustainable energy decision-making *Renew. Sustain. Energy Rev.* **13** 2263–78
[25] Klemm K, Marks W and Klemm A J 2000 Multicriteria optimisation of the building arrangement with application of numerical simulation *Build. Environ.* **35** 537–44
[26] Marks W 1997 Multicriteria optimisation of shape of energy-saving buildings *Build. Environ.* **32** 331–9
[27] Wright J A, Loosemore H A and Farmani R 2002 Optimization of building thermal design and control by multi-criterion genetic algorithm *Energy Build.* **34** 959–72
[28] Lu J and Ruan D 2007 *Multi-objective group decision making: methods, software and applications with fuzzy set techniques* vol 6(Imperial College Press)
[29] Kennedy M C, Ford E D, Singleton P, Finney M and Agee J K 2008 Informed multi-objective decision-making in environmental management using Pareto optimality *J. Appl. Ecol.* **45** 181–92

[30] Chew M Y L, De Silva N and Tan S S 2004 Maintainability of wet areas of non-residential buildings *Struct. Surv.* **22** 39–52

[31] Armacost R L, Componation P J, Mullens M A and Swart W W 1994 An AHP framework for prioritizing customer requirements in QFD: an industrialized housing application *IIE Trans.* **26** 72–9

[32] Chen C C, Chiang C M, Horng R S and Lee S K 2011 Multicriteria evaluation of the selection of green glazing material by using the analytic hierarchy process (AHP) *Advanced Materials Research* vol 224(Trans Tech Publ)pp 152–8

[33] Skibniewski M J and Chao L-C 1992 Evaluation of advanced construction technology with AHP method *J. Constr. Eng. Manag.* **118** 577–93

Identifying the relevant features of the National Digital Cadastral Database (NDCDB) for spatial analysis by using the Delphi Technique

N Z A Halim[1], S A Sulaiman[1], K Talib[1] and E G Ng[2]
[1]Centre of Studies for Surveying Science and Geomatics, FAPS, UiTM Shah Alam
[2]Department of Survey and Mapping Malaysia

*nurzurairah@gmail.com

Abstract. This paper explains the process carried out in identifying the relevant features of the National Digital Cadastral Database (NDCDB) for spatial analysis. The research was initially a part of a larger research exercise to identify the significance of NDCDB from the legal, technical, role and land-based analysis perspectives. The research methodology of applying the Delphi technique is substantially discussed in this paper. A heterogeneous panel of 14 experts was created to determine the importance of NDCDB from the technical relevance standpoint. Three statements describing the relevant features of NDCDB for spatial analysis were established after three rounds of consensus building. It highlighted the NDCDB's characteristics such as its spatial accuracy, functions, and criteria as a facilitating tool for spatial analysis. By recognising the relevant features of NDCDB for spatial analysis in this study, practical application of NDCDB for various analysis and purpose can be widely implemented.

1. Introduction

1.1. National Digital Cadastral Database

Since 2010, the cadastral surveyed information in Malaysia is kept digitally in the form of a survey accurate spatial database, namely the National Digital Cadastral Database (NDCDB). NDCDB is the improved version of the State Digital Cadastral Database (PDUK). PDUK was referenced to the respective ten state origins and was initially developed internally for the Department of Survey and Mapping Malaysia (DSMM). The coordinate system applied in PDUK was incompatible with GIS and satellite-based technology [1]. Data integration was also challenging and required a good understanding of map projections and transformation coordinates. Yusoff and Halim [2] describe the shortfalls of PDUK for spatial analysis were because of its low spatial accuracy of meter levels, topologically incorrect for spatial queries due to lack of connective an adjacent information and different set of spatial attributes were stored and sometimes duplicated in the same lot parcel. To overcome the shortcomings of PDUK, DSMM introduced the Geocentric Datum of Malaysia, 2000 (GDM2000) and adopted it to comprehensively readjust the nationwide cadastral survey data based on International Terrestrial Reference Frame (ITRF2000) and Geodetic Reference System 1980 (GRS80) reference ellipsoid. The survey accurate NDCDB was the result of the GDM2000 and the Coordinated Cadastral System (CCS) concept implementation. Its database structure became GIS format friendlier with the implementation of the cadastral fabrics readjustment and repopulation exercise. Today, the

NDCDB contains two types of data; the primary NDCDB – that has been verified and accepted as survey accurate by using absolute positioning survey and the 'relative NDCDB' – that has been verified and accepted as relatively correct by using the relative surveying methods and computation. The 'Relative NDCDB' is a temporary database and is the outcome of the non-seamless accuracy of NDCDB which was either due to the previously lower class survey, poor cadastral control network connections, and undeveloped, remote land or simply blunder errors. Nevertheless, continuous survey work are being carried out by DSMM to achieve the desired centimetre level spatial accuracy of NDCDB.

1.2. Why is NDCDB required for spatial analysis?
Spatial analysis enables one to answer complex spatial questions that include determining relationships, making predictions or finding the best location for planning, implementation or recovery purposes, thus eliminates risk and cost investments. However, a person is considered spatially-enabled to make sound decision-making when the basis of the spatial analysis which is the cadastral data are official, authentic, complete, comprehensive, updated and accessible [3]. NDCDB is primarily used in Malaysia for survey works, but not exclusively, for spatial analysis. Therefore, unused or incorrect usage of cadastral data (NDCDB) for spatial analysis may reduce the analysis's reliability when land and people's relationship is the concern. The significance of the cadastral data has been discussed extensively internationally, but identifying the relevant features of NDCDB for spatial analysis is limitedly studied. The situation has led to various perceptions and misinformation on NDCDB. Sometimes, underused. The benefits of factoring geolocation into spatial analytics are many, such as improved communication, better collaborations, informed decision making and increase investments. A strong statement to describe NDCDB's technical relevance is deem necessary to enable it.

1.3. Delphi Technique
The Delphi technique was chosen in this study to determine the relevant features of NDCDB for spatial analysis. The technique provides a platform where consensus by subject matter experts can be achieved on a topic despite the limited evidence, lack of precise information and prior research [4, 5]. Moreover, the Delphi technique is suitable for application whenever policies, plans, or ideas have to be based on informed judgment. The disadvantages associated with face-to-face meetings by using email and online questionnaire, can be avoided [4, 6] and elude influential result and inaccurate analysis.

2. Research Methodology
2.1. Questionnaire design, validity, and reliability
Questionnaires were used to collect data in this study. In designing the questionnaires for Round 1, open-ended close structured questions were formed based on the overall research framework, while close-ended close structured for Round 2 questionnaire was the result of the Round 1 qualitative analysis result. Waweru and Omwenga [8] mentioned a minimum of three participants or 10% of the overall sample is sufficient for qualitative pretesting. The questionnaires were assessed for validity by 4 experts and pretested for reliability by 3 mock Delphi participants. All experts in this study were in agreement that all of the six questions related to identifying the technical significance of NDCDB for spatial analysis are relevant, which resulted in the calculated Content validity index (I-CVI) value as 1.00. According to Polit and Beck [7], the experts must agree on the content validity of an item (I-CVI of 1.00) when the panel consists of five or fewer experts. Pretesting was carried out respectively for Round 1 and Round 2 questionnaires which provided valuable feedback that contributed to the overall refinement of the study.

2.2. Composition and panel size
To date, there are no standards that stipulate the panel selection criteria. Considering NDCDB has a direct association with the land, cadastre and GIS domain, a heterogeneous group of a panel of experts

was set up in this study to diversify member's background and reduce bias result. Initially, 18 potential members were carefully selected based on the criterions used by previous Delphi researchers [4, 9, 10], but the final totalled panel size was 14 and exceeded the recommended minimum panel size [11]. The Delphi participants can be summarized in three major groups: policy makers, interest group, and stakeholders. The profiles of the Delphi participants are as shown in Table 1.

Table 1. Profile of Delphi Heterogeneous Participants

Criteria/ Panel	Groups	Specialist	job-relevant experience	expert's certification or accreditation	current management level	labelled experts by social acclamation	expert's contribution inside and outside of the organisation
1	Policy Makers	Survey & Cadastre	>20 years	Professional body and professional board members	Decision maker	Yes	National and international level
2	Interest Group	Survey & Cadastre	>20 years	Professional body and professional board members	Decision maker	Yes	National and international level
3	Interest Group & Stakeholders	Survey & Cadastre	>20 years	Professional body and professional board members	Decision maker	Yes	National and international level
4	Policy Makers & Stakeholders	GIS & Civil Engineering	>20 years	Professional body	Decision maker	Yes	National and international level
5	Interest Group & Stakeholders	Survey & Cadastre	>20 years	Professional body	Decision maker	Yes	National and international level
6	Policy Makers	Survey & Cadastre	>20 years	Professional body and professional body council member	Decision maker	Yes	National and international level
7	Policy Makers	Land law, Survey & Cadastre	>20 years	Professional body	Decision maker	Yes	National and international level
8	Policy Makers & Stakeholders	Land & Cadastre	>20 years	Subject matter expert	Decision maker	Yes	National and international level
9*	Policy Makers & Stakeholders	Land & GIS	>20 years	Subject matter expert	Decision maker	Yes	National and international level
10	Stakeholders	Utility planning & GIS	> 20 years	Subject matter expert	Decision maker	Yes	National and international level
11	Policy Makers & Stakeholders	Planning& GIS	>20 years	Professional Body	Decision maker	Yes	National and international level
12	Policy Makers & Stakeholders	Land & Cadastre	>20 years	Professional Body	Decision maker	Yes	National and international level
13	Stakeholders	Planning & GIS	>20 years	Professional Body	Decision maker	Yes	National and international level
14	Interest Group & Stakeholders	Survey & Cadastre	>20 years	Professional body and professional board members	Decision maker	Yes	National and international level
15	Stakeholders	Cadastre & GIS	>20 years	Subject matter expert	Decision maker	Yes	National level

*Note *: the participant was involved during the 1st round only.*

2.3. *Consensus Building*

Holey, Feeley, Dixon and Whittaker [12] described the minimum round for consensus building can be two but, three rounds are common in most studies. However, if group consensus is desirable and the sample is heterogeneous, then Skulmoski, Hartman and Krahn [13] highlighted three or more rounds might be required. In this study, Round one is the foundation for identifying issues of the research area, while the following rounds two and three are consensus building rounds.

2.3.1. Round 1
An email inviting participation, explaining the study, outlining the Delphi process and requesting responses to the question was sent to the potential participants. Round 1 questionnaire was also attached for them to complete. The questions were focused on the participants' expertise views on the technical relevance or relevant features of NDCDB for spatial analysis. Upon agreeing to participate the Delphi process, participants were given seven days to respond to the questionnaire questions. A cut-off date was set to determine the timeline of Round 1. Only 15 participants as shown in Table 1 agreed to participate and returned the completed questionnaire through email or hand-delivered. Each response was of anonymity and was collated then sent via e-mail to each participant, instead of a face-to-face group discussion, to reduce bias results. They were instructed to review the response from other anonymous respondents and were able to revise their responses within seven days.

2.3.2. Round 1 Result
Responses from Round 1 were analysed by using the Thematic Analysis process. The qualitative data was firstly scrutinised, and emerging patterns or findings that were identified as nodes were processed using the Nvivo 11 Plus software. The nodes were listed under the Free Nodes theme. From the Free Nodes, a deductive approach was used to generate the Parent Node theme which is the technical significance or relevant features of NDCDB for spatial analysis. Similar ideas related Parent Node theme were clustered together into emerging sub-themes namely 'Survey accurate,' 'Homogenous dataset,' 'GIS-ready' and "Facilitating format.' Statements that best described the essence of the majority of opinions within each emerging sub-theme mentioned were generated and presented as the 'original statements; in Table 2. The 'original statements' provided the basis for Round 2 questionnaires.

Table 2. Statements on the relevant features of NDCDB's for spatial analysis

No	Original Statements	Rephrased Statements
1	The NDCDB has the spatial accuracy of centimeter level and is the most accurate land related database.	The NDCDB has the spatial accuracy of centimeter level *which is equivalent to land title registration requirement and therefore perceived to be the most spatially accurate database as far as the land-related database is the concerned.*
2	NDCB may not be seamlessly accurate, but the positional spatial accuracy of the NDCDB is survey accurate and is suitable for most user's spatial accuracy requirements and spatial analysis purposes.	*Despite not being seamlessly accurate where some plots are of higher accuracy,* the positional spatial accuracy of the NDCDB is survey accurate and is *sufficient* for most user's spatial accuracy requirements and spatial analysis purposes.
3	The criteria of NDCDB which are GIS-ready, survey accurate and homogenous dataset enables it as a facilitating tool for data integration, sharing, and convergence with other geospatial data or GIS-based application system and non-geospatial data.	No amendments.

2.3.3. Round 2
Participants were presented with an online questionnaire that included the 'original statements' shown in Table 2. A five-point Likert scale was used to gather participant's opinion and perception on the statements and linguistic scales from 5 points as being "strongly agree", 4 points for "slightly agree", 3 points for "less agree", 2 points as "disagree" and lastly 1 point as "strongly disagree". Five-point scales was preferred to transmit as much information and reduce stifle respondents. Space was provided for optional comments justifying their scale allocation decisions. All 15 original participants were offered via e-mail to participate in Round 2 and were requested to respond within seven days. Delphi process in Round 1, was repeated in Round 2. Only comments were added by the Round 2 timeline.

2.3.4. Round 2 Result
The number of participants allocating points to each statement in Round 2 was totalled as 14. 1 participant was transferred to a new agency and had to be withdrawn from the panel member. Likert scale points allocated were analysed with descriptive statistics using SPSS version 23 software. No statements were removed as the means values were higher than 4.00 as shown in Table 3. Additional comments received suggested minor amendments which were presented as 'rephrased statements' in Table 2, with the aim of moving towards group consensus and stability.

2.3.5. Round 3
The Round 2 methodology was repeated but with the rephrased Statement 1 and 2, and original Statement 3. The original and rephrased statements became the basis for Round 3 questionnaire. 14 participants were offered to participate in Round 3 and were requested to respond within three days.

2.3.6. Round 3 Result
The number of respondents received in Round 3 was totalled as 14 or 100% feedback. Analysis methodology in Round 2 was repeated in Round 3. The increase in mean values was found in the 'rephrased statements' as shown in Table 3 with means value are more than 4.70. The Median and Interquartile Range (IQR) are best statistical choices to measure central tendency and dispersion to calculate data scored on an ordinal scale in Delphi processes [14]. The median values in Round 3 were all 5.0 while the IQR values were within 0.00 to 1.00, which is suitable as consensus indicator for 5-unit scales [15]. The values showed consistency and high agreement, and therefore it was decided the Delphi process has come to a group consensus and stability which in result supports the decision that further Delphi rounds were unnecessary. Moreover, additional comments were not received in this round.

Table 3. Statistical analysis of Round 2 and 3

Statement	Round 2					Round 3					Mean Percent Change 100%
	N	IQR	Median	Mean	Std. Deviation	N	IQR	Median	Mean	Std. Deviation	
Statement 1	14	1.00	5.00	4.57	0.514	14	1.00	5.00	4.86	0.363	5.80%
Statement 2	14	1.00	5.00	4.64	0.497	14	0.00	5.00	4.79	0.426	3.00%
Statement 3	14	1.00	5.00	4.64	0.497	14	1.00	5.00	4.71	0.469	1.40%

3. Analysis
3.1. Statistical result analysis
Analysis of the Delphi technique results showed a change in participants' views towards consensus (agreement) and stability in Round 3. Therefore, the Delphi is concluded to have reach panel consensus and stability in Round 3 as indicated by a trend towards the following: i) the median for the 5-Point Likert scale is the highest measure (5 – "strongly agree") to all statements; ii) the IQR values are within 0.00 to 1.00 which indicate high agreement; iii) the increase of mean values in each round; iv) percent change between each statement from both rounds are less than 15% and is considered stable [14]; v) a decrease in comments in each round; and vi) the evolution of statements towards consensus.

3.2. Agreed statements on the relevant features of NDCDB for spatial analysis
The final and agreed statements describing the relevant features of NDCDB for spatial analysis by the Delphi panel as per Round 3 are as follows:

- **Statement 1:** The NDCDB has the spatial accuracy of centimetre level which is equivalent to land title registration requirement and therefore perceived to be the most spatially accurate database as far as the land-related database is the concerned.
- **Statement 2:** Despite not being seamlessly accurate where some plots are of higher accuracy, the positional spatial accuracy of the NDCDB is survey accurate and is sufficient for most user's spatial accuracy requirements and spatial analysis purposes.
- **Statement 3:** The criteria of NDCDB which are GIS-ready, survey accurate and homogenous dataset enables it as a facilitating tool for data integration, sharing, and convergence with other geospatial data or GIS-based application system and non-geospatial data.

4. Conclusion

This study has achieved its objective to establish statements that best describes the relevant features of NDCDB for spatial analysis based on the consensus of experts related to land, cadaster and GIS domain by optimising the Delphi technique. Agreement and stability of the consensus on the statements were achieved during Round 3. The agreed statements as stated in paragraph 3.2 of this paper have defined NDCDB to be the most accurate land based databases as far as GIS is concern, its spatial accuracy is sufficient for most spatial analysis purposes and its criteria or characteristics facilitates spatial analysis activities. The established statements could aid land surveyors, land administrators, and GIS users, as well as the general public, to recognize the importance of NDCDB and adopt it for the critical stance to spatial thinking. The notion of becoming spatially enabled signifies the importance of NDCDB in providing basic and crucial components to allow sound and transparent decision-making, but not limited to spatial analysis.

Acknowledgements

The authors would like to Universiti Teknologi MARA, DSMM, and all related agencies for providing valuable information for this study.

References

[1] Kadir, M., Desa, G., Ses, S., and Omar, K.: 'Studies towards the development of implementation plan of Coordinated Cadastral System for Malaysia', in Editor (Ed.)^(Eds.): 'Book Studies towards the development of implementation plan of Coordinated Cadastral System for Malaysia' (UTM, 2002, edn.), pp.

[2] Yusoff, M.Y.M., and Halim, N.Z.A.: 'Country Report: Unleashing the Full Potential of eKadaster on The Cadastral System of Malaysia ', in Editor (Ed.)^(Eds.): 'Book Country Report: Unleashing the Full Potential of eKadaster on The Cadastral System of Malaysia ' (United Nations - Economic and Social Council, 2009, edn.), pp.

[3] Steudler, D., and Rajabifard, A.: 'Key Elements for a Spatially Enabled Society': 'Spatially Enabled Society' (International Federation of Surveyors (FIG) & Global Spatial Data Infrastructure Asscociation (GSDI), 2012), pp. 68

[4] Avella, J.R.: 'Delphi Panels: Research Design, Procedures, Advantages, and Challenges', International Journal of Doctoral Studies, 2016, 11

[5] Yousuf, M.I.: 'Using experts' opinions through Delphi technique', Practical assessment, research & evaluation, 2007, 12, (4), pp. 1-8

[6] McGeary, J.: 'A critique of using the Delphi technique for assessing evaluation capability-building needs', Evaluation Journal of Australasia, 2009, 9, (1), pp. 31

[7] Polit, D.F., and Beck, C.T.: 'The content validity index: are you sure you know what's being reported? Critique and recommendations', Research in nursing & health, 2006, 29, (5), pp. 489-497

[8] Waweru, P.K., and Omwenga, J.: 'The Influence of Strategic Management Practices on Performance of Private Construction Firms in Kenya', International Journal of Scientific and

Research Publications, 2015, 5, (6)
[9] Warth, J., Heiko, A., and Darkow, I.-L.: 'A dissent-based approach for multi-stakeholder scenario development—the future of electric drive vehicles', Technological Forecasting and Social Change, 2013, 80, (4), pp. 566-583
[10] Markmann, C., Darkow, I.-L., and von der Gracht, H.: 'A Delphi-based risk analysis—Identifying and assessing future challenges for supply chain security in a multi-stakeholder environment', Technological Forecasting and Social Change, 2013, 80, (9), pp. 1815-1833
[11] Bueno, S., and Salmeron, J.L.: 'Fuzzy modeling Enterprise Resource Planning tool selection', Computer Standards & Interfaces, 2008, 30, (3), pp. 137-147
[12] Holey, E.A., Feeley, J.L., Dixon, J., and Whittaker, V.J.: 'An exploration of the use of simple statistics to measure consensus and stability in Delphi studies', BMC Medical Research Methodology, 2007, 7, pp. 52-52
[13] Skulmoski, G.J., Hartman, F.T., and Krahn, J.: 'The Delphi method for graduate research', Journal of information technology education, 2007, 6, pp. 1
[14] Heiko, A.: 'Consensus measurement in Delphi studies: review and implications for future quality assurance', Technological forecasting and social change, 2012, 79, (8), pp. 1525-1536
[15] Yang, Y.N.: 'Testing the Stability of Experts' Opinions between Successive Rounds of Delphi Studies', 2003

Comparative Study of Drift Compensation Methods for Environmental Gas Sensors

M Z Abidin[1], Arnis Asmat[1,2] and M N Hamidon[3]

[1] Faculty of Applied Sciences, Universiti Teknologi MARA, 40450 Shah Alam, Selangor, Malaysia
[2] Climate Change and Carbon Footprint Research Group, Universiti Teknologi MARA, 40450 Shah Alam, Selangor, Malaysia
[3] Institute of Advanced Technology, Universiti Putra Malaysia, 43400 Serdang, Selangor, Malaysia

E-mail: rnis_annis@salam.uitm.edu.my

Abstract. Most drift compensation attempts in environmental gas sensors are only emphasize on the "already-known" drift-causing parameter (i.e., ambient temperature, relative humidity) in compensating the sensor drift. Less consideration is taken to another parameter (i.e., baseline responses) that might have affected indirectly with the promotion of drift-causing parameter variable (in this context, is ambient temperature variable). In this study, the "indirect" drift-causing parameter (drifted baseline responses) has been taken into consideration in compensating the sensor drift caused by ambient temperature variable, by means of a proposed drift compensation method (named as RT-method). The effectiveness of this method in its efficacy of compensating drift was analysed and compared with the common method that used the "already-known" drift-causing parameter (named as T-method), using drift reduction percentage. From the results analysis, the RT-method has outperformed T-method in the drift reduction percentage, with its ability to reduce drift up to 64% rather than the T-method which only able to reduce up to 45% for TGS2600 sensor. It has proven that the inclusion of drifted baseline responses into drift compensation attempt would resulted to an improved drift compensation efficiency.

1. Introduction

Volatile organic compounds (VOCs) can be used as a precursor for the increasing of ozone and particulate matter in tropospheric. The main source of VOC is emitted from vehicle, traffic flows and fuel transport. Many VOCs have toxic characteristics which not only give great influence on climatic change [1], but also deteriorate air quality and threaten human health (i.e., benzene could lead to carcinogenic, while toluene may cause vertigo symptoms; for long term cumulative effect on health conditions) [2]. Miniaturized-devices (namely gas sensors) have been developed for monitoring purposes which giving better option on low at cost, continuous monitoring, compact and portable [3-9]. However, the existing sensor designs have a drawback when signal response cannot be fully optimized due to its sensitivity to surrounding environmental variables (i.e., environmental temperature and relative humidity) that has caused drift-effects [10-16]. In this paper, the study was aimed to compensate the sensor's signal drift using drift compensation model of T- and RT- methods. In compensating the signal drift, RT-method was emphasized on drifted baseline response utilization whereas T-method was used ambient temperature variable as compensating parameter.

2. Methodology

2.1. Data collection

Two Taguchi Gas Sensor (TGS2600 and TGS2602) sensors were used in this study and drifted responses have obtained under varied ambient temperature at 25, 30 and 35 °C with two different exposures; i) in exposure of clean air (baseline responses), and ii) in exposure of benzene at concentration of 0.5, 1, 5 and 10 ppm. Sensor resistance (R_S) was then calculated from the acquired responses (V_{OUT}) using equation:

$$R_S = \frac{V_C \times R_L}{V_{OUT}} - R_L \quad (1)$$

where V_C is the sensor's circuit voltage and R_L is the sensor's load resistor. Since there are two types of exposure (in clean air and in benzene), therefore R_S for these two exposures are denoted as R_a (for exposure in clean air) and R_g (for exposure in benzene).

The calculated R_a and R_g were then been pre-processed before further analysis can be proceed, in order to carefully select a number of parameter that are descriptive of the signal responses [17]. The pre-processing steps of relative baseline manipulation (R_a/R_g) and compression (using steady-state responses) were performed in order to retain the signal responses structure that carry relevant information (particularly information about the drift due to ambient temperature variable).

2.2. Drift Compensation

Linear regression model is used for the compensations of T- and RT- methods. The T-method is a one stage compensation method that compensates drift upon ambient temperature variable. Meanwhile, the RT-method include a two stages compensation method; which i) compensates drift upon drifted baseline response (R_a), and ii) compensates the compensated responses obtained from the first compensation stage (R_a-compensated) upon ambient temperature variable – as for final result standardisation, so that results for both T- and RT- methods are reflected into sensor responses that are unaffected with the change of ambient temperature values.

Both of these two compensation methods have taken sensor responses at ambient temperature of 25°C as the point of calibration to compensate the other responses that had been drifted at 30 and 35°C. The drifting model for both of these T- and RT- methods are as follows:

for T-method, $$F = a + bT \quad (2)$$

for RT-method (first compensation stage), $$F = a + bR_a \quad (3)$$

where F is the feature of drifted sensor responses (uncompensated) denoted as R_a/R_g-UNCOMP, a is constant, b is coefficient of slope line obtained from the drifting model, T is ambient temperature variable and R_a is drifted baseline responses.

A new feature of sensor responses (compensated) was derived based on the coefficient of b for each corresponding methods – F_{T25} for T-method (compensated from ambient temperature variable) and F_{Ra25} for first stage of RT-method (compensated from drifted baseline responses), respectively obtained using equation as follows;

$$F_{T25} = F - b(T - T_{25}) \quad (4)$$

$$F_{Ra25} = F - b(R_a - \overline{R_{a25}}) \quad (5)$$

where T_{25} is the ambient temperature of 25°C, and $\overline{R_{a25}}$ is the mean value of drifted baseline response at 25°C.

The second compensation stage of RT-method was performed using the R_{a25}-compensated responses (obtained from the previous compensation stage) through model equation as follows:

$$F_{Ra25} = a + bT \quad (6)$$

where F_{Ra25} is the R_{a25}-compensated responses, a is constant, b is coefficient of slope line obtained from the model and T is the ambient temperature variable.

The coefficient of b was used in obtaining a new feature of sensor responses, which is F_{RT25}, compensated from the dependence of ambient temperature variable, T, using equation:

$$F_{RT25} = F - b(T - T_{25}) \quad (7)$$

where T_{25} is the ambient temperature variable of 25°C.

From both of these T- and RT- compensation methods, the obtained F_{T25} were the final results of compensated sensor responses for T-method compensation (denoted as R_a/R_g-T25COMP), meanwhile the obtained F_{RT25} were the final results of compensated sensor responses for RT-method compensation (denoted as R_a/R_g-RT25COMP).

2.3 Compensation comparison

The compensation efficiency for both methods was quantified using a variability measure of coefficient of variation (CV). CV is defined as the ratio of the standard deviation (σ) to the mean (μ), showing the extent of drift in relation to the sensor response's population as follows;

$$CV = \frac{\sigma}{\mu} \tag{8}$$

Through this *CV*, the drift reduction percentage of compensated responses were measured using the equation:

$$CV_{red\%} = \left(\frac{CV_{UNCOMP} - CV_{xCOMP}}{CV_{UNCOMP}}\right) \times 100 \tag{9}$$

where $CV_{red\%}$ is the sought drift reduction percentage value, CV_{UNCOMP} is the *CV* value of uncompensated sensor responses (R_a/R_g-UNCOMP), and CV_{xCOMP} is the *CV* value of any compensated sensor responses (R_a/R_g-T25COMP or R_a/R_g-RT25COMP).

3. Results and Discussion

Fig. 1 shows the distribution of TGS2600 sensor responses before (UNCOMP) and after (T25COMP and RT25COMP) compensation based on the concentration of benzene exposure (0.5, 1, 5 and 10ppm).

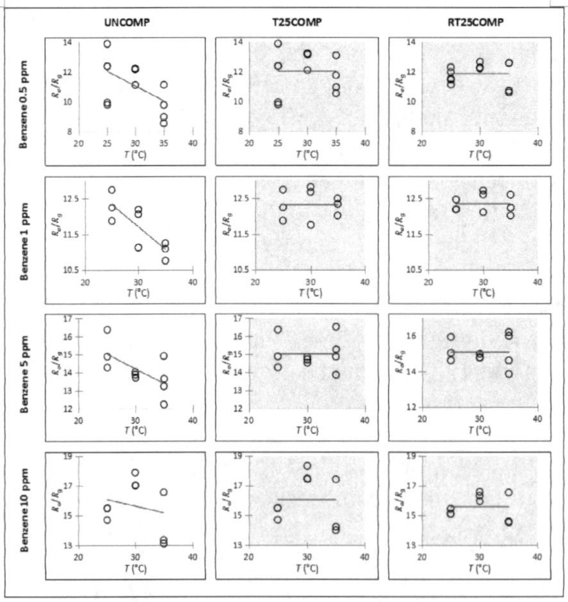

Figure 1. TGS2600 sensor responses distribution before (UNCOMP) and after (T25COMP and RT25COMP) compensation.

From the figure, in graphs for UNCOMP, results show negatively correlated between the drifted sensor responses and ambient temperature variable. As this drift was compensated, the distribution of the resulted compensated responses for T25COMP and RT25COMP has improved to calibrate around the sensor responses at ambient temperature of 25°C (as the point of calibration).

For example, in graphs for T25COMP, since the compensation's calibration point is based on ambient temperature variable at 25°C, thus the responses at 25°C were remain unchanged as its original UNCOMP responses, and responses at 30 and 35°C were compensated/calibrated to resemble this responses of 25°C. This compensation has resulted a "rotated" version of responses distribution (from the original UNCOMP) to be around the responses at 25°C (observed through the "rotated" trend line).

Meanwhile for RT25COMP, since the compensation was performed through two compensation stages, which main emphasize was given on the first one (compensation upon drifted baseline responses), the responses distribution can be seen differ from T25COMP. From this difference of responses distribution, obviously the span of compensated responses for RT25COMP has been reduced further than the T25COMP compensation (reduced variability). This drift reduction had been analyzed through the quantification of coefficient of variation (CV) value (as the form of variability or drift measure). Results of this drift reduction quantification are as shown in Table 1.

Table 1. Drift reduction percentage ($CV_{red\%}$) values for T25COMP and RT25COMP compensations.

Benzene Concentration (ppm)	T25COMP ($CV_{red\%}$)	RT25COMP ($CV_{red\%}$)
0.5 ppm	22%	58%
1 ppm	45%	64%
5 ppm	28%	37%
10 ppm	5%	52%

Table 1 shows the drift reduction percentage results for RT25COMP are much higher as compared to the drift reduction percentage by T25COMP. At the concentration of 0.5 ppm of benzene, the $CV_{red\%}$ for RT25COMP compensated responses is 58%, which is higher than the $CV_{red\%}$ for T25COMP compensated responses that is 22%. From the overall percentage result, has shown that the compensation using RT25COMP is able to reduce drift from the original UNCOMP responses with much higher degree than the compensation using T25COMP. This significant drift reduction ability of using RT25COMP from RT-method is congruent with the study in [18], in which utilization of baseline responses in compensating drift would results to a surprising drift-effects reduction. Thus, shows that RT-method is more effective in compensating drift compared to the compensation attempt using T-method.

4. Conclusion

The high degree of reduced drift in drift reduction percentage for RT-method indicates the role of baseline responses is significant in compensating the sensors drift. Thus, proves the inclusion of drifted baseline responses into drift compensation attempt able to give better solution to the drift issue in sensor responses, rather than using the "already-known" drift-causing parameter alone, as what commonly found in current practice. This improved compensation method would then contribute to enhanced prediction accuracy of gas concentration estimation in further gas sensing analysis.

Acknowledgement

The authors would like to thank Universiti Teknologi MARA (UiTM) to support the study under 600-IRMI/DANA 5/3BESTARI (0012/2016).

References

[1] Gómez M C, Durana N, Navazo M, Alonso L, García J A & Ilardia J L 2004 Application of validation data tests from an on-line volatile organic compound analyser to the detection of air pollution episodes in urban areas. *Analytica Chimica Acta* **524**(1–2 SPEC. ISS.) 41–49

[2] Chen X, Zhang G & Chen H 2010 Controlling Strategies and Technologies of Volatile Organic Compounds Pollution in Interior Air of Cars *2010 International Conference on Digital Manufacturing & Automation* 450–453

[3] Lee D D & Lee D S 2001 Environmental gas sensors *IEEE Sensors Journal* **1**(3) 214–224

[4] Tsujita W, Ishida H & Moriizumi T 2004 Dynamic gas sensor network for air pollution monitoring and its auto-calibration *Proceedings of IEEE Sensors* 56–59

[5] Conrad T, Reimann P & Schütze A 2007 A hierarchical strategy for under-ground early fire detection based on a T-cycled semiconductor gas sensor *Proceedings of IEEE Sensors* (2) 1221–1224

[6] Mabrook M & Hawkins P 2001 A rapidly-responding sensor for benzene, methanol and ethanol vapours based on films of titanium dioxide dispersed in a polymer operating at room temperature *Sensors and Actuators B: Chemical* **75**(3) 197–202
[7] Ren H 2001 Current Voltage Characteristics of a Semiconductor Metal Oxide Sensor (University of Maine)
[8] Tinoco A V 2006 Improving the performace of micro-machined metal oxide gas sensors: Ptimization of th temperature modulation mode via pseudo-random sequences (June)
[9] Ueno Y, Horiuchi T, Niwa O, Zhou H, Yamada T & Honma I 2003 Portable automatic BTX measurement system with microfluidic device using mesoporous silicate adsorbent with nano-sized pores **95** 282–286
[10] Abidin M Z, Asmat A & Hamidon M N 2013 Identification of initial drift in semiconductor gas sensors caused by temperature variation *Proceedings of 2013 IEEE 9th International Colloquium on Signal Processing and its Applications* (Kuala Lumpur) 285–288
[11] Abidin M Z, Asmat A & Hamidon M N 2014 Temperature drift identification in semiconductor gas sensors *Proceedings of 2014 IEEE Conference on Systems, Process and Control* (Kuala Lumpur) 63-67
[12] Hossein-Babaei F & Ghafarinia V 2010 Compensation for the drift-like terms caused by environmental fluctuations in the responses of chemoresistive gas sensors *Sensors and Actuators B: Chemical* **143**(2) 641–648
[13] Marco S & Gutierrez-Galvez A 2012 Signal and data processing for machine olfaction and chemical sensing: A review *IEEE Sensors Journal* **12**(11) 3189–3214
[14] Vergara A, Vembu S, Ayhan T, Ryan M A, Homer M L & Huerta R 2012 Chemical gas sensor drift compensation using classifier ensembles *Sensors and Actuators B: Chemical* **166–167** 320–329
[15] Wang C, Yin L, Zhang L, Xiang D & Gao R 2010 Metal oxide gas sensors: Sensitivity and influencing factors *Sensors* **10**(3) 2088–2106
[16] Korotcenkov G & Cho B K 2011 Instability of metal oxide-based conductometric gas sensors and approaches to stability improvement (short survey) *Sensors and Actuators B: Chemical* **156**(2) 527–538
[17] Gutierrez-Osuna R 2002 Pattern analysis for machine olfaction: A review *IEEE Sensors Journal* **2**(3) 189–202
[18] Macías M, Agudo J, Manso A, Orellana C, Velasco H & Caballero R 2014 Improving Short Term Instability for Quantitative Analyses with Portable Electronic Noses *Sensors* **14**(6) 10514–10526

IOP Conf. Series: Earth and Environmental Science **117** (2018) 012032 doi:10.1088/1755-1315/117/1/012032

Ammonia-nitrogen and Phosphate Reduction by Bio-Filter using Factorial Design

Norhafezah Kasmuri[1]*, Nur Ashikin Mat Damin[1], Megawati Omar[2]

[1]Faculty of Civil Engineering, Universiti Teknologi Mara 40450 Shah Alam, Selangor, Malaysia

[2]Academy of Language Studies, Universiti Teknologi MARA, Shah Alam, 40450, Selangor, Malaysia

*norhafezahkasmuri@salam.uitm.edu.my

Abstract. Untreated landfill leachate is known to have endangered the environment. As such new treatment must be sought to ensure its cost-effective and sustainable treatment. Thus this paper reports the effectiveness of bio-filter to remove pollutants. In this research, the reduction of nutrients concentration was evaluated in two conditions: using bio-filter and without bio-filter. Synthetic wastewater was used in the batch culture. It was conducted within 21 days in the initial mediums of 100 mg/L ammonia-nitrogen. The nitrification medium consisted of 100 mg/L of ammonia-nitrogen while the nitrite assay had none. The petri dish experiment was also conducted to observe the existence of any colony. The results showed 22% of ammonia-nitrogen reduction and 33% phosphate in the nitrification medium with the bio-filter. The outcome showed that the bio-filter was capable to reduce the concentration of pollutants by retaining the slow growing bacteria (AOB and NOB) on the plastic carrier surface. The factorial design was applied to study the effect of the initial ammonia-nitrogen concentration and duration on nitrite-nitrogen removal. Finally, a regression equation was produced to predict the rate of nitrite-nitrogen removal without conducting extended experiments and to reduce the number of trials experiment.

1. Introduction

The increase of population in urban areas has created industrial areas all over the world. This increase in turn leads to the surge of municipal and industrial solid waste [1]. To treat these municipal solid wastes (MSW), sanitary landfill is employed by city authorities in many countries despite the existence of many other treatment methods [2]. Sanitary landfill method continues to be used due to its relatively simple procedure and low operational cost [1]. Landfill leachate generated from landfills must be treated to meet the standard as it creates social and environmental issues [3]. The leachate contains various pollutants and high concentration of organic compound, nutrients, heavy metal, chemicals and inorganic salts [4].

Therefore, to reduce the nitrogen in order to improve the treatment of treated landfill leachate, wastewater requires a treatment process that involves biological oxidation which converts ammonia-nitrogen (NH_3-N) into nitrate-nitrogen (NO_3-N), which is called nitrification. Then, it is followed by the process of biological reduction known as de-nitrification which converts NO_3-N into nitrogen gas (N_2) [5]. The nitrification initiated from oxidation of NH_3-N to nitrite-nitrogen (NO_2-N) by several genera of autotrophic nitrifying bacteria example; *Nitrosomonas* for AOB. Then, NO_2-N is oxidized to NO_3-N which much less toxic by several other genera of nitrifying bacteria, which is *Nitrobacter* as NOB. In addition, the importance of biological phosphorus removal is chosen because of the lower operation cost and environmental impacts [6].

At present, bio-filter packing medium has received considerable attention as an effective way to remove pollutants and ammonia nitrogen in landfill leachate. In the same vein, [7] studied the reduction of ammonia-nitrogen by bio-filters. The biological process of a bio-filter system to remove pollutants has been an alternative due to its low cost and maintenance compared to other treatment. The efficiency to remove pollutants is mainly from its physicochemical and microbiological properties. The bio-filter had been found to be practical in treating landfill leachate in previous reports [8]. Hence, the objective of the experiment was to reduce the ammonia-nitrogen and phosphate, and to produce nitrite-nitrogen and nitrate-nitrogen. This study was also to develop a regression equation for nitrite-nitrogen production for batch culture experiment as a tool to predict which in turn will limit the number of trial experiment in bio-filter system.

2. Material & Methods

This study employed three types of synthetic media. They were to be mixed to produce two types of media called nitrification and nitrite assay media [9]. The synthetic medium was employed instead of real as the later contained a variety of pollutants.

To note, this experiment took longer retention time as the nitrifying bacteria (AOB and NOB) are slow growers [9]. Two sets of flasks were prepared where each set contained two flasks. The two flasks were labeled Flask 1 and Flask 2. Flask 1 contained a bio-filter while Flask 2 did not. All flasks were injected with 20 mL of leachate for AOB and NOB. After the experiments, both retained media were used for agar plate experiments. The agar and nutrient agar plates were prepared to measure the presence of AOB and NOB colonies.

The formulations of media were chosen based on the previous study conducted by other researchers [7, 9]. Medium A was the formulation of ammonia-nitrogen solution to determine the ammonia-nitrogen reduction while Medium B was the formulation of nitrite-nitrogen solution, of which no ammonia-nitrogen was added. However, Medium C was the formulation of phosphate solution to reduce phosphate. Phosphate is also the nutrient requirement for autotrophic nitrifying bacteria to grow [6].

2.1. Nitrification Medium

The formulation of Medium A included $(NH_4)_2SO_4$, 5 g and 1000 mL of distilled water. The solution was sterilized at 121°C for 15 minutes. The Medium C included a sterilized stock of Na_2HPO_4, 3.5 g, KH_2PO_4, 0.7 g, $MgSO_4.7H_2O$, 0.1 g, $FeCl_3.6H_2O$, 0.014 g, $CaCl_2.2H_2O$, 0.18 g and 1000 mL of distilled water. It was prepared separately from the basal medium. The stock solution of Medium C was added aseptically to give the final nitrification medium volume of 500 mL [7].

2.2. Nitrite Assay Medium

Medium B contained $NaNO_2$, 0.5 g, Na_2HPO_4, 13.5 g; $MgSO_4.7H_2O$, 0.1g, $FeCl_3.6H_2O$, 0.014 g, $CaCl_2.2H_2O$, 0.18 g and 1000 mL of distilled water and sterilized at 121°C for 15 minutes. A sterilized Medium C of Na_2HPO_4, 3.5 g, KH_2PO_4, 0.7 g, $MgSO_4.7H_2O$, 0.1 g, $FeCl_3.6H_2O$, 0.014 g, $CaCl_2.2H_2O$, 0.18 g and 1000 mL of distilled water was prepared separately from the basal medium. It was added aseptically to give the final nitrite assay volume of 500 mL [7]. The difference between nitrification and nitrite assay medium was their chemical measurements.

2.3. Batch Culture Experiment

Flask 1 contained bio-filter while Flask 2 did not. Ten (10) pieces of bio-filters were used and inserted in Flask 1 for both media. Then, 20 mL of leachate was injected into each flask. The flasks then were placed under aeration to encourage the growth of bacteria. The samples were taken daily to record the concentrations of NH_3-N, NO_2-N, NO_3-N and PO_4^{3-} for 21 days.

2.4. Ammonia-agar plate

Ammonia-nitrogen agar was prepared using the formulation of 14 g of agar; 250 mL of Medium A and 250 mL of Medium C to reach a total volume of 500 mL. The solution was stirred using the glass rod until the agar dissolved completely. Afterwards, the solution was sterilized by autoclaving at 121°C for 15 minutes.

2.5. Nitrite-agar Plate

The formulation of the nitrite-agar plate was 14 g of agar, 250 mL of Medium B and 250 mL of Medium C. The solution was stirred by a glass rod to mix it well, then, autoclaved at 121°C for 15 minutes.

2.6. Nutrient-agar Plate

The nutrient agar plate contained 28 g nutrient agar and 1000 mL distilled water. Similarly, the solution was stirred, sterilized and autoclaved at 121°C for 15 minutes.

2.7. Factorial Design by Design of Experiment (DOE)

The relationship between the two factor variables and the response for the nitrite-nitrogen reduction was analyzed by a full general factorial design by the DOE [10]. The general factorial regression is used to predict the future percentage of nitrate-nitrogen removal. The multilevel factorial design used in this experiment was the one having at least one factor with more than two levels [10].

The two factors of the initial ammonia-nitrogen concentration and duration were denoted $\varkappa_1$ and $\varkappa_2$, respectively. The initial concentration consisted of two levels which were 0 and 100 mg/L, coded as -1 and 1. While the level of the duration that were referred to coded values -1, 0 and 1, used in this part of experiment were 8, 15 and 21 days. The values of each level mentioned were based on the results obtained from the batch experiment. Table 1 shows the summary of two factors with two levels of the initial ammonia-nitrogen concentration and three levels of the duration discussed earlier.

Table 1. Coded and Real Values for Multilevel Factorial Design of Initial Concentration and Time

Factor variables	Variables	Unit	Response		
			-1	0	1
Initial Ammonia Concentration	$\varkappa_1$	mg/L	0	-	100
Duration	$\varkappa_2$	days	0	18	21
Initial Ammonia Concentration	$\varkappa_1$	mg/L	0	-	100

The results showed the coefficient parameters which were estimated using factorial regression analysis of the software Minitab 17 [10]. Table 2 displays the standard order, run order and design matrix of the two factors with multilevel factorial designs. The standard order is sometimes known as the non-randomized order which contain the common arrangement of factorial design [10]. On the other hand, the run order is identified as the randomized order, which refers to the sequence of the experimental run. This dissimilarity is done to minimize the effect of unexplained variability in the observed response due to systematic errors [11]. In this research, the randomized order was set to avoid any systematic errors.

Table 2. Experimental Design Matrix by Factorial Design

Standard Order[a]	Run Order[b]	Initial Concentration of Ammonia-nitrogen		Duration	
		Coded values	Real values (mg/L)	Coded values	Real values (days)
4	1	1	100	-1	1
6	2	1	100	1	21
5	3	1	100	0	18
2	4	-1	0	0	18
3	5	-1	0	1	21
1	6	-1	0	-1	1

[a] Nonrandomized order
[b] Randomized order

3. Results and Discussion

3.1. Statistical Analysis by Factorial Design

The factorial design in this experiment consisted of two factors. They were the initial ammonia-nitrogen concentration and the duration, denoted as $\varkappa_1$ and $\varkappa_2$, respectively. The two levels of the initial ammonia-nitrogen concentrations were 0 and 100 mg/L while the three durations were 0, 18 and 21 days. The combination of these factors and the levels outlined a total of 6 runs experimental matrix, shown in column 2 in Table 3. Coded (±1) and real levels of two factors are given on columns 3 - 6 in Table 3. The last column shows the percentage nitrite-nitrogen removal obtained in each run. The objective of using factorial design was to predict future removal of nutrients and to investigate the effect of each factor on a response. The response is recognized in the amount of the nitrite-nitrogen being removed as presented in Table 3.

Factorial Design Analysis Figure 1 shows the study of the factors affecting the nitrite-nitrogen removal. The mean of the response variable for various levels of each factor is represented by the points in the plot. The initial ammonia-nitrogen concentration of 100 mg/L has a high percentage of nitrite-nitrogen removal. The duration of 21 days also has a high percentage of nitrite-nitrogen removal which indicated that when the duration is increases the percentage of removal also tend to be increased. The horizontal reference line drawn at the grand mean of the response data shows the mean processing time for all runs. In this study, the time interval has a significant effect on the nitrite-nitrogen removal. The mean of the nitrite-nitrogen removal increases in polynomial manner by increasing the duration as well as increasing the concentration (Figure 1). The two mentioned effects contributed to enhance the nitrite-nitrogen removal by the bio-filter.

IOP Conf. Series: Earth and Environmental Science **117** (2018) 012032 doi:10.1088/1755-1315/117/1/012032

Table 3. Two Factors with Two Levels of Initial Concentrations and Three Levels of Duration in Factorial Design

Standard Order[a]	Run Order[b]	Factors				Response
		Initial Ammonia -nitrogen Concentration ($\varkappa_1$)		Duration ($\varkappa_2$)		% Nitrite-nitrogen Removal
		Coded values	Real Values (mg/L)	Coded values	Real Values (Days)	
4	1	1	100	-1	0	0
6	2	1	100	1	21	37.5
5	3	1	100	0	18	25
2	4	-1	0	0	18	8
3	5	-1	0	1	21	24
1	6	-1	0	-1	0	0

[a] Nonrandomized order [b] Randomized order

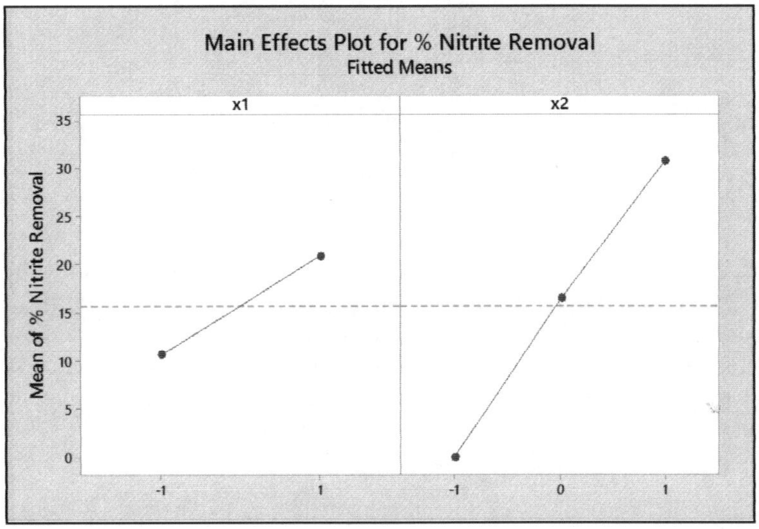

Figure 1. Main Effects Plot of the Nitrite Percentage Removal from the Minitab 17' Software

The Minitab 17's software developed the regression equations by the main effect plot (Minitab, 2017). Other studies reported that the main effects of operation time and current density have significant effect on the nitrite-nitrogen removal. The regression model which explains the removal efficiency is optimized to find the maximum level of removal [12]. Meanwhile in this research, the duration of time is the significance factor for the regression model for percentage removal of nitrite-nitrogen in developing the regression equation.

3.1.1. Factorial Design Analysis Factorial Regression Analysis: Percentage of Nitrite-nitrogen Removal Versus $\varkappa_1$ and $\varkappa_2$ The summary report presented in Figure 2 demonstrates that the model is statistically significant with the duration, $\varkappa_1$ at the 95% confidence level since p-value were less than alpha (α) of 0.10. The R^2 is high showing the data good fit.

IOP Conf. Series: Earth and Environmental Science **117** (2018) 012032 doi:10.1088/1755-1315/117/1/012032

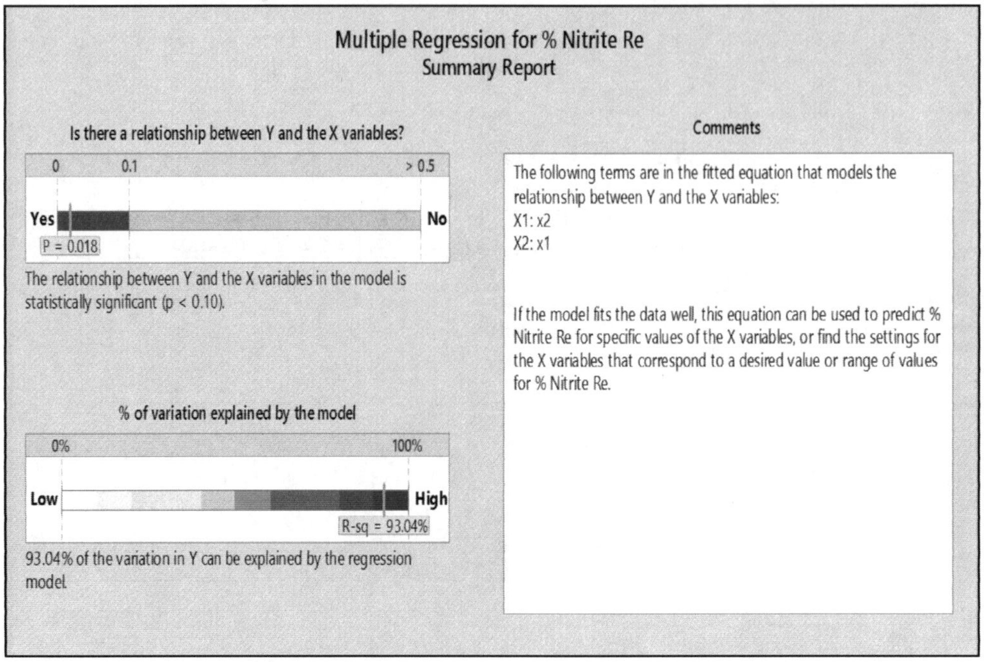

Figure 2. Summary Report from the Minitab 17' Software

The positive correlation R = 0.96 indicates that when $\varkappa_1$ increases, the removal of nitrite-nitrogen increases. The R^2 obtained in the experiment was 0.9304, which indicates a good agreement between the response (% nitrite-nitrogen removal) and the factor variables (duration). The factor (duration) with their square interactions ($P=0.018 < 0.10$) was significant at the 95% confidence level.

The factorial regression model obtained from the Minitab 17' software, as shown in Equations 1 and 2, can be used to predict the long term removal of nitrite-nitrogen without conducting any extended experiment. This in turn can reduce the number of experimental trials traditionally required to assess multiple parameters and their interactions.

For $\varkappa_1 = 0$ mg/L,

$$\text{\% Nitrite-nitrogen Removal} = 10.67 + 15.37\,\varkappa_2 \quad (1)$$

For $\varkappa_2 = 100$ mg/L,

$$\text{\% Nitrite-nitrogen Removal} = 20.83 + 15.37\,\varkappa_2 \quad (2)$$

where: $\varkappa_1$: initial ammonia concentration
$\varkappa_2$: duration

4. Conclusion

In conclusion, the bio-filters are potential as reliable supporting media for microbial population to grow in order to remove pollutants in leachates. It is also an economical means to develop more advanced biological treatment system. Not only beneficial in terms of cost, bio-filter is also sustainable. In this case, the duration taken to reduce pollutants is found to have a positive influence on the response and it is classified significantly able to remove the nitrite-nitrogen. The regression equation to remove nitrite-nitrogen can be studied deeper by future researchers by including more parameters to enhance the regression equation. This may help reduce reliance on the use of sanitary landfills to treat municipal solid wastes, leachates and pollutants.

Acknowledgments
The authors gratefully acknowledged the financial support, Lestari Grant (600-IRMI/DANA 5/3/LESTARI (0013/2016), provided by RMI.

References
[1] Jumaah M A, Othman M R and Yusop M R 2016 Characterization of Leachate from Jeram Sanitary Landfill-Malaysia International Journal of Chem.Tech. Research **9 08** pp 571-574
[2] Fernandes A, Pacheco M J, Ciríaco L and Lopes A 2015 Review on the electrochemical processes for the treatment of sanitary landfill leachates: Present and future Applied Catalysis B: Environmental **176–177** pp 183–200
[3] Zhang D, Vahala R, Wang Y and Smets B 2016 Microbes in biological processes for municipal landfill leachate treatment: Community, function and interaction. International Biodeterioration & Biodegradation **113** pp 88-96
[4] Yuan Q, Jia H and Poveda M 2016 Study on the effect of landfill leachate on nutrient removal from municipal wastewater Journal of Environmental Sciences pp 153-158
[5] Ge S, Wang S, Yang X, Qiu S, Li B and Peng Y 2015 Detection of nitrifiers and evaluation of partial nitrification for wastewater treatment: A review Chemosphere **140** pp 85–98
[6] Yadav D, Pruthi V and Kumar P 2016 Enhanced biological phosphorus removal in aerated stirred tankreactor using aerobic bacterial consortium Journal of Water Process Engineering **13** pp 61–69
[7] Kasmuri N, Lovitt R W and Omar M 2016 Reducing Ammonia-Nitrogen by Bio-Filters 5th International Technical Conference Kota Kinabalu pp 545-551
[8] Xie B, Lv B Y, Hua C, Liang S B, Tang Y and Lu J 2010 Landfill leachate pollutant removal performance of a novel biofilter packed with mixture medium Bioresource Technology **101** pp 7754–7760
[9] Reddy C, Beveridge T, Breznak J, Marzluf G, Schmidt T and Synder L 2007 Methods for General and Molecular Microbiology 3rd ed Washington D C: ASM Press
[10] Minitab, I 2017 Bizit Systems (M) Sdn Bhd Retrieved from http://support.minitab.com/en us/minitab/17/topic-library/modeling-statistics/doe/factorial-designs/factorial-designs/
[11] Zhu C and Liu X 2013 Optimization of extraction process of crude polysaccharides from Pomegranate peel by response surface methodology Journal of CarbohydratePolymers **92** pp 1197-1201
[12] Ghanim, A 2013 Application of Response Surface Methodology to Optimize Nitrate Removal from Wastewater by Electrocoagulation International Journal of Scientific & Engineering Research **4** pp10

A quantitative analysis study on the implementation of partnering in the design and build construction project

F M Halil[1], N M Nasir[1], A S Shukur[1], H Hashim[1]
[1]Faculty of Architecture, Planning and Surveying, Universiti Teknologi MARA, 40450 Shah Alam, Selangor, Malaysia

* fammuha@gmail.com

Abstract. Design and Build construction project involved the biggest scale of the cost of investment as compared to the traditional approach. In Design and Build, the client hires a design professional that will design according to the client's need and specification. This research aim is to explore the concept of partnering implementation practiced in the design and build procurement approach. Therefore, the selection of design professionals such as Contractors and consultants in the project is crucial to ensure the successful project completion on time, cost, and quality. The methodology adopted using quantitative approach. Administration of the questionnaire was distributed to the public client by using postal survey. Outcomes of the results, the public clients agreed that project management capabilities and commitment to budget as a crucial element of partnering from the design professional in design and build construction project.

1. Introduction

Cartlidge [1] defines partnering as cooperation and teamwork, openness and honesty, trust, equity and equality between the parties in the construction project. This concept of the relationship depend on the mutual trust established between two parties. The trust occurs between client and contractor after the first project partnering successfully completed on time, cost and quality. Therefore, certain client maintains the relationship and the next project was given to the same contractor in client's development, it's called strategic partnering. [2] Anita & Fellows explained strategic partnering as the parties create a longer term relationship through a cycle of projects for which contracts are usually negotiated. Agreed by Bennet & Jayes [3] strategic partnering occurs when two or more company use partnering for a long term basis to undertake more than one construction project, or some continuing construction activity [3]. The client initiates to maintain the relationship with a similar contractor in project due to the trust has been established between both parties. Reported by Cartlidge [1] most of the clients across all sectors have been adopting partnering in response to the proven long-term benefits that can be achieved through this approach and achieved savings up to 40% on costs and 70% on time by using partnering approaches. Therefore, through a good services reputation showed by the contractor in the construction project, the continuity project was established between client and contractor.

1.1 Partnering

In the construction industry, partnering refers to a long-term mutual agreement between two companies to achieve an unusually high degree of cooperation to accomplish main objectives. This mutual agreement allows the two parties to work more effectively and efficiently. Both parties received a benefit such as increase workload stability and reduced overheads cost. The long-term relationship is considered critical to partnering success because it generates an atmosphere that lends itself to problem-solving, frees each party from constant re-evaluation and permits 'lessons learned' to be passed from one project to the next [4]. The crucial need of this kind of collaboration to be practiced in the construction project is to ensure the successful completion of the project on time, cost, and quality. Through this approach of partnering that encourages openness and trust between the parties to a contract. In the design and build project is a very risky project that involves risk in design liability and disputes among parties. Therefore, the selection of partner involved in design and build project is crucial. Through partnering the selection of an appropriate partner can reduce the risk of delay and dispute in the construction project. This type of selection of partner in design and build project requires that contractors have well experienced and have a good knowledge in design and strong project management team to ensure the project complete on time, between the cost, stipulated and good quality. The identification of good contractors can be evaluated by the clients through previously completed projects received and based on this situation, continuous projects only will be given to the contractors that completed projects as mentioned above. The benefits of partnering received in design and build as follow;

a) Reduce risks of cost overruns and delays due to better time and cost management;
b) Increase concern for and commitment to the user needs as a basis for design;
c) Effective overall project management in design and build project;
d) Open communication with familiar parties involved;
e) increased degree of trust and parties;
f) Financially successful projects completion on time, cost and quality.

1.2 Design and build

Design and build system of procurement are not new in the Malaysian construction industry. This kind of procurement method mostly practice in the public sector project as compare to the private sector. Turner [5] described design and build as the procurement practice where one organisation is responsible to the client for both design and construction. The elements of the design and build system consist of the following;

a. Establishing the need to build
b. Establishing the client's requirement
c. Selecting and inviting tenderers to bid
d. The contractor preparing their proposals for design, time and cost
e. Evaluation and acceptance of a tender which then becomes a contract
f. Design and construction of the works

From the above, showed that in design and build procurement, the client need to have in-house skills to ensure the client's requirement is clearly identified. Scriven [4] explained that the intention of design and build contract is to place the entire responsibility for the construction to the contractor, if any problems arise in project, the contractor is responsible, subject to specific exceptions. Others benefits achieved through this procurement approach in terms of well management of cost control, completion period is faster, less dispute and others. As compare to the adversarial relationship practice in traditional approach, it is more difficult to ensure value for money due to the un-familiar contractor involved in the construction project. The primary reason for applying the design–build delivery method is to potentially shorten the project's duration [6]. Design–build in the public sector is mostly used for schools, hospital, government office or highways to achieve a targeted completion date that cannot be achieved through traditional methods.

1.3 Partnering concept in Design and Build Procurement approach

Partnering occurs in design and builds throughout continuity of working repeatedly by the contractor for the same clients. Normally, the clients were only searching reputable contractor to be involved in the design and build the project from successful previous project completion. [1] stated that partners must have confidence in each other's organization, and each organization needs to have confidence in its own team, which means careful selection of the people involved in the development of the client's project. Figure 1.0 show the concept of partnering practice in the design and build procurement approach.

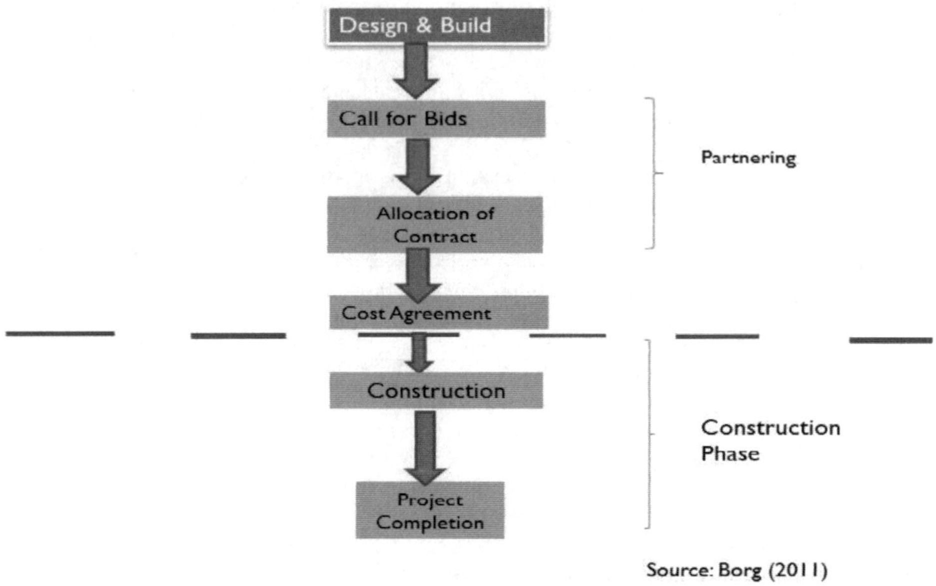

Figure 1. The concept of Partnering in the Design and Build

Partnering occurs during the selection process of the contractor in client's development project. Therefore, the client should look into who are the reputable contractor not only focus on the financial background but on the aspect of trusting people involved in client's project. The maximum benefits of partnering are realized when all parties in the chain of partners cooperate.

2. Research methodology

Research methodology is a logical way to solve a problem and involves the procedure by which researchers carry out their work for the purpose of describing, explaining and predicting phenomena are called research methodology [7]. In this research, the method adopted in this study used is a quantitative approach. Quantitative analysis is a scientific approach to managerial decision making whereby raw data are processed and manipulated resulting in meaningful information. The illustration of this method employed as follow;

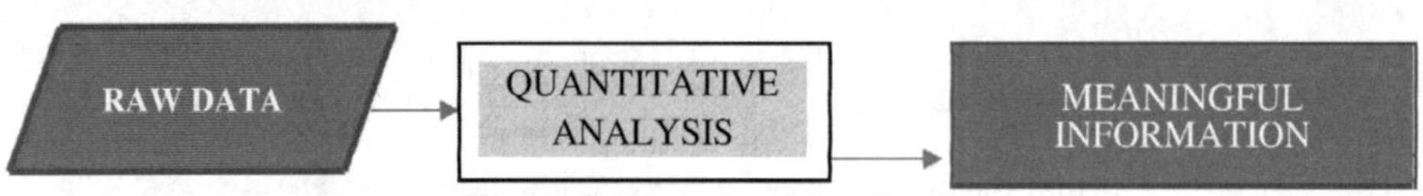

Figure 2. Method employed

Therefore, the crucial need to use this method due to the result received represent the whole design and build project practice in the public work department.

2.1 Data collection

Data collection is the process of gathering information which then forms the database for determining what is needed. The process is essential to ensure that information regarding scope, quality, time and cost is gathered in a way which provides easy access for the researcher to fulfill the research need [8]. This research is to explore the concept of partnering implementation practiced in the design and build procurement approach.

2.2 Sampling

In this research, the method use for sampling is purposive sample. The respondent involve are from the public client. This selection was based on data gathered from the public client, which included the following ministries and departments:

a) Ministry of Finance
b) Ministry of Home Affairs
c) Ministry of Youth and Sport
d) Ministry of Health
e) Ministry Rural and Regional Development
f) Ministry of Women and Family Development
g) Ministry of Transportation
h) Ministry of Agriculture
i) Ministry of Human Resources
j) Prime Minister's Department
k) Ministry of Education (Four Departments)
l) Ministry of Science & Technology & Innovation
m) Ministry of Defence
n) KWP-Perbadanan Putrajaya, DBKL
o) Ministry of Works (Nine Departments)

2.3 Response rate

Administration of the questionnaire was distributed to the public client as listed above by using postal questionnaire. All of this clients involve in the design and build project in the construction industry. By using the purposive sample, only 26 questionnaires was distributed to each of these 26 ministries and department. At the early stage, the response rate is low which is only one respondent replied. However, the researcher took the alternative approach to send a questionnaire through by hand. Lastly, all of the respondents replied to the questionnaire and 100% received from a total of 26 ministries department.

3. Result

The result of the study was measured for the element of partnering need by the client and type of evaluation consider by the client to evaluate the performance of contractor using partnering approach in the design and build project.

Table 1 shows the elements of partnering practice between client and contractor. The result indicates that the client revealed that project management capability at mean 4.27 at highest rank among others factor. It shows that in design and build project, project management capability as primary factor client looking for partnering in design and build project. This is due to the client consider financial and technical risk in project management capability to ensure the selected contractor is capable to manage a design and build project. Follow by good communication at mean 4.26 and timely project goals at mean 4.25. In design and build project communication between both parties is crucial to ensure the project achieved is timely project goals.

Table 1. Elements of partnering with contractors in design and build projects

Factors	Mean	Rank
1. Image and reputation; satisfying client needs	4.09	8
2. Innovative ways of using technology in construction	4.00	9
3. Timely project goals	4.25	3
4. Good communication	4.26	2
5. Project management capability	4.27	1
6. Commitment	4.18	6
7. Early involvement of contractors at the design stage	4.22	5
8. Trust	4.14	7
9. Learning culture/exchange of knowledge	4.23	4

Table 2 shows the evaluation of performance contractor in design and build project by the public client. The result indicates the commitment to budget at mean 4.90, in design and build project the client looking for the design professional that commit to the budget proposed. Other than that, supply chain management at mean 4.86 is considered by the client to ensure successful implementation of design and build construction project. Therefore, satisfaction with the project outcomes received through the design professional selected which practice a systematic approach of supply chain management in the organization is considered.

Table 2. Evaluation Performance of Contractors by Public Client

Performance Measure	Mean	Rank
1. Good progress report	4.57	3
2. Key performance indicator	3.95	5
3. Environment and sustainability	4.33	4
4. Project management capability	3.86	6
5.Commitment to budget	4.90	1
6. Quality	3.82	7
7. Supply chain management	4.86	2

4. Conclusion

The findings indicate that to secure in design and build construction project from delay and cost overrun, the importance of client to carry out the relationship with an appropriate design professional through partnering approach. The evaluation of the performance of the previous project by the design professional creates trust between both parties. Therefore, outcomes of the results, the public clients stated that project management capabilities and commitment to budget as a crucial element of partnering from the design professional in design and build construction project. These contribute to the successful implementation of a high-risk project in design and build construction approach.

References

[1] Cartlidge Duncan,(2009). Quantity Surveyor's Pocket Book. 1st Edition Elsevier.

[2] Cox, A. & Thompson, I. (1997). 'Fit for purpose' contractual relations: Determining a theoretical framework for construction projects. *European Journal of Purchasing and Supply Management, 3*(3), 127–135.

[3] Bennett, J. & Jayes, S. (Eds.). (1998). The seven pillars of partnering: A guide to second generation partnering. London: Thomas Telford.

[4] Scriven, J., Pritchard, N. & Delmon, J. (Eds.). (1999). *A construction guide to major construction projects*. London: Sweet & Maxwell.

[5] Turner A. (1997). Building Procurement. 2nd Edition Macmillan

[6] Cho, K., Hyun, C., Koo, K. & Hong, T. (2010). A partnering process model for public-sector fast-track design-build projects in Korea. *Journal of Management in Engineering, 26*(1), 19–29.

[7] S.Rajasekar,P.Philominathan & V.Chinnathambi (2013). *Research Methodology.*

[8] Wiersma, W. & Jurs, S. G. (Eds.). (2005). *Research Methods in Education* (8th ed.). Boston, MA: Pearson

External shading devices for energy efficient building

M S Shahdan[1], S S Ahmad[1*] and M A Hussin[1]
[1]Faculty of Architecture, Planning and Surveying, Universiti Teknologi MARA Selangor, Puncak Alam Campus, 42300 Puncak Alam, Selangor, Malaysia

*sabrin63@gmail.com

Abstract. External shading devices on a building façade is an important passive design strategy as they reduce solar radiation. Although studies have proven the benefits of external shading devices, many are designed solely for aesthetic purposes without fully considering its high potential to reduce solar radiation and glare. Furthermore, explorations into shading devices by the design team are mostly left too late in the design development phases. Hence, the paper looks into the effectiveness of external shading devices on a building towards more energy efficient building. The study aims to analyse the effects of various configurations of external shading devices towards the energy consumption of a case study building based on computer simulations. This study uses Building Information Modelling (BIM) through Autodesk Revit software as simulation tool. The constant variables for the simulation are the orientation of the building, types of glazing used by the building and the internal loads of the building. Whereas, the manipulated variable is the types of shading device used. The data were sorted according to the categories and translated into a chart. Analysis of the findings indicate that shading devices with different configurations show significant results in the energy consumption and the best configuration is the egg-crate shading devices. The study recommends that the consideration for shading device as a passive design strategy needs to be developed at the early stage of the building design.

1. Introduction

Energy efficiency is now recognised as one of the fastest and suitable approaches to reduce any energy related emissions associated with, air pollution, global warming and climate change [1]. Buildings are expected to achieve energy efficient status by most nations, which could be achieved through active or passive design strategies. The biggest percentage of energy consumption for any building in hot-humid climate is for air conditioning followed by artificial lighting [2]. Besides enhancing the aesthetic appearance to the buildings, shading devices on the building facades help to minimise energy consumption in the building. Several researchers have conducted studies on improvements and impact of shading devices on the building energy consumption where several types of passive design strategies can be implemented.

Numerous strategies can be applied to minimise energy consumption in a building such as better OTTV values for walls and external shading devices. A study in Hong Kong reported that an energy efficient envelope design could save as much as 35% of total the cooling demands [3]. Another study in Malaysia confirmed that high-rise office buildings use 46% of total energy use for cooling [4]. External shading devices can either be made of building materials or living plants and could reduce the indoor cooling load by 30% [5]. This study also confirmed that shading devices are more effective to reduce cooling energy load compared with the use of high performance glazing [5]. Even for plant based external shading, a study in Penang with vertical green system showed a façade surface temperature reduction of up to 6.4°C [6].

Shading devices, which is a component of a building envelope, performs a crucial role to give positive influences towards energy efficiency in buildings [7, 8]. Generally, shading devices are used to protect inner spaces from direct solar gain through openings, windows and large glazed surfaces.

Different climatic region needs different configuration of shading devices. Buildings in hot-humid climate need to reduce solar radiation and sunlight penetration into the building. On the other hand, for cold climate, it is critical to let the sunlight enter into the building or the envelope material absorb the solar radiation to keep the warmth within the building especially in winter.

2. Problem Statement

Shading of buildings as a means to reduce energy consumption needs careful consideration. Numerous studies explored the use of shading devices, for example between 10% and 11.3% of energy savings can be expected from the benefit of external shading devices in hot and humid climates [9, 10]. However, most shading devices are designed for aesthetic purposes rather than their energy savings potentials. Designers mostly ignore findings on the importance of shading devices as they leave the analysis too late in the design development phases. Shading devices need careful consideration at the early design stage especially for facades with high window to wall ratio [11, 12].

This paper aims to study the effectiveness of external shading devices for more energy efficient building in the hot-humid climate of Malaysia. This paper analyses the effects of various configurations of external shading devices towards the energy consumption of a school building based BIM simulations. The significance of this research is to understand through how the various configurations of external shading devices can contribute to lessen cooling energy consumption.

3. Methodology

This paper employs the quantitative approach for data collection through Autodesk Revit Building Information Modelling (BIM) simulations. Published articles on studies into shading devices and energy savings potentials have explored other simulation tools besides Revit [13, 14, 15, 16] with encouraging results. Simulations were conducted to gather accurate values of the energy consumption of a case study building based on various configurations of external shading devices. Through the simulations outputs, the readings collected through the simulation are reliable primary data collection. The detailed information of the methodology is discussed in this section.

3.1. Building Information Modelling (BIM) Simulation through Revit

To determine the effectiveness of the external shading devices, a strategy was introduced based on Building Information Modelling (BIM) simulations using the validated Revit software. BIM is a smart 3D modelling process that provides an integrated and comprehensive architectural, engineering, and construction (AEC) tools for more proficient planning, design, construction, and operation management of buildings and infrastructure for architects, engineers and construction managers [17]. This technology enhances decision-making and performance across the building and infrastructure lifecycle.

The case study selected for the simulation is a 3-storey school building located in Shah Alam, Malaysia. Table 1 shows the total floor areas of air-conditioned and non-air-conditioned spaces. The ground floor houses several classrooms, conference rooms, cafeteria and other mechanical and technical rooms. The first floor has several classrooms, drafting room, library, computer labs, and administration office while the second-floor houses classrooms, media room and administration office (Figure 1-3).

For this simulation, seven shading device models are considered as variants, which include the basic forms of shading devices that are typically used in buildings. The behaviours of the shading provided by these models demonstrate the influence of each configuration of shading device on the cooling loads of the building, which would lead to the understanding of the devices efficiency.

IOP Conf. Series: Earth and Environmental Science **117** (2018) 012034 doi:10.1088/1755-1315/117/1/012034

Table 1. Total air-conditioned and non-air-conditioned floor areas.

Floor level	Total A/C floor area (m2)	Total non A/C floor area (m2)
Ground floor	1386.08	130.02
First floor	1320.35	334.84
Second floor	1462.22	203.97
Total	4168.65	668.83

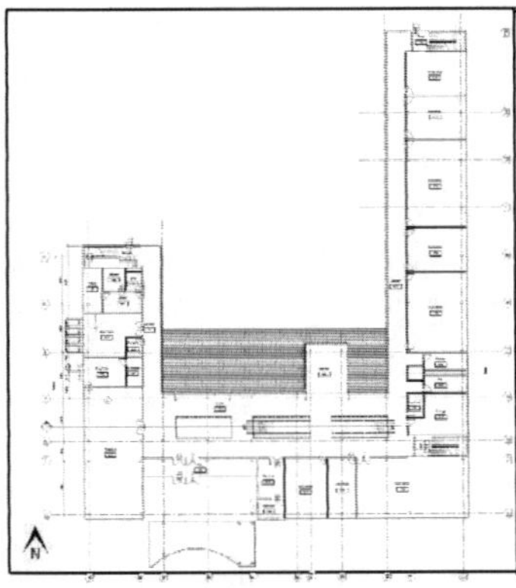

Figure 1. Ground Floor Plan

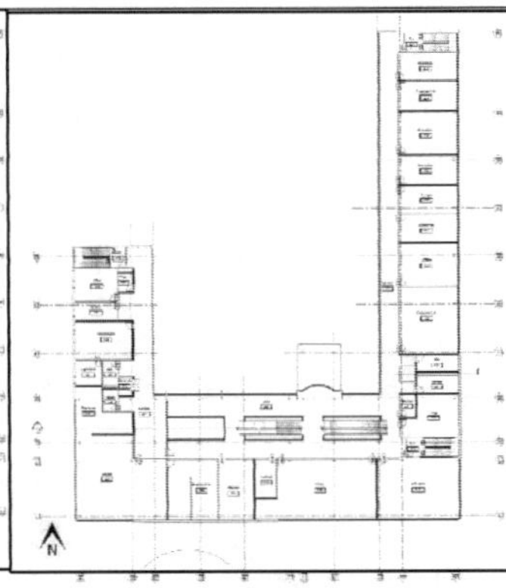

Figure 2. First Floor Plan

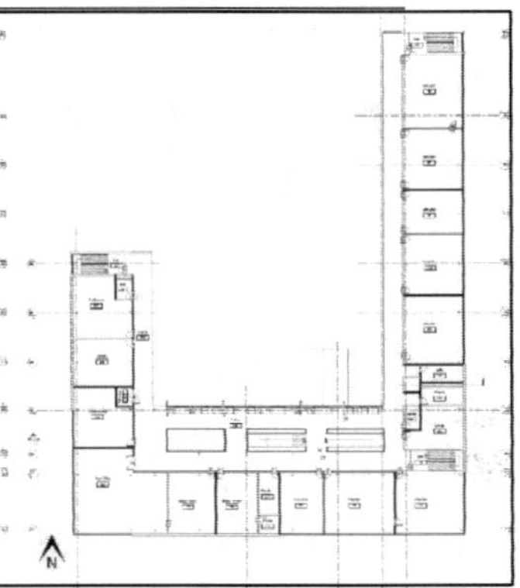

Figure 3. Second Floor Plan

As shown in Table 2, the options of external shading devices are:

i. No shading device
ii. Horizontal single panel shading device
iii. Horizontal double panel shading device
iv. Horizontal inclined double panel shading device
v. Horizontal louvers shading device
vi. Vertical fins shading device
vii. Vertical slanted fins shading device
viii. Egg-crate shading device

The depth of the size and the material of these shading devices were fixed for the simulation as different types of material and sizes of the shading could cause inconsistencies in the result.

Table 2. Eight types of external shading devices used in the simulations.

No shading device (option1)	Horizontal single panel (option 2)	Horizontal double panel (option 3)	Horizontal inclined double panel (option 4)
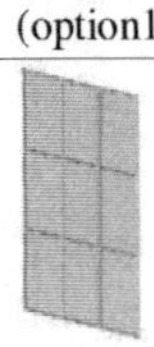	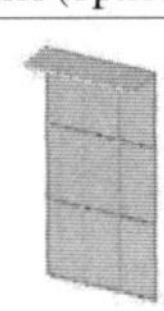	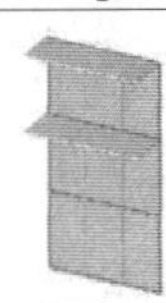	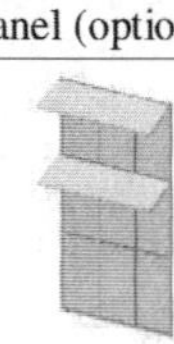
Horizontal louvers (option 5)	Vertical fins (option 6)	Vertical slanted fins (option 7)	Egg-crate (option 8)

IOP Conf. Series: Earth and Environmental Science **117** (2018) 012034 doi:10.1088/1755-1315/117/1/012034

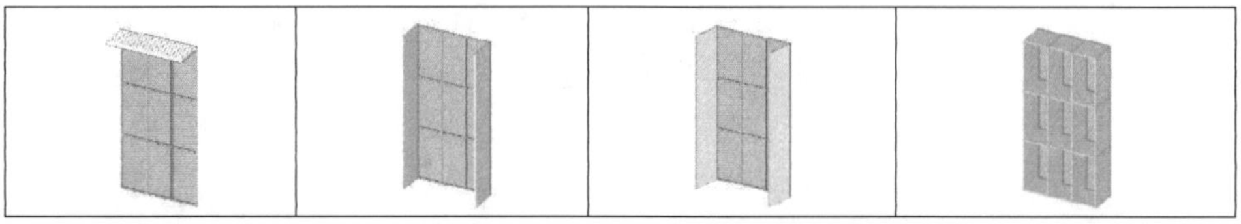

In this simulation, several variables were selected to ensure the most effective result. The following variables remain constant to make sure that the results are accurate and reliable on all categories:

i. Internal loads of the building
ii. Orientation of the building
iii. Type of glazing

The dependent variables are the types of shading devices. The efficiency of the external shading devices was determined according to their capability to reduce the cooling load of the building in comparison with the base case condition without any shading on the facade. The less energy needed to reduce the cooling load indicates the more efficient external shading device.

3.2. Simulation Process

The simulation process involves the application of several steps to ensure that the process is clearly well thought and reliable results can be achieved. The details of the process for the first variable, which is the base case (walls without any shading devices) then followed by other types of shading devices, are as follows:

Step 1: Model/draw of the building, which includes the floor plans, elevation and section in scale of 1:500. The total floor area for air-conditioned space is 4168.64 m_2 and the total floor area for non-air-conditioned space is 668.83 m_2.

Step 2: Choose the orientation, location, date and weather data. The orientation of the building is as shown in Figures 1-3 and the building is located in Shah Alam, Malaysia and the month for the simulation is March.

Step 3: Choose the building materials from the library or add materials. These materials must be constant for all simulation since different types of material might affect the results.

Step 4: Run the simulation for energy efficiency based on façade without any shading device (Table 2 option 1) as base case study. Record the results.

Step 5: Apply the second type of shading device (Table 2 option 2) onto the building and record the results.

Step 6: Repeat step 5 with applying different types of shading devices (Table 2 options 3 –8) onto the building and record the results.

Step 7: Repeat steps 2 – 6 with different month, namely September.

Step 8: Tabulate the results and translate them into a bar chart. Discuss and analyse the results.

Table 3. Parameters and conditions used in the simulation.

Month of simulation	March and September
Orientation	North-South
Total floor area	4837.48 m^2
Storey	Three storeys
Shadin g configurations	Eight types of shading configuration
Simulation for	Monthly energy consumption for cooling load
Location	Shah Alam, Selangor

4. Results

The results from the simulation through Revit software are collected and interpreted into a table and bar charts for ease of reference and comparisons. The results are then analysed to get the findings and later concluded. Some recommendations are introduced based on the findings at the end of the report. Table 4 shows all the results for simulations on the types of shading devices applied on the wall facades.

Table 4. Electricity consumption for cooling system for the school building in March and September 2017

Types of shading devices	Electricity used for cooling (kWh) in March	Electricity used for cooling (kWh) in Sept	% of reduction in Sept from March
None (base case)	45,337	42,898	5.38
Horizontal single panel	43,882	41,392	8.70
Horizontal double panel	33,433	31,491	30.54
Horizontal inclined double panel	32,204	29,542	34.84
Horizontal louvers	40,725	38,399	15.30
Vertical fins	41,808	39,259	13.41
Vertical slanted fins	40,000	37,476	17.34
Egg-crate	25,475	22,834	49.63

Table 4 highlights the comparison of cooling energy consumption for the school in March and September. Table 4 shows the most effective shading device for March is the egg-crate shading device as it reduces the energy consumption for cooling from 45,337.67 kWh to 25,475.75 kWh. In March, the total energy saved by installation of egg-crate shading devices is 19,861.92 kWh. The second most efficient shading device for March is the horizontal inclined double panel, while the least efficient shading device is horizontal single panel, which only save about 1,455.09 kWh of electricity.

Table 4 also indicates the cooling energy used for the school in September where the egg-crate shading device is still the most effective shading device with a 49.63% reduction in cooling energy consumption from the March base case of no shading (Table 4). The least effective is the horizontal single-panel shading device with only an 8.7% reduction in cooling energy consumption from the March base case of no shading installed.

Generally, from Figure 4 we can see a similar pattern of energy consumption for March and September, although all devices perform incrementally better in September. From both months of simulations, the most effective shading device is the egg-crate shading device since it consumes the lowest energy for cooling system for the school.

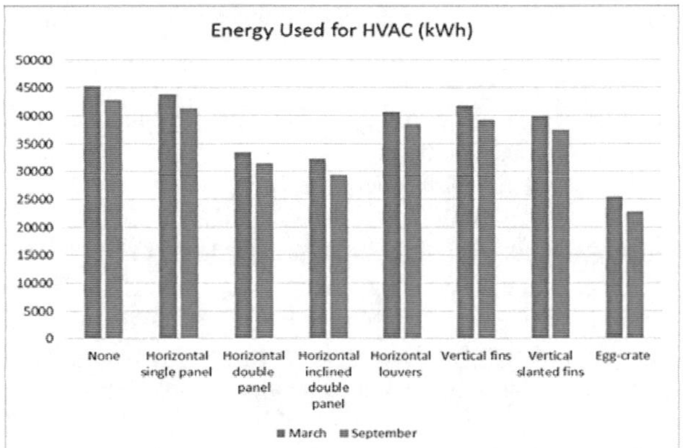

Figure 4. Comparison of energy used for cooling for March and September

The results of this study are in agreement with published work on shading devices. A simulation study in Milan during summer showed a 20% reduction in energy consumption [13] when shading devices were installed on the building. A simulation study in South Korea on external shading devices also revealed that the experimental shading device with angled slats gave the most efficient performance and offers better views from indoors [14]. The effects of louvered shading devices on building energy consumption were studied across Mexico, Cairo, Lisbon, Madrid and London [15] using TRYNSYS and EES simulation tools. The results showed that the external louvered shading devices provided comfortable indoor conditions and led to significant energy savings compared with a building without shading devices. However, the significance of this study is that the exploration into egg crate shading devices proved to be more effective than other types of shading devices studied earlier.

5. Conclusion

In conclusion, this paper has achieved its objective which is to analyse the effects of various configurations of external shading devices towards the energy consumption of the building based on computer simulations. The BIM simulation clearly shows that the egg-crate shading device is the best solution for hot-humid climate area where it combines both horizontal and vertical shading devices. For hot-humid climate region, it is crucial to keep the building shaded from sunlight to minimise heat penetration into the building. All the tested shading devices give impact on the energy consumption of the building according to the types of shading device installed onto the building facades. This study can be further explored by varying the orientations and the building form, the size or depth of the shading devices as well as the materials of the shading devices as different types of materials have different thermal insulation. As a recommendation, it is important to take into consideration the design criteria of each shading devices prior to its application on the façade of the building. External shading devices will act as a passive design strategy for the building thus, planning for this at the early phase of the design is very important.

Acknowledgements
The funding for the publication of this paper is made possible by UiTM Bestari research grant scheme 600-IRMI/DANA 5/3/BESTARI (040/2017).

References

[1] Dincer, I 1998 Energy and environmental impacts: present and future perspectives. *Energ. Source.* **20**(4-5) 427-53
[2] Iqbal I, Al-Homoud MS 2007 Parametric analysis of alternative energy conservation measures in an office building in hot and humid climate. *Build. Environ.* **42**(5) 2166-77
[3] Chan KT, Chow WK 1998 Energy impact of commercial-building envelopes in the sub-tropical climate *Appl. Energ.* **60**(1) 21–39
[4] Lau AK, Salleh E, Lim CH, Sulaiman MY 2016 Potential of shading devices and glazing configurations on cooling energy savings for high-rise office buildings in hothumid climates: the case of Malaysia. *Int. J. Sust. Built Environ.* **5**(2) 387-99
[5] Balaras CA, Droutsa K, Argiriou AA, Asimakopoulos DN 2000 Potential for energy conservation in apartment buildings. *Energy. Buildings.* **31**(2) 143-54
[6] Basher HS, Ahmad SS, Rahman AMA, Qamaruz-Zaman N 2016 The use of edible vertical greenery system to improve thermal performance in tropical climate. *J. Mech. Eng.* **13**(1) 58-
[7] Nielsen MV, Svendsen S, Jensen LB 2011 Quantifying the potential of automated dynamic solar shading in office buildings through integrated simulations of energy and daylight. *Sol. Energ.* **85**(5) 757-68

[8] Ali-Toudert F, Mayer H 2006 Numerical study on the effects of aspect ratio and orientation of an urban street canyon on outdoor thermal comfort in hot and dry climate. *Build. Environ. 41(2) 94-108*
[9] Wong NH, Li S 2007 A study of the effectiveness of passive climate control in naturally ventilated residential buildings in Singapore *Build. Environ.* **42** 1395-1405
[10] Yu J, Yang C, Tian L 2008 Low-energy envelope design of residential building in hot summerand cold winter zone in China. *Energ. Buildings.* **40**(8) 1536-46
[11] Cellai G, Carletti C, Sciurpi F, Secchi S 2014 Transparent building envelope: windows and shading devices typologies for energy efficiency refurbishments. In *Build. Refurb. for Energ. Perform. 2014* 61-118 Springer International Publishing
[12] Echenagucia TM, Capozzoli A, Cascone Y, Sassone M 2015 The early design stage of a building envelope: Multi-objective search through heating, cooling and lighting energy performance analysis. *Appl. Energ*. **154** 577-91
[13] Bellia L, De Falco F, Minichiello F 2013 Effects of solar shading devices on energy requirements of standalone office buildings for Italian climates. *Applied Thermal Engineering. 54*(1) 190-201
[14] Kim G, Lim HS, Lim TS, Schaefer L, Kim JT 2012 Comparative advantage of an exterior shading device in thermal performance for residential buildings. *Energy and buildings, 46* 105-111
[15] Palmero-Marrero AI, Oliveira, AC 2010 Effect of louver shading devices on building energy requirements. *Applied Energy. 87*(6) 2040-2049
[16] Kirimtat A, Koyunbaba BK, Chatzikonstantinou I, Sariyildiz S 2016 Review of simulation modeling for shading devices in buildings. *Renewable and Sustainable Energy Reviews. 53* 23-49.
[17] Eastman CM, Eastman C, Teicholz P, Sacks R 2011 *BIM handbook: A guide to building information modeling for owners, managers, designers, engineers and contractors.* John Wiley & Sons

IOP Conf. Series: Earth and Environmental Science **117** (2018) 012035 doi:10.1088/1755-1315/117/1/012035

Comparative study of landslides susceptibility mapping methods: Multi-Criteria Decision Making (MCDM) and Artificial Neural Network (ANN)

Salleh S. A.[1], Abd Rahman A. S. A[1], Othman A. N[1] and Wan Mohd W. M. N[1]

[1]Applied Remote Sensing and Geospatial Research Group
Centre of Studies for Surveying Science and Geomatics, Faculty of Architecture, Planning and Surveying, Universiti Teknologi MARA, 40400 Shah Alam, Selangor.

*aekbal@salam.uitm.edu.my

Abstract. As different approach produces different results, it is crucial to determine the methods that are accurate in order to perform analysis towards the event. This research aim is to compare the Rank Reciprocal (MCDM) and Artificial Neural Network (ANN) analysis techniques in determining susceptible zones of landslide hazard. The study is based on data obtained from various sources such as local authority; Dewan Bandaraya Kuala Lumpur (DBKL), Jabatan Kerja Raya (JKR) and other agencies. The data were analysed and processed using Arc GIS. The results were compared by quantifying the risk ranking and area differential. It was also compared with the zonation map classified by DBKL. The results suggested that ANN method gives better accuracy compared to MCDM with 18.18% higher accuracy assessment of the MCDM approach. This indicated that ANN provides more reliable results and it is probably due to its ability to learn from the environment thus portraying realistic and accurate result.

1. Introduction

The landslide is one of the natural disasters that are significantly related to our country, multiple cases are reported throughout the entire country. The landslide is defined as the movement of a mass of rock, debris, or earth down a slope [1]. In a simple word, it is basically a type of mass wasting that relates to downslope movement of mass such as soil or rock with the influence of gravity.

It is usually triggered by the movement of the slope that forces the down-slope movement with the mass and surface structure that exist on the slope [2]. Other than that, climate changes also play a role and impact in triggering landslides. Other factors related in triggering the occurrence of landslides in an earthquake, heavy rainfall and other geological causes [3]. Each time landslide occurs, it changes the structure and surface cover of the area. Besides that, it causes a massive loss in terms of property damage, effect on resources and biodiversity and also it may lead to injury and death [4].

Back in 1993, a landslide event occurred at Highland Towers affecting hundreds of people killing more than 50 people. Based on the studies done by the agencies involved such as Ampang Jaya Municipal council (MPAJ) and Jabatan Kerja Raya (JKR), it is found that one of the factors that contribute to the landslide events is the presence of rainfall besides issues with the inadequate constructed drainage. During the month of November to January, rainfall season is the most intense period that leads towards landslides, mudslides and floods. During this period, flood and landslide occurred at a rapid rate and risk area such as Kuala Lumpur is vulnerable to flash flood and landslide event. The existence of rainfall triggered the movement of geology at the location, making it loosen and eventually fall [5].

Landslide is not something new and mitigation action needs to be taken in order to prevent it from occurring again. Mapping the risk area is crucial in determining susceptible zones to help in decision making for urban planning and management. One of the popular methods in mapping susceptible zones is by using Multi-Criteria Decision Making (MCDM) and also Artificial Neural Network (ANN). The approach taken by the government is the establishment of the National Slope Master Plan (NSMP) that provides detailed elements in a comprehensive and effective way to describe the national policy. It includes policy, strategy and also action plan in reducing the risk of landslide events. It focuses on nationwide scopes that started back in 2009 till 2023 [6].

1.1. Kuala Lumpur geographical characteristics

Kuala Lumpur is the city centre of Malaysia which located in the centre state of Selangor. It is also known as Klang Valley and surrounded by various types of topographic features such as Titiwangsa Mountains in the east side, several ranges on the north and south while Malacca Straits on its west. The city has a tropical rainforest climate which is warm and sunny alongside abundant rainfall. The rainfall season usually occurs during northeast monsoon period of October to March. Due to monsoon season, Kuala Lumpur is frequently struck by flood and landslide occurrence especially during intense rainfall.

2. Methodology

The conceptual framework of the study is summarised in Figure 1. It consists of six phases beginning with the process of determining of issues, literature review, and selection of criteria, data collection, data processing and analysis. Each of the phases describes the process in achieving the objectives of the study.

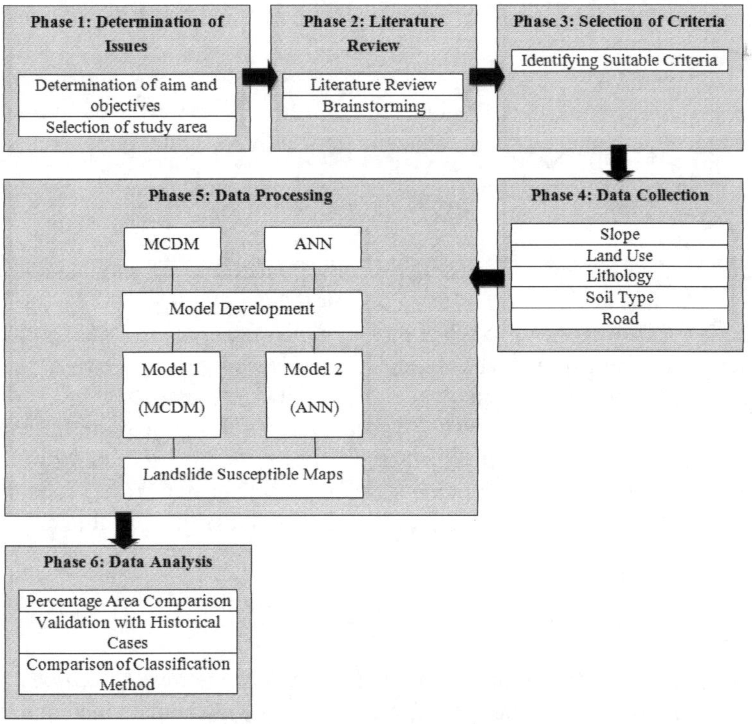

Figure 1. Conceptual framework

IOP Conf. Series: Earth and Environmental Science **117** (2018) 012035 doi:10.1088/1755-1315/117/1/012035

2.1. Data Collection

Based on discussion with various experts and government agencies, 5 selected criteria were used, slope, land use, lithology, soil type and proximity to road. The criteria involve in this study is collected from Kuala Lumpur City Council (DBKL). The slope is generated from the DEM surface that is based on LiDAR data while other layers acquired are in vector form. Besides that, previous location and historical data is captured based on landslide occurrence and previous study. Land use data are classified into six classes such as agriculture, grassland, forestry, rubber, unused land and urban area. While for lithology layer, it consists of five types of lithology that exist in the study area (i.e. acid intrusive, limestone, phylite, slate, shale and sandstone, schist, schist, phylite, slate and limestone). The soil type in the study area consists of several types of soil such as mined land, munchong, rengam, serdang, steepland, telemong and urban land. The proximity to the road is categorized into five classes, each with an interval of 40 meters (0-40m, 40-80m, 80-120m, 120-160m, 160-200m).

2.2. Data Processing

2.2.1. Multi Criteria Decision Making (MCDM) Technique. In MCDM, there are various techniques that can be used in assigning weightage to the criteria. One of the techniques is by ranking method, Rank Reciprocal (RR). In RR, it uses the technique of deriving the weighted value from normalize reciprocals of a criterion rank. It uses a formula to calculate each of the criteria weight. The formula of Rank Reciprocal is shown below, where wi is the weight value for each criteria and rj is the rank of each criteria [7].

$$wi = [1/rj\,] \,/\, \Sigma\, (1/\,rk) \tag{1}$$

Based on interviews and discussions with expert, the score for each of the criteria is obtained. From the score, weight is calculated based on formula to perform a model. The model for landslide susceptible zones is shown in Equation 2.

$$LHZ = (0.438 * s\_slp) + (0.109 * s\_rd) + (0.146 * s\_lith) + (0.219 * s\_lu) + (0.088 * s\_st) \tag{2}$$

Where *s_slp* is the standardized score for slope sub critera, *s_rd* is the standardized score for proximity to road sub criteria, *s_lith* is the standardized score for lithology sub criteria, *s_lu* is the standardized score for land use sub criteria and *s_st* is the standardized score for soil type sub criteria.

2.2.2. Artificial Neural Networks (ANN). ANN uses experience and training sites to learn from and produce results. The key element of the process is the structure of information processing system [8]. The methods are configured basically for a specific purpose and application such as pattern recognition or data classification. It learns from the biological systems and performs adjustments to the connection that exist between neurons. The model adopts the changes in the structure based on the information then flows through the network during its learning phase. ANN requires a learning algorithm to perform the model network [9]. Training sites are essential, the training site method consists of two, supervised and unsupervised. For supervised training method, the input and output of the study are provided to the network. It will then process both data and compare the output against the desired outputs. From the comparison process, an error is determined and propagated back through the system. After that, the system will then readjust the weight assigned to the inputs until the weight is reliable.

3. Results and Discussions.

The landslide susceptible maps from both methods are shown in Figure 2. The results are overlaid with historical cases and it is classed into five classes (i.e. no risk, low risk, medium risk, high risk and very high risk). From the result, the total area was compared.

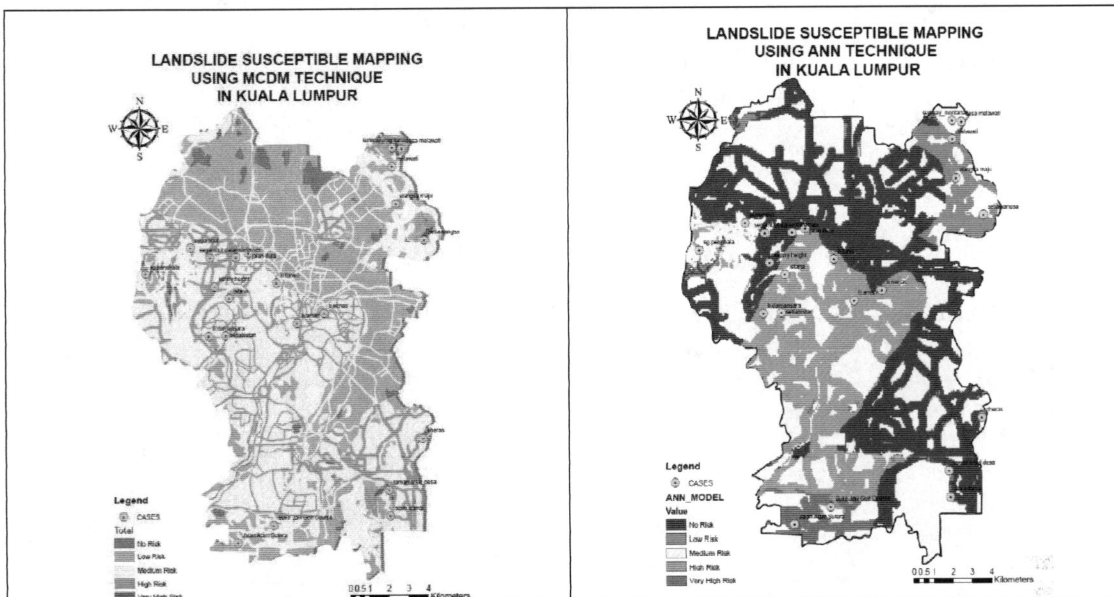

Landslide Susceptible Mapping Using MCDM Method	Landslide Susceptible Mapping Using ANN Method

Figure 2. Landslide susceptible mapping using MCDM and ANN methods.

3.1. Comparison of Risk Area between MCDM and ANN

Table 1 explains the area covered for all risk classes using both methods. There are differences in risk area coverage between MCDM and ANN methods. For example, in medium risk, MCDM covers a total of 115.57 square kilometres while ANN method covers only 5.97 square kilometres.

Table 1. Comparison of Area for MCDM and ANN Methods

Risk Ranking	MCDM [Area (sq km)]	ANN [Area (sq km)]
No Risk	10.438702	87.144128
Low Risk	80.391987	11.542029
Medium Risk	115.574903	5.974561
High Risk	36.929774	49.259483
Very High Risk	1.213424	6.205911

3.2. Accuracy Assessment between MCDM and ANN with Historical Cases

Table 2 below shows the comparison of results between both methods in terms of historical cases. The result from both methods is compared to see the differences of zoning for the previous cases location. The classification method used for both result is Natural Break classification [10]. It shows that MCDM technique has the accuracy of 50% comparing to historical cases while ANN technique scores 68.18% of accuracy comparing to historical cases.

Table 2. Comparison of Zoning Classification for MCDM and ANN Method

Name of Cases	Zoning	MCDM Zoning	ANN Zoning
Bukit Tunku	High	Medium	High
Cheras	Very High	Very High	Very High
Bukit Damansara	High	Medium	High
Bukit Nenas	High	Medium	High
Segambut	High	High	High

Name of Cases	Zoning	MCDM Zoning	ANN Zoning
Segambut Dalam	Low	High	Low
Setiabistari	High	Medium	High
Setiawangsa	Medium	Medium	Medium
Taman Melawati	Medium	Medium	Low
Wangsa Maju	High	High	Medium
Kenny Height	Low	High	Low
Jalan Duta	High	Low	High
Taman Orkid Desa	Very High	Very High	Very High
Istana Negara	Medium	Medium	-
Bukit Aman	High	High	High
Sunway Montana	Low	Very High	-
Dutamas	High	High	High
Alam Damai	Medium	Medium	No Risk
Sungai Penchala	Very High	Very High	High
Desa Melawati	Low	High	-
Jalan Alam Sutera	Very High	Low	Very High
Bukit Jalil Golf Course	High	Medium	High
	Percentage	50%	68.18 %

Table 3 below shows the comparison of results by classification methods with historical cases. The result from both methods is compared to see the differences of zoning for the previous cases location. The classification method used for both result is Quantile classification technique. From the result, it shows that MCDM technique has the accuracy of 54.54% comparing to historical cases while ANN technique scores 68.18% of accuracy comparing to historical cases.

Table 3. Comparison of Zoning Classification for MCDM and ANN Method using Quantile Classification Technique

Name of Cases	Zoning	MCDM Zoning	ANN Zoning
Bukit Tunku	High	High	High
Cheras	Very High	Very High	Very High
Bukit Damansara	High	Medium	High
Bukit Nenas	High	Medium	High
Segambut	High	High	High
Segambut Dalam	Low	High	Low
Setiabistari	High	Medium	Very High
Setiawangsa	Medium	Medium	Medium
Taman Melawati	Medium	Medium	Medium
Wangsa Maju	High	High	Medium
Kenny Height	Low	High	Low
Jalan Duta	High	No Risk	High
Taman Orkid Desa	Very High	Very High	Very High
Istana Negara	Medium	Medium	-
Bukit Aman	High	High	High
Sunway Montana	Low	Very High	-
Dutamas	High	High	Low
Alam Damai	Medium	Medium	Low
Sungai Penchala	Very High	Very High	Very High
Desa Melawati	Low	High	-
Jalan Alam Sutera	Very High	Low	Very High
Bukit Jalil Golf Course	High	Medium	High

Name of Cases	Zoning	MCDM Zoning	ANN Zoning
	Percentage	54.54 %	68.18 %

4. Conclusion and Recommendations

Both MCDM and ANN can produce reliable results in mapping susceptible zones for landslide event. Each method has its own uniqueness in processing data to achieve the aim and objective of the study. The results indicated that ANN method can provide more reliable as it portrays much realistic information through its result based on its ability to learn from the environment as compared to MCDM where the result is generalized at certain area. Through the process of accuracy assessment, ANN produces better accuracy with 68.18% as compared to the MCDM method with 50% accuracy. It was suggested that accurate data such as land use data will produce more accurate results. Other than that, comparing results from both techniques with other methods such as Fuzzy Membership in order to validate the results obtained. Perhaps, by having more historical cases, it will also help in determining the training sites for ANN method while it can be used to validate the model and the result accurately.

Acknowledgments

The authors would like to thank Kuala Lumpur City Council (DBKL) for providing various data and valuable advice for this research, also to the Malaysia Meteorological Department that provides reports and documents in completing this study.

References

[1] Cruden, D. M. (1991). A simple definition of a landslide. *Bulletin of the International Association of Engineering Geology, 43*(1), 27-29. doi:10.1007/bf02590167

[2] Hungr, O., Leroueil, S., & Picarelli, L. (2013). The Varnes classification of landslide types, an update. *Landslides, 11*(2), 167-194. doi:10.1007/s10346-013-0436-y

[3] Lee, S., & Talib, J. A. (2005). Probabilistic landslide susceptibility and factor effect analysis. *Environmental Geology, 47*(7), 982-990. doi:10.1007/s00254-005-1228-z

[4] Davies, T. (2015). Landslide Hazards, Risks, and Disasters *Landslide Hazards, Risks and Disasters* (pp. 1-16): Elsevier BV

[5] Lora, M., Camporese, M., Troch, P. A., & Salandin, P. (2016). Rainfall-triggered shallow landslides: infiltration dynamics in a physical hillslope model. *Hydrological Processes, 30*(18), 3239-3251. doi:10.1002/hyp.10829

[6] Jaapar, A. R. B. *A framework of a national slope safety system for Malaysia.* The University of Hong Kong Libraries. Retrieved from http://dx.doi.org/10.5353/th_b3738460

[7] Murray, A. T. (2002). GIS and Multicriteria Decision Analysis, by Jacek Malczewski, 1999. *Geographical Analysis, 34*(1), 91-92. doi:10.1111/j.1538-4632.2002.tb01077.x

[8] Lee, S., Ryu, J.-H., Won, J.-S., & Park, H.-J. (2004). Determination and application of the weights for landslide susceptibility mapping using an artificial neural network. *Engineering Geology, 71*(3-4), 289-302. doi:10.1016/s0013-7952(03)00142-x

[9] Lin, K., Simossis, V. A., Taylor, W. R., & Heringa, J. (2004). A simple and fast secondary structure prediction method using hidden neural networks. *Bioinformatics, 21*(2), 152-159. doi:10.1093/bioinformatics/bth487

[10] Othman, A. N., Mohd, W. M. N. W., & Noraini, S. (2014). Accuracy assessment of landslide prediction models. *IOP Conference Series: Earth and Environmental Science, 18*, 012078. doi:10.1088/1755-1315/18/1/012078

Identifying critical success factors (CSFs) of Facilities Management (FM) in non-low cost high-rise residential buildings

F M Dahlan[1], **A Zainuddin**[1]
[1]Faculty of Architecture, Planning & Surveying, Universiti Teknologi MARA, Seri Iskandar Campus, Seri Iskandar, 32610, Perak, Malaysia

*farha221@perak.uitm.edu.my

Abstract. Critical success factors (CSFs) are important key areas of activity that must be performed well in any Facilities Management (FM) organisation to achieve its missions, objectives or goals. Before implementing CSFs, an FM organisation must identify the key areas where things must be done properly to enable the business to flourish. Although many performance measurements in FM organisation have been discussed in previous research, not much research has been done on CSFs from the perspective of FM business in non-low cost high-rise residential buildings. The purpose of this study is to develop a methodology in developing the CSFs group and CSFs for FM organisation in non-low cost residential buildings. This research will involve three (3) phases of research strategy to achieve the objective of this research.

1. Introduction

High-rise residential building living is unlike that in a freestanding house. The actions, activities and attitudes of residents, residents' association and the management can have significant living and environmental impacts on the satisfaction and enjoyment of others [1]. The main difference in high-rise living as compared to that in landed property is the need for residents to set up a Management Corporation (MC), the residents' association, as enacted by the Strata Management Act 2013, to regulate and manage all the facilities and amenities together with the common area [2] such as the parking area, lighting and other common facilities that are shared, which were provided by the developer. Maintenance service charges payment which is collected from the residents, is the sources of fund for the MC to smoothen the activities of managing, maintaining and operating the building, services and facilities to ensure its optimal performance over its life cycle [3].

The major issue in high rise residential buildings is the dispute between the residents and the management of the MC [4][5]. Lai [6], consistent with Che Ani et al. [7] and Rahman et al. [8] claimed that the dispute was due to the existence of a management gap when there is a significant difference in the satisfaction levels among respondents regarding various components of facility management including finances, maintenance, and resident aspects, as well as regarding the efficiency of the management. Che Ani et al. [7] emphasized the dispute between residents and MC that occur due to the existence of a gap between FM service expectation and FM service quality received and those problems are unresolved regardless of the size and owners of the housing project. This gap results in the failure to keep up the payment, leading to insufficiency of funds, which in turn, affects the FM capacity to provide good service quality [9].

The failure to keep up payment by residents had led to building quality deficiencies [10] [11], rendered unreliable cost estimates [11] [10] and compromising safety and health issues [12]. To ensure that the buildings remain in good condition and to avoid the potential risk of accidents, excellent and efficient FM is needed [13]. The Deming Theory of Management 1982 claimed that organisations should improve the system of service constantly. This theory also highlighted that in any product or service, the quality should be considered in the first place. According to Rochart [15], critical success factors (CSFs) are the few key areas that must go right and receive constant and careful attention from the management for the business to flourish. The author claimed that if results in these areas are not adequate, the organization's effort for the period will be less than desired. Previous research on CSFs of FM in non-low cost high-rise residential buildings in Malaysia, categorized with the built up area of 800 square feet to more than 1000 square feet [14] are still scarce. Therefore, the purpose of this research is to develop a methodology for the development of CSFs groups and CSFs which is an extension of earlier study by Che Ani et al. [7].

2. Literature Review

This research attempts to identify the CSFs group and CSFs from the perspective of FM in non-low cost high-rise residential buildings. Therefore, the Deming Theory of Management 1982, underpinned this research. This theory highlights product quality as the mainstay. In terms of service quality, an organization should focus on identifying quality which serves as key factors as their CSFs to be successful [15]. Previous research on CSFs efficiency of FM in non-low cost high-rise residential buildings in Malaysia are still scarce.

In Hong Kong, using the Structural Equation Modelling (SEM), Hui and Zheng [16] identified that the service quality delivered to the residents significantly affects management quality instead of safety consciousness, security level, fire services equipment, and also professional knowledge of staff. Meanwhile, Lai and Yik [17] revealed that the result from an interview conducted on residents in Hong Kong showed that the cost effectiveness of FM services is the significant factor for evaluating the FM services delivered by the company. In addition, their findings also discovered that the quality of facilities, environmental quality and construction quality have a significant factor for achieving a quality and sustainability environment. Further study done by Lai [18] in three (3) different residential estates (public, semi-public and private) found that the highest service performance and cost levels was the private estate followed by the semi-public estate and public estate. However, the cost effectiveness was found reversed between the estates. Another study by Lam [30] on the perspective of outsourcing FM organisations found that instead of the transaction monitoring factor, market competition was also found as the fundamental economic forces for efficient FM in Hong Kong.

Previous studies in the Malaysian context done by Che Ani et al.[7], Mohd Ali et al.[21] and Ali et al.[22] discovered the financial factor as the crucial factor for efficient FM in high-rise residential buildings. Che Ani et al.[7] identified financial effectiveness, financial sources, financial allocation and financial expenses as the factors that need to be highlighted the most by FM organisations towards efficient FM. Another research by Mohd Ali et al. [21], who took sampling from flats, apartments and condominiums in Klang Valley revealed that the maintenance aspect and the financial factor influence the management capabilities. However, another study done by Douglas [25] focusing on outsourcing Malaysian FM companies in various types of property, revealed that flexibility strategy, integration of goals and companies strategy were the factors that can sustain the business of FM organisations.

Studies on CSFs of efficient FM focusing on non-low cost high-rise residential buildings had not been discussed a great deal and they were not comprehensive where the researchers only focused on the financial factor, human factor and economic factor. However, limited studies provide a comprehensive approach as the concept of the FM in organizations as defined by Douglas [25] theoretically lies in three elements, namely process, place and people. Therefore, the financial, human and economic factors are insufficient to achieve a more holistic approach towards attaining efficient FM. Those CSFs groups include; financial, customer base, internal process, learning & growth as well as design & construction quality.

In order to identify the CSFs group and CSFs, this research will involve three (3) phases of methodology to achieve the objective of this research.

Table 1. Critical Success Factors of Facilities Management

No	Critical Success Factors (CSFs)	AUTHORS
1	Financial	
	• Cost efficiency/ value for money	Kalumbu, Mutingi, & Mbohwa[19], Yongtao, Liyin, Craig, Weisheng, and Michael [20], Tucker, Turley, and Holgate [23]
2	Customer	
	• Communication between organization and customers	Zushi and Sohal [24], Kalumbu, Mutingi, & Mbohwa [19], Yongtao, Liyin, Craig, Weisheng, and Michael [20]
	• Relationship with customers	Zushi and Sohal [24], Kalumbu, Mutingi, & Mbohwa [19], Yongtao, Liyin, Craig, Weisheng, and Michael [20]
	• Customer expectation	Yongtao, Liyin, Craig, Weisheng, and Michael [20], Ganisen, Mohammed, Jawahr Nesan, Kanniyapan [12]
	• Reliability of service	Yongtao, Liyin, Craig, Weisheng, and Michael [20]
	• Quality standards/certification	Kalumbu, Mutingi, & Mbohwa [19], Yongtao, Liyin, Craig, Weisheng, and Michael [20]
	• Quality resources	Zushi and Sohal [24], Kalumbu, Mutingi, & Mbohwa [19], Yongtao, Liyin, Craig, Weisheng, and Michael [20]
	• Responsiveness to incidents	Yongtao, Liyin, Craig, Weisheng, and Michael [20], Ganisen, Mohammed, Jawahr Nesan, Kanniyapan [12]
3	Internal Process	
	• Top management support	Zushi and Sohal [24], Kalumbu, Mutingi, & Mbohwa [19], Yongtao, Liyin, Craig, Weisheng, and Michael [20]
	• Clear policy, strategy & planning	Zushi and Sohal [24], Kalumbu, Mutingi, & Mbohwa [19]Yongtao, Liyin, Craig, Weisheng, and Michael [20], Tucker, Turley, and Holgate [23], Ganisen, Mohammed, Jawahr Nesan, Kanniyapan [12]
	• Experience in maintenance business/familiarity with maintenance & other maintenance plan	Kalumbu, Mutingi, & Mbohwa [19], Yongtao, Liyin, Craig, Weisheng, and Michael [20]
	• Adequate resources/dedicated resources	Zushi and Sohal [24], Ganisen, Mohammed, Jawahr Nesan, Kanniyapan [12]
	• Appointment of capable manager	Zushi and Sohal [24], Ganisen, Mohammed, Jawahr Nesan, Kanniyapan [12]
	• Staff qualification & experience	Yongtao, Liyin, Craig, Weisheng, and Michael [20], Ganisen, Mohammed, Jawahr Nesan, Kanniyapan [12]
	• Reference to industry guideline/standards	Zushi and Sohal [24]
	• General training & awareness for suppliers & other stakeholders	Zushi and Sohal [24]
	• Necessity & usage of audits/monitor performance/quality assessment	Zushi and Sohal [24], Yongtao, Liyin, Craig, Weisheng, and Michael [20], Tucker, Turley, and Holgate [23], Ganisen, Mohammed, Jawahr Nesan, Kanniyapan [12],Abdul Mutalib, M. Sapri, and I. S. Mohammad [26]
	• Document control system (hard or soft version)/ IT &Technology	Zushi and Sohal [24], Kalumbu, Mutingi, & Mbohwa [19], Yongtao, Liyin, Craig, Weisheng, and Michael [20], Ganisen, Mohammed, Jawahr Nesan, Kanniyapan [12],Abdul Mutalib, M. Sapri, and I. S. Mohammad [26]
	• Company reputation/certification	Yongtao, Liyin, Craig, Weisheng, and Michael [20]
	• Member of professional organizations	
	• Working condition/effective working practice/good communication	Yongtao, Liyin, Craig, Weisheng, and Michael [20], Tucker, Turley, and Holgate [23], Ganisen, Mohammed, Jawahr Nesan, Kanniyapan [12]

IOP Conf. Series: Earth and Environmental Science **117** (2018) 012036 doi:10.1088/1755-1315/117/1/012036

	• Contract and risk management	Yongtao, Liyin, Craig, Weisheng, and Michael [20]
	• Lesson learnt	Ganisen, Mohammed, Jawahr Nesan, Kanniyapan [12]
4	Learning & Growth	
	• Employee training	Zushi and Sohal [24], Kalumbu, Mutingi, & Mbohwa [19], Yongtao, Liyin, Craig, Weisheng, and Michael [20], Tucker, Turley, and Holgate [17], Ganisen, Mohammed, Jawahr Nesan, Kanniyapan [12]
	• Cultural changes/innovativeness	Zushi and Sohal [24], Yongtao, Liyin, Craig, Weisheng, and Michael [20], Tucker, Turley, and Holgate [23], Ganisen, Mohammed, Jawahr Nesan, Kanniyapan [12], Abdul Mutalib, M. Sapri, and I. S. Mohammad [26]
	• Learning from other organisations' experiences and benchmarking	Zushi and Sohal [24]
5	Design & Construction Defects	
	• Defective construction materials	Waziri [28]
	• Poor supervision	Waziri [28]
	• Defects due to specification	Waziri [28]
	• Poor quality control on site	Waziri [28]
	• Incompetent workforce	Waziri [28]
	• Architectural defects	Waziri [28]
	• Use of new & untested materials	Waziri [28]
	• Incompetent workforce for construction	Waziri [28]

3. Methodology

This research will involve three (3) phases of methodology to achieve the objective of this research. The flow chart of the research strategy is shown in Figure 1.

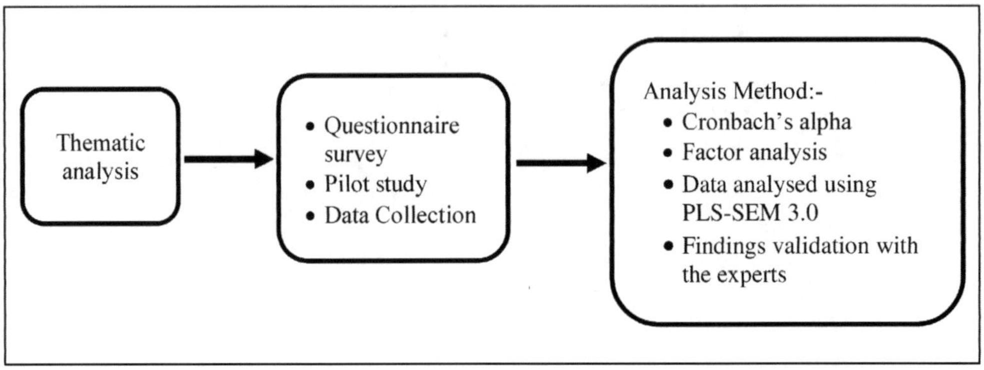

Figure 1. Research Strategy Flow Chart

4. Result and Findings

This research highlights issues related to the dispute between the residents and the management of the MC. Leading journals between 2013 and 2017 have been analyzed under these keywords: success factors, critical success factors, facilities management and high-rise residential building.

The aim of this research is to develop a CSFs structural modeling for efficient FM in non-low cost high rise residential buildings, involving three (3) phases of research strategy. From this theoretical framework, five (5) group factors and 34 factors have been identified. The results are shown in Table 1.

5. Conclusion

This research is intended to narrow the gap in the efficiency of FM in non-low cost high-rise residential buildings by developing CSFs structural modelling. There are 5 CSF groups and thirty four (34) CSFs that have been analysed. The CSFs group and CSFs can be seen to help in producing successful competitive performance for the non-low cost high-rise residential buildings in the area of FM. This research is also intended to provide a useful guideline and can be used as a benchmark for the efficiency of FM especially for non-low cost high-rise residential buildings that will enhance the current practice.

References

[1] N. R. Muhamad Ariff, "Multi-owner low cost houding in Malaysia," Deakin University, 2011.

[2] Z. N. Musa, A. G. Sarip, W. N. A. W. A. Aziz, N. R. Hanif, Z. Al-Sadat, and P. A. Tedong, "Issues in Managing Vertical Residential Building in Malaysia: Homeowners Perspective," in The Asia Pasific Network for Housing Research (APNHR), 2015, pp. 1–8.

[3] O. A. Abdul Lateef, "Building maintenance management in Malaysia," J. Build. Apprais., vol. 4, no. 3, pp. 207–214, 2009.

[4] T. L. Ta, "Managing High-Rise Residential Building in Malaysia: Where Are We?," Naprec Conf. Inspen, pp. 1–25, 2006.

[5] F. Kamarudin, "The Development of an efective and efficient dispute resolution processes for Strata Scheme Disputes in Peninsular Malaysia," Queensland University of Technology, 2014.

[6] J. H. K. Lai, "Investigating the Quality of FM Services in Residential Buildings," in CIB World

Building Congress : Building a Better World, Salford, UK, 10-13 May 2010, 2010, pp. 13–25.
[7] A. I. Che-Ani, N. M. Tawil, A. Sairi, N. A. G. Abdullah, M. M. Tahir, and M. Surat, "Facility management indicators for high-rise residential property in Malaysia," WSEAS Trans. Environ. Dev., 2010.
[8] M. S. Rahman, B. Hussain, A. N. M. M. Uddin, and N. Islam, "Exploring residents' satisfaction of facilities provided by private apartment companies," Asia Pacific Management Review, vol. 20, no. 3, pp. 130–140, 2015.
[9] Jamila H., Strata Title In Malaysia. Selangor. Pelanduk Publications (M) Pte. Ltd., 1994.
[10] S. Ganisen, L. J. Nesan, I. S. Mohammad, A. H. Mohammed, and G. Kanniyapan, "Facility Management Variables That Influence Sustainability of Building Facilities," J. Teknol., vol. 10, pp. 27–38, 2015.
[11] M. F. Mohamed Isa and M. Usmen, "Improving University Facilities Services Using Lean Six Sigma: A Case Study," J. Facil. Manag., vol. 13, no. 1, pp. 70–84, 2015.
[12] S. Ganisen, A. H. Mohammed, L. J. Nesan, and G. Kanniyapan, "Critical Success Factors for Low Cost Housing Building Maintenance," J. Teknol., vol. 2, no. 2015, pp. 31–40, 2015.
[13] N. V. Siegert, What to do with public rental housing? Challenges and Option. Washington DC: Inter-American Development Bank., 2004.
[14] K. A. Rahman, A. Hariri, A. M. Leman, M. Z. M. Yusof, and M. N. M. Najib, "Energy Consumption in Residential Building : The Effect of Appliances and Human Behaviour," vol. 20018, 2017.
[15] J. F. Rockart, "Chief executives define their own date needs," Harvard Bus. Rev., vol. March-Apri, pp. 81–93, 1979.
[16] E. C. M. Hui and X. Zheng, "Measuring customer satisfaction of FM service in housing sector," Facilities, vol. 28, no. 5/6, pp. 306–320, Apr. 2010.
[17] J. H. K. Lai and F. W. H. Yik, "An analytical method to evaluate facility management services for residential buildings," Build. Environ., 2011.
[18] Lai, J.H.K., "Analytical assessment and comparison of facilities management services for residential estates," Int. J. Strateg. Prop. Manag., vol. 16, no. 3, pp. 236–253, 2012.
[19] R. Kalumbu, M. Mutingi, and C. Mbohwa, "Critical Success Factors for Developing Building Maintenance Strategies : A Case of Namibia," in Proceedings of the 2016 IEEE IEEM, 2016, pp. 1402–1406.
[20] T. Yongtao, S. Liyin, L. Craig, L. Weisheng, and C. H. Y. Michael, "Critical success factors for building maintenance business : a Hong Kong case study," Facilities, vol. 32, no. 5/6, pp. 208–225, 2014.
[21] H. Mohd Ali et al., "The Development of Property Management Capability Model (PMCM) for Effective Service Delivery In Stratified Residential Property," in 12th NAPREC Conference Proceedings, 2016.
[22] A.-S. S. Ali, S.-N. N. Kamaruzzaman, R. Sulaiman, Y. Cheong Peng, and Y. C. Peng, "Factors affecting housing maintenance cost in Malaysia," J. Facil. Manag., vol. 8, no. 4, pp. 285–298, 2010.
[23] M. Tucker, M. Turley, and S. Holgate, "Critical success factors of an effective repairs and maintenance service for social housing in the UK," 2014.
[24] A. Zushi and A. S. Sohal, "Adoption and maintenance of environmental management systems: Critical success factors," Manag. Environ. Qual. An Int. J., vol. 15, no. 4, pp. 399–419, 2004.
[25] M. U. Douglas, "Finding the niche towards performance excellence: A study of facilities management firms in Malaysia," J. Facil. Manag., vol. 14, no. 7, pp. 330–349, 2016.
[26] A. F. Abdul Mutalib, M. Sapri, and I. S. Mohammad, "Facility management knowledge development in Malaysia: added value in hospirality managerial competency," Facilites, vol. 33, no. 1/2, 2015.
[27] H. Mohd Ali et al., "The Development of Property Management Capability Model (PMCM) for Effective Service Delivery In Stratified Residential Property," in 12th NAPREC Conference

Proceedings, 2016.
[28] B. S. Waziri, "Design and construction defects influencing residential building maintenance in Nigeria," Jordan J. Civ. Eng., vol. 10, no. 3, pp. 313–323, 2016.
[29] M. U. Douglas, "Finding the niche towards performance excellence: A study of facilities management firms in Malaysia," J. Facil. Manag., vol. 14, no. 7, pp. 330–349, 2016.
[30] T. Y. M. Lam, "Economic perspective on outsourcing of property management services," J. Prop. Manag., vol. 30, no. 4, pp. 138–332, 2013.

Mitigating delay and non-payment in the Malaysian construction industry

N Mohamad[1], A S Suman[1], H Harun[1], H Hashim[1]

[1]Centre of Quantity Surveying Studies, Faculty of Architecture, Planning and Surveying, Univeristi Teknologi MARA, 40450 Shah Alam, Selangor, Malaysia

*norsi564@salam.uitm.edu.my

Abstract. Construction industry is one of the industries that have contributed towards the rapid growth of development and economics in Malaysia. However, the industry is inundated with delay and non-payment issues between the two parties in contract that is the clients and contractors Even though there are contractual and administrative provisions in the standard forms of contract in Malaysia regarding payments, delay and non-payment issues still occur between them. The aim of the study is to develop measures to mitigate delay and non-payment issues between contractors and clients in the Malaysian construction industry. Questionnaire survey was conducted with clients and contractors in Klang Valley. Results from data analysis identified significant measures to mitigate delay and non-payment issues between contractors and clients which include contractors should submit their progress work invoicing with adequate documents; contractors should follow up constantly with client regarding payment; proper understanding of requirements with regards to payment; mutual discussion of problems with client to address problems in a timely manner and proper use of payment provisions in the standard form of contract. This study is significant to contractors and clients and to other construction players in order to reduce and minimise delay and non-payment issues for the growth of economy in the Malaysian construction industry.

1. Introduction

The practice of efficient and timely payment in construction project is a significant factor for a project's success [1]. According to [2], payment is the crucial lifeblood of the construction industry. It is needed to pay for preliminaries, materials, labour, plant, subcontractors' work and general overheads during the progress of the work [3].

Major problems faced by construction industry are late payment and non-payment. Previous researches have shown that payment disputes that arise are due to default by clients, and contractors are victimized by this scenario. According to [4] in the Malaysian construction industry, not getting paid or payments have been delayed by the clients purposely are common contractor's arguments. [5] cited from [6] stated that this is may be due to certain clients that like to delay the contractor's payment for his own benefit such as to fund their other projects. Another reason for delay and non-payment is due to client's poor financial management [7]. On the other hand, delay or non-payment can also be due to the fault by the contractors. A common issue is when contractors fail to agree with the valuation of work at site [8]. Other issues form the contractor which can cause delays in payment are errors in submitting claim documents, insufficient document submitted by contractor for claims and delay in submitting the claims itself [1].

As pointed out by [9] construction payment can have domino effects since the whole chain of payment flow is affected by delayed payment by the client. [4] highlighted the consequences from the failure by contractors receiving regularly and timely payment from clients are i) project delays, ii) reduced profitability and, iii) the company may go into liquidation in the worst case scenario. This is agreed by [10] where late payments by clients affect the cash flow of a company which can affect the overall progress at site and may eventually lead to company's insolvency. Hence, to ensure sustainable performance for both parties, the aim of this study is to develop mitigation measures to reduce payment issues in the Malaysian construction industry.

2. Literature Review

What is payment? It plainly means the money given or to be given to a person in return for the goods sold or services delivered. Payment in construction is the monetary consideration for the contractor's performance or work done [4]. Payment is further defined by Suman (2016) as what needs to be paid by the client within a stipulated time - which is the value of work done by the contractor at project site and/or material by the supplier which is delivered to site as specified in the contract. Failure in timely and regular payment from clients can disrupt the contractor's overall progress of work at site. Liability on delay or non-payment should be distributed equally to both contractors and clients. Therefore, mitigation measures must be taken by both parties in construction projects.

Previous researches have highlighted measures that both contractors and clients can take to mitigate late payment issues. It is crucial that payment provisions in the standard form of contract are followed and used properly [12]. The standard forms of construction contract used in Malaysia clearly state the client's responsibility to pay the contractors, which is stipulated within a certain time frame or a specific time given. For instance, the PWD 203A (2010) Standard Form of Contract used for public projects funded by the Government, states 30 days from the date of the issuance of the certificate as the period for honouring the certificate. As for private projects, the PAM (2006) Standard Form of Contract is normally used. This contract states 21 days after the issuance of the certificate as the period of honouring certificate. Period of honouring certificate means the period where payment should be made to the contractor. Hence, what is considered as late or delayed payment is any payment made later than the time frame or period of honouring the payment certificate as stipulated in the provisions of the contract [4]. According to [10], in order to benefit the contractors and clients, one of the measures to address late or non-payment issues between both parties is to engage in mutual discussion between them. In addition, payment terms or clauses in the contract should be properly understood by the contractor especially for particular projects as clients sometimes have particular requirements that contractors should follow in order to receive payment.

According to [1] in order to avoid delay in payment, contractors have to make sure that documentation for payment claim is complete with adequate documents before submitting to the clients. Contractors should also liaise with clients constantly regarding payment, in other words, the contractors should communicate with the clients to keep informed of their payment status [10].

3. Research methodology

To achieve the aim of study, the measures investigated are to mitigate delay and non-payment for payments issues which arise during the construction stage. They are administrative measures to encourage both parties to take precautionary steps before payment issues are brought up to Alternative Dispute Resolution (ADR) techniques such as mediation or adjudication provided in the contract, or statutory Construction Industry Payment Adjudication Act (CIPAA), or arbitration and even to litigation.

The populations for this study are clients and G7 contractors of construction projects in Klang Valley. Since construction projects are diversified in nature, the clients will include public (for public projects) and private clients (for private projects). Public clients are represented by local authorities and Public Work Department (PWD) Malaysia; while, private clients are represented by developers. Contractors registered with Construction Industry Development Board (CIDB) under G7 are considered appropriate to represent the contractors as they are in the CIDB's highest classification based on financial capabilities of contractor to tender for projects, hence payment matters will be of the most significant. The respondents are from both the contractors' and clients' organisations to ensure fairness and unbiased response; and also viability of the research since payment are transacted between them. Both points of view must be taking into consideration in order to obtain significant results from the study. The limitation of the study is that data collection is conducted within Klang Valley area. It is justifiable to use Klang Valley as locality for this study due to active construction development that takes place in Klang Valley as compared to other locations.

3.1. Research process

The research process is divided into three (3) phases; Phase1 - Secondary data collection, Phase 2 - Pilot survey and Phase 3 - Primary data collection.

3.1.1. Phase 1: Secondary data collection. Secondary data is important to build up the literature review and questions in the questionnaire for this study. The secondary data is obtained from: i) published printed sources – which include the Standard Form of Building Contracts, government and statutory body circulars, standards and other publications; Acts, other construction industry related publications and books not available on-line, ii) published electronic sources - this include articles in journals, conference proceedings, updates on news related links and others alike.

3.1.2. Phase 2: Pilot survey. The pilot survey is used to pre-test the research instrument. In order to ensure the validity of the research, this research instrument is tested first - before the final questionnaire survey is constructed and distributed to the target respondents. The pilot survey consisting of 15 questions was tested to five selected respondents. The respondents for the pilot survey are from the same background to the targeted population for this study. The demography of the pilot survey respondents is shown in Table 1.

Table 1. Demography of pilot survey respondents

Respondent	Organisation	Profession	Position	Working experience	Projects completed
1	Contractor	Quantity surveyor	Senior executive	More than 10 years	More than 10 projects
2	Contractor	Engineer	Executive	1 year - 5 years	Less than 5 projects
3	Contractor	Quantity surveyor	Executive	1 year - 5 years	Less than 5 projects
4	Private client	Quantity surveyor	Manager	More than 10 years	More than 10 projects
5	Contractor	Engineer	Manager	1 year - 5 years	Less than 10 projects

The results from the pilot survey were tested using Cronbach's coefficient alpha (α) to determine the consistency and reliability of the questions. Based on Cronbach's coefficient alpha (α) rule of thumb where the acceptable values range from 0.70 to 0.95 (Cohen, 2010), the results of the reliability test on Cronbach's coefficient alpha (α) is 0.860 (good) This shows that the questions in the pilot survey are consistent and reliable. Therefore, no significant changes were made for the questionnaire survey. However, based on the response and recommendation from the respondents from the pilot survey, the questions was reworded and rephrased to improve the understanding and reliability of the questions before distribution to target population for questionnaire survey.

3.1.3 Phase 3: Questionnaire survey. This research adopted a quantitative research design. Data is collected through a structured questionnaire survey. 120 questionnaire surveys were distributed through direct walk in and e-mail to the organisations of G7 contractors, public and private clients, in Klang Valley. From the 120 samples, 48 respondents were received through e-mail and another 36 from 'walk in office' making a total number of 84 respondents, which is a 70% response rate, which is considered creditable.

4. Data analysis and discussion

Data from questionnaire survey was analysed using SPSS to determine the frequency and mean. The reliability of data is tested by using Cronbach's coefficient alpha (α). Then, data collected were interpreted in the form of tabulations.

4.1. Demography

Table 2 shows the demography of the respondents. Contractors make up the majority of the respondents (53.60% = 45), then the public clients (23.80% = 20) and private clients (22.60% = 19) Contractors make up the most respondents as compared to others. The contractors' perceptions toward payment issues are significant as they are normally the casualty of the payment issue in the construction project. The data collected has also taken into consideration the clients' perception in order to improve viability and the fairness of the results obtained for this study.

Table 2. Demography of respondents

Category		Frequency	Percentage (%)
Type of organisation	Contractor	45	53.60
	Public client	20	23.80
	Private client	19	22.60
	Total	**84**	**100.00**
Working experience	Less than 1 year	1	1.20
	1 year – 5 years	44	52.40
	5 years – 10 years	19	22.60
	More than 10 years	20	23.80
	Total	**84**	**100.00**
Projects completed	1 project	14	16.70
	2- 5 projects	33	39.30
	6- 10 projects	11	13.00
	More than 10 projects	26	31.00
	Total	**84**	**100.00**

Majority of the respondents' working experience is 1-5 years (52.40% = 44 no.), followed by more than 10 years (23.80% = 20), 5-10 years (22.60% = 19) and less than 1 year (1.20% = 1). The working experience of the respondents gives significant impact to the study as respondents who have more experience in the construction industry will have more exposure and understanding on payment issues. This improves the reliability and validity of the findings for this study. Majority of the projects completed by the respondents are: 2-5 projects (39.30% = 33), followed by more than 10 projects (31.00% = 26), 1 project (16.70% =14) and 6- 10 projects (13% = 11). This shows that most of the respondents have a sound experience in number of projects completed, reflecting their experience on payment issues which enhances the viability and reliability of date collected.

4.2 Measures to mitigate delay and non-payments issues between clients and contractors
4.2.1 Reliability test. The result of Cronbach's coefficient alpha (α) on measures to mitigate delay and non-payments issues between clients and contractors is equal to 0.768 (acceptable). Hence, the data collected for the questionnaires survey is reliable and viable. This shows that the findings for this study have significant value as the reliability of data collected is acceptable.

4.2.2. *Measures to mitigate delay and non-payment issues in construction industry - Descriptive analysis.* The measures are divided into three (3) different categories; i) measures by client and contractor (both parties), ii) measures by client, iii) measures by contractor. Table 3 shows the overall ranking of measures to mitigate delay and non-payment issues.

Table 3. Overall ranking of measures to mitigate delay and non-payment issues

Rank	Measure by	Measures	Mean
1	Contractor	The latest progress work invoicing submitted must be with adequate documents	1.58
2	Both parties	Understanding the terms or clauses of payment in the project	1.89
3	Both parties	In order address the problem in a timely manner, mutual discussion of problems with client	1.91
4	Contractor	The payment matter should be followed up constantly with the client	1.92
5	Both parties	Payment provision in the standard form of contract is applied properly	1.94
6	Both parties	Statutory enactment must be implemented to deal with payment in construction industry	2.01
7	Both parties	People mentality and attitude must be changed in order to achieve timely payments	2.03
8	Client	Financial management must be implemented to mitigate cash flow problems	2.09
9	Client	Setting an established time frame for payment	2.15
10	Both parties	The effect of payments issues on the project progress must be understood by all parties and personnel involved	2.17
11	Contractor	Right for contractors to suspend work in the event of late or non-payments to be established	2.19
12	Both parties	The provision of 'pay when paid clause' to be introduced in the construction contract	2.21
14	Contractor	Right of contractors to charge interest on late payment to be established	2.33
15	Client	The limit of amount that client can hold the money to be established	2.35

The measures to mitigate delay and non-payment are ranked in ascending order. This is based on Likert scale i.e. strongly agree (mean = $0.00 \leq a < 1.00$), agree (mean $1.01 \leq a < 2.00$), undecided ($2.01 \leq a < 3.00$), disagree ($3.01 \leq a < 4.00$) and strongly disagree ($4.01 \leq a < 5.00$) of the respondents' perception on the variable. This discussion highlights five (5) measures (in the categories of strongly agree and agree; mean = $a < 2.00$) to mitigate delay and non-payment issues. As shown in Table 3, two (2) of the measures are by contractor, and other three (3) by both parties, the contractor and client. The two (2) measures by the contractor are: the latest progress work invoicing submitted must be with adequate documents (rank 1, mean = 1.58); and the payment matter should be followed up constantly with the client (rank 4, mean = 1.92). The other three (3) measures by both the client and contractor are; understanding the terms or clauses of payment in the project (rank 2, mean = 1.89); in order address the problem in a timely manner, mutual discussion of problems with client (rank 3, mean = 1.91); payment provision in the standard form of contract is applied properly (rank 5, mean = 1.94). Hence, it can be said that the contractor himself and both the contractor and client must cooperate together to mitigate delay and non-payment issues. The advantages of cooperation between both parties will initiate an efficient process of certification by the client and releasing payment to the contractor leading to a timely project completion.

5. Conclusion

This study focuses on measures that clients and contractors can implement towards mitigating delay and non-payment issues in the Malaysian construction industry. They are administrative measures encouraging both parties to take preventative steps during the construction stage before payment issues are bought up to Alternative Dispute resolution (ADR) techniques such as mediation or adjudication, or the statutory CIPAA, arbitration or even to litigation.

This study is significant since insights from both parties in construction contracts in the industry, that is, the contractor and the client are considered, which provides fairness and reliability on the measures identified. It is also useful to other construction players in the endeavour to reduce payment disputes in the Malaysian construction industry. Having timely payment to the contractor provides efficient project management of a construction project, hence, the project can complete on time, within budget and achieving the quality specified. This domino effect will contour a rapid growth of development in the construction sector resulting in economic growth in Malaysia.

References

[1] Hasmori, M. F., Ismail, I., & Said, I. 2012. *Issues of Late and Non-Payment Amon Contractors in Malaysia.* 3rd International Conference on Business and Economic Research\ (3rd ICBER 2012)\) Proceeding 12-13 March 2012, Bandung, Indonesia.

[2] Naseem, N.A. 2006. *A Construction Industry Payment and Adjudication Act - Reducing Payment-Default and Increasing Dispute Resolution Efficiency.* International Forum Construction Industry Payment and Adjudication Act Kuala Lumpur: CIDB and ISM.

[3] Odeyinka, H.A. and Yusif, A, 1998. *The causes and effects of construction delays on completion cost of housing projects in Nigeria,*" Journal of Financial Management of Property and Construction, 2(3), p 31- 44.

[4] Judi S., and Abdul Rashid, K. 2010. *Contractor's Right of Action for Late or Non-Payment by the Employer.* Journal of Surveying, Construction & Property, 1(1), p 65-95.

[5] Ramachandra, T. and Rotimi, J.O.B., 2015.*Causes of Payment Problems in the New Zealand Construction Industry*, Construction Economics and Building, 15(1), p 43-55.

[6] Kenley, R., 2002. *Milking the cash cow.The New Zealand Herald.* [online] 17 August. Available at: http://www.nzherald.co.nz/business/news/article.

[7] Kaka, A. P. and Price A.D.F 1991.*Net cash flow models: are they reliable*? Construction Management Economics 9, p 292-308.

[8] Odeh, A.M. and Battaineh, H.T., 2002. C*auses of construction delay: Traditional contracts.*International Journal of Project Management, 20(1), p 67-73.

[9] Mohamad, M. I., Nekooie, M. A., &Kamaruddin, N. C. 2012.The *Adequacy of Contractual Provisions in Managing Construction Failure in Malaysia.* European Journal of Business Management. 4: p 22–38.

[10] Ye, K.M. and Rahman, A.H., 2010. *Risk of late payment in the Malaysian construction industry.* International Scholarly and Scientific Research and Innovation, 4(5), p 81- 9.

[11] Suman, A. S. 2016, Mitigating delay and non-payment issues in the Malaysian Construction Industry, Masters thesis, Universiti Tekmologi MARA, Selangor.

[12] Ramachandra, T. and Rotimi, J.O.B., 2012. *Construction payment delays and losses: Perceptions of New Zealand*, 18th Annual Conference, Wellington, New Zealand, September 2012,p 24-26

The effect of building facade reflectivity on urban dwellers in tropics.

N M Ishak[1], W N Hien[2], H S Jenatabadi[3] , M Ignatius[2], R. Yaman[1]

[1] Faculty of Architecture, Planning and Surveying, Universiti Teknologi MARA, Shah Alam, Selangor, Malaysia
[2] Department of Building, National University of Singapore, Singapore
[3] Department of Science and Technology Studies, Faculty of Science, University Malaya.

*norish@salam.uitm.edu.my

Abstract. With the rapid growth and use of modern architecture practices for high-rise buildings, highly reflective materials have been adopted extensively for aesthetical reasons. However, outdoor glare from highly reflective facades might cause thermal and visual problems towards the occupants of neighbouring buildings and outdoor dwellers, particularly pedestrians. In tropical countries, this negative impact can be greater due to the higher solar radiation received throughout the year. At the present, there are few building guidelines limiting outdoor glare, or daylight reflectance from a building facade. This study aims to introduce a framework for outdoor glare studies that focus on perceived glare from highly reflective facades by pedestrians in Singapore. The introduced framework includes age, glare time, glare duration, avoidance and sensitiveness. For this study, the survey is carried out with the application of Structural Equation Modelling (SEM). This paper is helpful for planners, designers, and engineers to estimate the sensitivity of pedestrians' discomfort glare and towards the creation of sustainable architecture in Singapore.

1. Introduction

The scarcity of land in urban areas of some tropical regions has created a construction demand for high rise buildings in an increasingly dense manner, creating simultaneously benefits and problems to the urban sustainability of these regions. To understand the concept of being sustainable, it is essential to understand the behaviour and properties of building materials. For example, the use of materials with lower albedo and impermeable surfaces will result in more heat being absorbed from the sun's radiation, thereby raising the temperature in urban areas [1].

Building facades have been extensively designed and built with highly transparent and reflective materials that receive much attention in R&D. However, these materials may cause severe outdoor glare problems and further increase the surface temperature of the ground and neighbouring building [2]–[8]. The reflectance value of some facade materials can go up to 40%, which can impose a severe impact towards pedestrians and road users. When sunlight hits a reflective surface of a building, it will reflect, absorb and transmit the radiation simultaneously. Furthermore, the amount of light reflected will depend on the surface and type of material used, the building geometry and the facade form [8]–[11]. A prominent case depicting this issue is the Disney Concert Hall in Los Angeles [2], where the use of shimmering stainless steel materials combined with concave facade geometries had worsened the reflected glare to the surrounding areas. Other related cases that shown similar incidences are the "Walkie Talkie" building, 20 Fenchurch Street in London [12] and the Vdara Hotel in Las Vegas [13]. Both buildings have concave facades that concentrate and reflect incident solar radiation, creating a "death ray" when the sunlight hit the glass facade at certain angles. This intense reflected sunlight is focused into a small areas on the surrounding, generating more urban heat. Therefore, it can significantly affect the urban microclimate, influencing both the visual and outdoor thermal comfort of the inhabitants.

Several models have been developed to quantify interior glare inside buildings [14]–[16] , but less attention has be given to external glare. While some countries have developed guidelines to limit outdoor glare/daylight reflectance from building facades, no clear standard or parameters have been established. For example, some Australian cities have developed prescriptive measures to limit glare, which includes normal reflectance limits of than 20% for glass or all facade elements [17]–[20]. These guidelines are also followed by Singapore [21] and Hong Kong [22] on limiting external reflectance. Outdoor glare may also cause a problems to road safety if it has not been properly estimated during the design stage. The two types of glare that have been studied and evaluated the most are discomfort and disability glare, however both types are mostly assessed for visual performance inside buildings, compared with outdoor dwellers, where the impact of glare concerning pedestrian can be further intensified if they look directly at the source.

In tropical urban regions, glare and heat problems can have a more significant effect on urban outdoor dwellers, due to the greater incidence of solar radiation compared to temperate climate regions. Thus, there is a necessity to have reccomendations in form of guidelines/standards so that architectural designs do not come at the cost of comfort and safety. In a tropical country such as Singapore, most of the new high rise buildings have been designed with 100% fully glazed or translucent materials. Some buildings have been built using high reflective curtain walling materials similar to glass (Figure 1). These materials have caused visual and the thermal discomfort on the surrounding outdoor environment. The Building Construction Authority (BCA) of Singapore reported that there were an increasing number of feedbacks on discomfort associated with reflected glare from neighbouring buildings [21]. Currently, BCA has updated the regulations on daylight reflectance for materials used on the building facade. The new regulations have changed from a prescriptive based of less than 20% daylight reflectance to a performance based method that requires the external surface of a building to be designed and constructed without affecting the surrounding areas.

Figure 1. Outdoor glare from reflected sunlight in Singapore.

Regression modelling, factor analysis, and correlation analysis are the most familiar statistical techniques applied to analyse road users' glare. Nevertheless, for the estimation of dependent or output variables, there are some concerns when using regression analysis. The multicollinearity of independent variables is the main preventable issue of regression analysis. Furthermore, this issue has an effect on separate predictors but no impact on predictive power. In other words, in multiple regression techniques, there may be a well-fitted model, but correlated predictors may not produce trustable results for research model coefficients. Moreover, regression modelling does not facilitate estimating causal and indirect effects in a single and integrated equation.

In the last decades, Structural Equation Modelling (SEM) is an improved modelling technique, has been applied as a better statistical method of analysing normal and causal relationships with lack of multicollinearity among latent (construct) and measurement variables. Hence, the purpose of the current paper is to introduce a new framework which can provide an overall evaluation of observed and non-observed (latent) research variables by the combination of age, glare time, glare duration, avoidance and sensitiveness with SEM technique.

2. Methods

SEM is a robust method to hypothesize any kinds of relations and interactions among research variables in a single causal framework. Based on the objective of the study with capabilty of SEM, the present figure 2 illustrates the conceptual framework of the study.

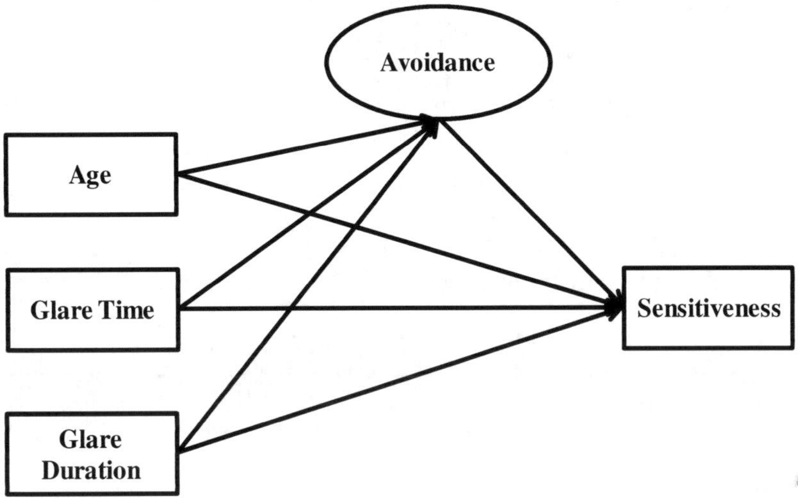

Figure 2. Research Framework

The squares (or rectangles; ▭) and circles (or ellipses; ⬭) represent observed and unobserved (latent) variables respectively. In the above figure 2, age, glare time, and glare duration act as independent variables, avoidance is a mediator, and sensitiveness is the dependent variable. In figure 2, age, glare time, and glare duration have direct and indirect impact on sensitiveness.

3. Sampling

The stratified sampling method has been chosen for the survey. The population surveyed was divided into four different locations (areas). These are respectively Marina Boulevard, Robinson Road, Chulia Street, and Shenton Way located in the city of Singapore. For each zone, 60 questionnaires were distributed. Therefore, the sample size is equal to 240. Data collection was performed between February 2016 and April 2016. The classical methods of standard SEM analysis are focused on the sample covariance matrix.

4. Data Analysis

The research framework developed can be observed in Figure 3. Prior to study the influence of independent variables on the dependent variable (sensitiveness), six model fitting tests were applied, to evaluate the consistency between the predicted and observed data matrix in the research equation. These indices were respectively Goodness of Fit Index (GFI), Relative Fit Index (RFI), Incremental Fit Index (IFI), Tucker Lewis Index (TLI), Comparative Fit Index (CFI), and Normed Fit Index (NFI).

For an acceptable model fitting, the value of the indices should be above 0.9. Figure 3 represents the model fitting results based on the SEM application. The results confirm that the defined indices values are within acceptable ranges. Therefore, the model fitting of the research framework is accepted.

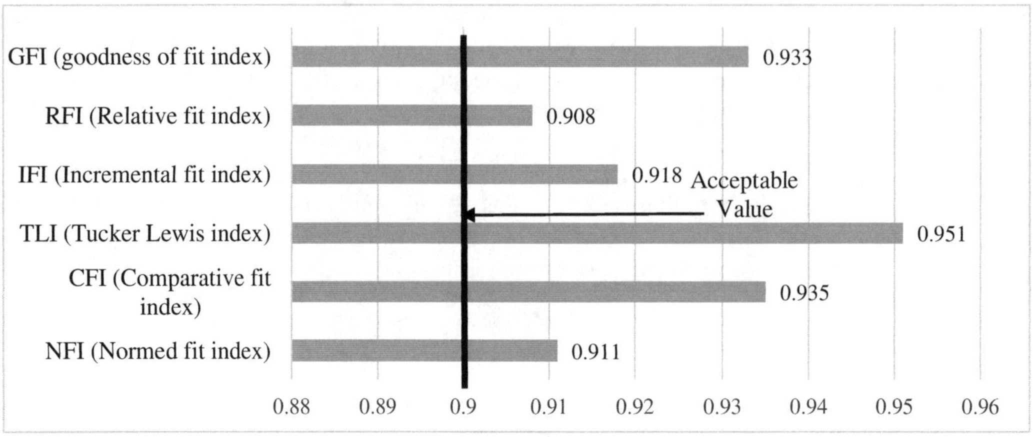

Figure 3. Model fit analysis

The structural model for the research framework is determined and shown in Table 1. There are seven relations in the research model. Their impact was evaluated using the AMOS software.

Table 1. The output of the structural model.

Path	Standardized Coefficient	C.R.	*p*
Age → Avoidance	0.56	7.329	<0.001
Age → Sensitiveness	0.44	5.451	<0.001
Glare Time → Avoidance	0.23	3.142	<0.001
Glare Time → Sensitiveness	0.29	3.881	<0.001
Glare Duration → Avoidance	0.11	1.612	0.113
Glare Duration → Sensitiveness	0.18	2.312	0.031
Avoidance → Sensitiveness	-0.33	−4.113	<0.001

5. Discussion

The aim of this study is to examine a multi-factorial model for the relationship between pedestrians and outdoor glare issues by applying SEM. The research framework contains four measurable variables (age, glare time, glare duration, and sensitiveness) and one latent variable (avoidance). The sensitiveness is the main dependent variable; the age of the road users, glare time, and glare duration status are the main independent variables, whereas avoidance is considered the mediator between the dependent and independent variables. As shown in Table 1 the impact of pedestrian age on avoidance and sensitiveness has the highest significance compared to glare time and glare duration on dependent and mediator variables. Glare time also has a major impact on avoidance and sensitiveness. Glare duration has a significant impact on sensitiveness. Interestingly, avoidance has a significant negative impact on sensitiveness. According to the findings, R^2 of the introduced model is 0.71, which means that 71% of a sensitiveness variation is dependent on age, glare time, glare duration, and avoidance.

The current study is limited by the cross-sectional data analysis, which does not allow influential temporal connections between research variables. The main suggestion is that this research should be carried out with longitudinal data, which can provide more robust and precise data analysis. The computation system in SEM has the main concern with normality assumption. In some data structures, the tested multivariate normality is not accepted based on normality tests. Thus, several scholars [23],[24]rely on Bayesian approach in SEM which is confirmed to have the capability to overcome the non-normality issues.

6. Conclusion

The main contribution of this study is about the application of SEM in estimating the perceived outdoor glare from highly reflective building facade by pedestrians in the city of Singapore. By applying SEM, a framework for outdoor glare evaluation that can fill in the modelling gaps that exist in regression analysis is presented. For a better sustainable development, planners, designers and engineers should understand the relationship between microclimate and glare holistically. It is hoped that this complicated relationship and predictive evaluate of glare should not be a reason to embolden the government initiatives for promoting sustainable design.

References

[1] Taha H. Urban climates and heat islands: albedo, evapotranspiration, and anthropogenic heat. Energy Build. 1997;25(2):99–103.
[2] Schiler M, Valmont E. Microclimatic Impact: Glare around the Walt Disney Concert Hall. Proc Sol World Congr … [Internet]. 2005 [cited 2014 Nov 11]; Available from: http://www.researchgate.net/publication/228838745_Microclimatic_Impact_Glare_around_the_Walt_Disney_Concert_Hall/file/5046352330b7f80b36.pdf
[3] Schiler M. Examples of Glare Remediation Techniques Four buildings. Plea2009. 2009;(June).
[4] Shih N-J, Huang Y-S. A study of reflection glare in Taipei. Build Res Inf [Internet]. 2001 Jan [cited 2014 Aug 14];29(1):30–9. Available from: http://www.tandfonline.com/doi/abs/10.1080/09613210150208778
[5] Harries A, Brunelli G, Rizos I. London 2012 Velodrome – integrating advanced simulation into the design process. J Build Perform Simul [Internet]. 2012 Aug 31 [cited 2013 Aug 23];(August 2013):1–19. Available from: http://www.tandfonline.com/doi/abs/10.1080/19401493.2012.718796
[6] Gil Vl-R. Evaluation of solar glare from reflective facades : A general method. Light Res Technol. 2015;1–18.
[7] Danks R, Good J. Urban Scale Simulations of Solar Reflections in the Built Environment : Methodology & Validation. Urban Scale Simulations Sol Reflections Built Environ Methodol Valid [Internet]. 2016;19–26. Available from: https://www.researchgate.net/publication/303312340_Urban_Scale_Simulations_of_Solar_Reflections_in_the_Built_Environment_Methodology_Validation
[8] Suk JY, Schiler M, Kensek K. Reflectivity and specularity of building envelopes: how materiality in architecture affects human visual comfort Reflectivity and specularity of building envelopes: how materiality in architecture affects human visual comfort. Archit Sci Rev [Internet]. 2017 [cited 2017 Sep 4];60(4):256–65. Available from: http://www.tandfonline.com/action/journalInformation?journalCode=tasr20
[9] Danks R, Good J, Sinclair R. Assessing reflected sunlight from building facades: A literature review and proposed criteria. Build Environ [Internet]. 2016;103:193–202. Available from: http://dx.doi.org/10.1016/j.buildenv.2016.04.017
[10] Suk JY, Schiler M, Kensek K. Development of new daylight glare analysis methodology using absolute glare factor and relative glare factor. Energy Build [Internet]. 2013 Sep [cited 2014 Aug 14];64:113–22. Available from: http://linkinghub.elsevier.com/retrieve/pii/S0378778813002569
[11] Schiler M. Examples of Glare Remediation Techniques Four buildings. In: PLEA2009 - 26th Conference on Passive and Low Energy Architecture, Quebec City, Canada, 22-24 June 2009 Examples. 2009.
[12] The Daily Telegraph. London skyscraper's "deathray" reflection is melting cars, burning

businesses, but also cooking eggs [Internet]. National Post. 2013 [cited 2017 Jul 20]. Available from: http://nationalpost.com/news/london-skyscrapers-deathray-reflection-is-melting-cars-burning-businesses-but-also-cooking-eggs
[13] Garfield L. The Vdara "death ray hotel" is still burning people in Las Vegas - Business Insider [Internet]. [cited 2017 Sep 2]. Available from: http://www.businessinsider.com/the-vdara-death-ray-hotel-is-still-burning-people-in-las-vegas-2016-6/?IR=T
[14] Araji MT, Boubekri M. Window Sizing Procedures based on Vertical Illuminance and Degree of Discomfort Glare in Buildings Interiors. Archit Sci Rev. 2008;51(3):252–62.
[15] Bessoudo M, Tzempelikos a., Athienitis a. K, Zmeureanu R. Indoor thermal environmental conditions near glazed facades with shading devices – Part I: Experiments and building thermal model. Build Environ [Internet]. 2010 Nov [cited 2014 Aug 7];45(11):2506–16. Available from: http://linkinghub.elsevier.com/retrieve/pii/S0360132310001514
[16] Yamin Garreton JA, Rodriguez RG, Ruiz A, Pattini AE. Degree of eye opening: A new discomfort glare indicator. Build Environ. 2015;88:142–50.
[17] City of Sydney, Central Sydney Development Control Plan. 2012; Available from: http://www.cityofsydney.nsw.gov.au/development/planning-controls/development-control-plans
[18] Streets D. Solar Light Reflectivity Analysis Sydney Water Corporate Headquarters. 2006;
[19] Bay H. Solar Light Reflectivity Analysis for the proposed development known as Site 4B , Sydney Olympic Park ,. 2007;1–23.
[20] Rofail A, Dowdle B, Perry J, Macdonald CM. Reflectivity impact on occupants of neighbouring properties. In: International Conference on Building Envelope Systems and Technology, 2004 Sydney, Australia [Internet]. 2004. p. 1–5. Available from: http://www.windtech.com.au/wp-content/uploads/2012/08/14.pdf
[21] Sreedharan S. BCA to update rule on use of reflective materials in buildings [Internet]. 2013 [cited 2017 Sep 3]. Available from: http://www.todayonline.com/singapore/bca-update-rule-use-reflective-materials-buildings
[22] Department HKB. Guidelines on Design and Construction Requirements for Energy Efficient of Residential Buildings. Vol. Hong Kong. 2014.
[23] Scheines R, Hoijtink H. Bayesian estimation and testing of structural equation models. 1999 [cited 2017 Sep 9];64(1):37–52. Available from: https://link-springer-com.libproxy1.nus.edu.sg/content/pdf/10.1007%2FBF02294318.pdf
[24] Ansari A, Jedidi K, Dube L. Heterogeneous factor analysis models: a bayesian approach. 2002 [cited 2017 Sep 9];67(1):49–78. Available from: https://link-springer-com.libproxy1.nus.edu.sg/content/pdf/10.1007%2FBF02294709.pdf

Local Stakeholder Perception on Community Participation in Marine Protected Area Management: A Q-Method Approach

P S Megat Jamual Fawaeed[1] and M S Daim[2]

[1]Centre of Biodiversity & Sustainable Development, UITM Shah Alam, Selangor
[2]Faculty of Architecture Planning & Surveying, UITM, Shah Alam, Selangor

*shahirah_puteri265@yahoo.com

Abstract. Local stakeholder involvement in Marine Protected Area (MPA) management can bring to a successful MPA. Generally, participatory research in marine protected area management is exploring the relationship between marine protected area management approach adopted by the management agencies and the level of participation of local stakeholder whom reside within the marine protected areas. However, the scenario of local community participation in MPA management in Malaysia seems discouraging and does not align with the International Aichi Biodiversity Target 2020. In order to achieve the International Aichi Biodiversity Target 2020, this paper attempts to explore the methodology on participatory research towards the local stakeholder of Pulau Perhentian Marine Park (PPMP), Terengganu, Malaysia. A Q-methodology is used to investigate the perspective of local stakeholder who represents different stances on the issues , by having participants rank and sort a series of statements by comply quantitative and qualitative method in collecting the data. A structured questionnaire will be employed across this study by means of face-to-face interview. In total, 210 respondents from Kampung Pasir Hantu are randomly selected. Meanwhile, a workshop with the agency (Department of Marine Park) had been held to discuss about the issues faces on behalf of management that manage the PPMP. Using the Q-method, researcher acknowledged wise viewpoints, reflecting how different stakeholders' perception and opinion about community participation with highlights the current level of community participation in MPA. Thus, this paper describes the phases involved in this study, methodology and analysis used in making a conclusion .

1. Introduction

The relationship between MPA and its local stakeholder is one of the most common issues in the field of MPA. This hate love relationship is laden with MPA management principal and policy adopted by most the management agencies mixed with the attitude, values and norms of the local community whom reside within the marine protected areas [11]. However recently, the importance of local stakeholder involvement in MPA management is being acknowledged by many management authorities as conflict resolution approach. It creates a win-win situation which later creates beneficial relationships to all parties involved in a MPA especially local stakeholder reside there.

MPAs are well known throughout the world as a tool for managing fisheries, conserving species and habitats, maintaining ecosystem functioning and resilience, preserving biodiversity and protecting manifold of human values related with the ocean [14]. While the myriad of human values and biodiversity are essential to conserving, conflict between conservationists and people desperate in accessing the resources had caused criticism of conservation organizations and their activities globally [7]. Moreover, a solution is needed in ways of seeking for transforming conflicts between MPA and local stakeholder into mutually beneficial relationships. Instead maintaining MPA by force, it should be done by good approach towards community because it is important of considering community livelihoods, particularly when 'no-take' MPAs was employed, as well as governance and management for the success of MPAs [13].

Therefore, this study reviews with the vital of participatory approach in marine protected area management. Particularly, there is a long overdue issue that always being concern is local stakeholder versus MPA management. In order to explore this issue, one MPA in the state of Terengganu, Malaysia- Pulau Perhentian Marine Park (PPMP) is selected. Precisely, this study aims to analyse the participatory approach in MPA management; to analyse the level of local stakeholder participation in PPMP; to analyse benefits that local stakeholder get from their participation in PPMP management; to analyse the support from authorities for encourage local stakeholder participation; and to develop a community participation model for PPMP.

2. Research Design

A plan for gathering the data required to achieve the research objectives must be devised once a decision is made to proceed with the next step of research [12]. The function of research design is to obtain relevant evidence in achieving objective with answering the initial question as unambiguously as possible [2]. A participatory research have evolved from qualitative method approaches with objective to reflect, explore and propagate the views, concerns, feelings and experiences of research participants from their own perspectives [16]. According to J. Bergold and S. Thomas, the objective of the inquiry and the research questions in participatory research was develop out with merging the two perspectives; science and practice whereas the results benefit both sides from the research process [8]. Thus, the Q-method had been chosen to be attempt in this research corresponding to the aims of the method which is act as cornerstone for the systematic study of subjectivity, a person's viewpoint, opinion, beliefs, and attitude [4].

2.1. Q-Method: An attempt in participatory research

Q-methodology is an approach to research at the interface between qualitative and quantitative methods where it is used particularly for the acquisition of complex views, attitudes and value orientations from a subjective perspective [3]. The Q-methodology is equivalent to this study because this study want to identify local stakeholder viewpoints on community participation, where it analyse the view of benefit, level of participation and support for participation from local stakeholder perception. Q-Method employs qualitative judgements of the researcher in defining the problem, developing statements to identify the participants' perspectives and selecting participants and used quantitative options of analysis.

Basically, data in Q are collected within five steps. The first step is definition of the concourse which refers to collection of all the possible statements the respondents can make about the subject at hand or in other meaning is the raw material [16]. In other meaning, it is the hub of information about research topic (in this study: the hub of available information on the local community participation in MPA) where is the concourse are all the data from interviewing local stakeholder, participant observation, and scientific literature such as papers, journals, and books from which the 'Q set' is taken. Secondly, develop the Q-set which is the subset of statement is drawn from the concourse which based on the literature review, to be presented to the participants [9]. All the selected statements included in the Q set is based on the concourse that basically indicate the diversity of discourses apropos the local community participation in MPA. In this study, the Q set consists of 61 statements (Table 2) with key themes within the opinion on local community participation in MPA had been expressed in the concourse (dependency on resources, issues, concern, opportunity, and benefit). To avoid potential biases, researcher used a 'structured approach' [9]. Next is selection of the P-set or Q participants which is usually smaller than the Q-set. In this study, it will be referred to the village committee (JKK) of Kampung Pasir Hantu and the agency involved (Department of Marine Park Malaysia) who represent the array of perspectives and opinions concerning the local community participation in MPA. Then, Q sorting which is the respondent is instructed to rank the statement according to their preference. This process shows the participants' subjective meaning to the statements [10]. The condition of instruction is normally the person's point of view regarding the issue.

Then, participants' sorts are being analysed using correlation and factor analysis to show clusters of subjectivity [10]. Thus, this methodology is suitable for this study because this process will reveals group of stakeholder perspective and the similarities and differences in viewpoint over the issue. On the other hands, it helps to create a platform for meaningful discussion between different stakeholder groups through stakeholders' perspective [10] for participatory research.

2.2. The main study

For the purpose of this study, one marine protected area (PPMP) is selected by the researchers as it will represent a continuance of the Department of Marine Park Malaysia effort in managing its 42 marine protected areas in peninsular Malaysia. The management authority of PPMP needs to increase their effort to encourage community participation in MPA management. This case study will show that effective MPA management can be achieved when the local stakeholder is a part of management and conservation initiative in MPA by involving them in decision-making process and conservation programme. The study will explore the wider perspective of PPMP from geographical locations, villages, chalet and resort within the Pulau Perhentian considering the size and various communities lived and use the resources within and around this reserve. The study will substantiate that it is essential to identifying benefits for the community to ensure a successful participatory process. A case study help researcher link the actions of individual people to social structure and process [19] in this case the local stakeholder participation and the marine protected area management.

2.3. Sampling design

The target population for the study is the local community whom were the stakeholders or play character to the community involvement initiative in PPMP. The only village in PPMP Kampung Pasir Hantu is selected for this study (Table 1).

Table 1. Villager's ethnicity, household and resident in Kampung Pasir Hantu

Household	Resident	Ethnicity	Total Population
450	210 door	Bumiputera : 2,160	2,163
		Others : 3	

From the data above, household is the total number of family live in Kampung Pasir Hantu. Meanwhile, resident is total number of house built in the village. In order to give ample power detecting relationships, the "5times" heuristics and the "10 times" heuristics is used as the guide to minimum sample size since this study will use SEM-Partial Least Square as method for analysis [11] and 210 sample was considered as good sample in fulfil representative criteria.

2.4. Data collection and instrument

This study employed the survey approach with a face-to-face interview with local stakeholders using structured questionnaire. This is a better approach in obtaining representative samples and produce the fewest number of constraints in terms of questionnaire design and questionnaire construction [11] such as constraints in wording aspects, the measurement scale used, and the layout design of the questionnaire [15]. The questionnaire consists of six sections (Table 2) which mostly using Likert Scale as measurement.

Table 2. Arrangement of the questionnaire

S ection	Questions Classification	Number of items
I	Respondent's background	7 items
II	Dependency on resources	14 items
III	Issues, concerns and opportunities	5 items
IV	Expectation benefits	30 items
V	Level of participation	7 items
VI	Support for community participation	5 items

The questionnaire had been design as open-ended structured. The questionnaire had been divided into six sections according it purpose to achieve the objective of the study. Respondent background is essential in order to know their profile demography. Then, the data from dependency on resources are important to evaluate how important is it the marine park towards stakeholder. The paramount data is from the issues, concerns and opportunities from their perceptions which help to identify marine park issues and improve the management. Besides, the expectations benefit can encourage for more conservation programme to protect the marine park values. Meanwhile, the section fifth and six is to measure the level of stakeholder participation in MPA management and if there any support for encouraging stakeholder participation. Henceforth, Q-method is very practical to use in this study in order to get the data systematically, precise and efficient.

3. Analysis

This study will use several statistical techniques including factor analysis and reliability analysis, descriptive statistics, and partial least square for analyzing the data, hypotheses testing and reaching conclusions. These statistical techniques are used to test the local stakeholder dependency on resources, benefits that local community get from the MPA, level of local stakeholder participation in MPA management, and support for local stakeholder actively participate in MPA management.

3.1. Modelling local stakeholder participant: Partial Least Square SEM (PLS-SEM)

PLS-SEM models are path models which some variables may be effects of others while still because for variables later in the hypothesized causal sequence where is an alternative model to covariance - based structural equation modelling [1, 4]. Primarily, Partial Least Square SEM (PLS-SEM) is used to establish theories in exploratory research by focusing on explaining the variance in the dependent variables when examining the model [6]. The rules of thumb for choosing PLS-SEM as method of analysis had been presents by [6]. The first rule is deciding the aim of the research is either to predict key target constructs or identifying key "driver" constructs. In this research, the aim is to predict the influence of social benefits, economic benefits, attitudinal benefits and legal/institutional benefits on community participation in MPA management. These beneficial are the "drivers" to community participation. A PLS-SEM is preferred method since the aim of this study is not theory confirmation. PLS-SEM can handle both reflective and formative measurement models as well as single-item measures without additional requirements or constraints [6]. In this study, reflective measurement model are being involved whereas the indicators refers to the functions of the latent construct and changes in the latent constructs are reflected in changes in the indicators variables [5]. This study involves many constructs and many indicators. As mentioned, PSL-SEM is capable of estimating very complex models to using it as the method of analysing the data is appropriate [6].

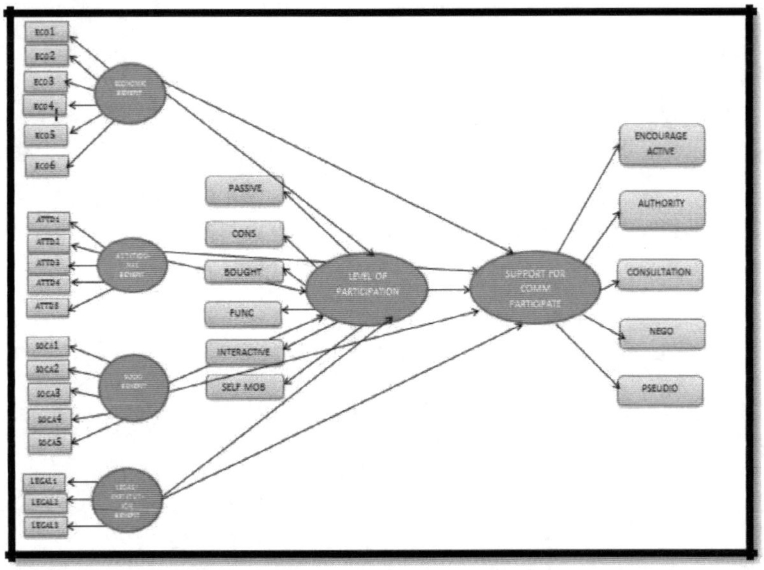

Figure 1. Hypothetical Model of the study

4. Conclusion

The purpose of this paper is to introduce Q methodology in the participatory research. The reason Q method had been chosen, because of focusing to highlighting the synergies between the approach and current theoretical concerns in participatory research. This is an appropriate method was said by Brown (1998), when the application are endless, then the term 'subjectivity' itself is 'in vogue' which welcoming the Q methodology in participatory research.

Acknowledgements

The authors would like to extend their greatest appreciation to all participants including the Management Plan for Pulau Payar and Pulau Perhentian project team members, staff from agency Department of Marine Park and the local stakeholder at Pulau Perhentian for giving their valuable time, cooperation and support. The authors also acknowledge the Department of Marine Park, Nature Resources and Environment GOV grant 100-IRMI/GOV 16/6/2 (009/2017) to give funding for this research.

References

[1] A. Monecke and F. Leisch (2010). *SemPLS: Structured Equation Modeling Using Partial Least Squares.*

[2] David A. de Vaus (2001). *Research Design in Social Research.* California, Sage Publication Inc.

[3] F. H. Müller and Elisabeth Kals (2004). *The Q Method: An Innovative Method for Collecting Subjective Attitudes and Opinions.*

[4] G. David Garson (2016). *Partial Least Squares: Regression & Structural Equation Models.* Statistical Associates Publishing.

[5] Hair, J. F., C. M. Ringle, et al. (2011). "PLS-SEM: Indeed a Silver Bullet." *Journal of Marketing Theory and Practice vol. 19(no. 2):* pp. 139-151.

[6] Hair, J. F., G. T. M. Hult, et al. (2017). A Primer On Partial Least Squares Structural Equation Modeling (PLS-SEM). California, Sage Publications Inc.

[7] J. Beaumont (1997). Community Participation in the Establishment and Management of Marine Protected Areas: A Review of Selected International Experience. *South African Journal of Marine Science.*

[8] J. Bergold and S. Thomas (2012). Participatory Methods: A Methodological Approach in Motion. *Forum: Qualitative Social Research. Vol. 13.*

[9] J. Hugé, et al. (2016). *Mapping discourses usin Q methodology in Matang Mangrove Forest,*

IOP Conf. Series: Earth and Environmental Science **117** (2018) 012039 doi:10.1088/1755-1315/117/1/012039

Malaysia.

[10] Melanie Kincaid (2011). *A Q-Methodology Study: Stakeholder Perspectives on the Future Management* of Capitol Lake, Olympia, Washington.

[11] M. S. Daim and J. Abdullah (2013). *Community Participation in Protected Area Management in Peninsular Malaysia: A Participatory Research.*

[12] Mukesh Kumar et al. (2013). *Business Research Methods.* Oxford Fajar Sdn. Bhd.

[13] N. J. Bennett and Philip Dearden (2013). *Why Local People Do Not Support Conservation: Community Perceptions of Marine Protected Area Livelihood Impacts, Governace and Management in Thailand.*

[14] N. J. Bennett and Philip Dearden (2014). *From Measuring Outcomes to Providing Inputs: Governance, Management, and Local Development for More Effective Marine Protected Areas.*

[15] Sekaran, U. (2006). *Research methods for business: A skill building app roach.* New Delhi, Wiley.

[16] Stevan R. Brown (1993). *A Primer on Q Methodology.*

[17] Swain J. and French S. (2004). *Researching Together: A participatory approach.*

[18] Van Exel NJA, G de Graaf (2005). *Q methodology: A sneak preview.*

[19] W. Lawrence Neuman(2003). *Social Research Methods.* Pearson Education Inc.

A preliminary investigation on the effects of characteristics and contractual behaviour on civil engineering project performance

W N W Ismail[1], H Adnan[1], N Yusuwan[1], M Maisham[1], A A Hassan[1]
[1]Faculty of Archithecture, Planning and Surveying, Universiti Teknologi MARA, Perak Campus, Seri Iskandar, Perak, Malaysia

*Wanno134@perak.uitm.edu.my

Abstract: The significant role of civil engineering project is not only to make the lives of people easier and secure but also to trigger the economic growth by providing infrastructure facilities as well as job opportunities. As it is dominantly initiated by government sectors, performance of the civil engineering projects is always observed. This study aims to investigate the characteristics of civil engineering project and the contractual behavior of the key participants and how do these two factors affect civil engineering projects performance. Literature reviews, content analysis and questionnaires survey were conducted to undertake the research. A total of 50 questionnaires were distributed and 10 questionnaires were returned, resulting in a 20% response rate. The research unveiled that performance of civil engineering projects are influenced greatly by the ability to handle the unpredictable character of the civil engineering projects and adequate behavioral management. Apart from that, balancing the factors with high quality of workmanship, avoidance or well managed conflicts and high satisfaction level will ensure performance in projects.

1. Introduction

Civil engineering structures such as roads, highways, dams, bridges, airports, ports etc. are the bloodlines to a progress of a country. Infrastructures improve transportation, communication and services, thus serve the country's development by underpinning the economic and linking social activities. A well-structured nation helps to generate higher economic activities and attracting local and foreign investment. Civil engineering structures have always been a traditional responsibility of the public sector. Expectation for delivery of work on time, satisfactory quality and responsive customer services is putting the public sector responsible for the works under the microscope.

In literature, there are many researchers who focus their study on the general building construction. Very few studies can be found research on the characteristics of civil engineering projects. Coincidentally, it is found that the research recently done by [1 – 3] and [4] are focusing on civil engineering projects. However, research by [1] is focusing more on the management of risk in major infrastructure projects, meanwhile [2] narrow down their research on the case of cost overrun in Dutch transport infrastructure projects. Meanwhile, [4] focus on the impact of irreversibility and uncertainty on the timing of civil engineering projects. None of the studies have established the characteristics of civil engineering projects. Hence, other than difficult in management and implementation, the characteristics of civil engineering projects which has numerous attribute that different from general building project also can lead to undesired behavior of project participants where eventually affect the performance of civil engineering projects.

On the other hand, the impacts of contractual behavior of the key participants on civil engineering performance also important to be explored. In fact, [5] and [6] found that most of the problems in construction projects were related to the contractual behavior of individual key participants in the project. The contractual behavior of the key participants i.e delay in paying interim payment, late in giving possession, architect's behavior, adversarial relationship, poor communication are among other things that affect the project performance [7] and [6]. Unfortunately, none of the study done by the mentioned authors differentiates by the types of project and mostly focus on the general building projects. The findings for the civil engineering project might be different from the general building project due to their distinctive characteristics. This is due to the assumption that the more uncertainty element exist in the project, the more unpredictable behavior of the key participants will occur [8]. Hence, it is important to investigate the common contractual behaviours of key participants since they are among the critical factors that will determine the success and the failure of civil engineering projects.

2. Literature review

[9 – 11] are among the researchers have explored and identified the characteristics of different types of construction projects. However, most of them have done the study focusing on the building projects. For general building project [11] explain the project characteristics based on a precise cost estimate before contract signing; time reductions; tight project milestones or deadlines; cost savings; project budget; ability to define the project scope; project size; complexity. From the perspectives of civil engineering works, [10] have determine the project characteristics based on contract size; tender type; bid ratio ; percentage difference between awarded bid and estimate; extra project cost; number of bidders; size of contractor; project complexity; type of design and supervision; experience of contractors; project regional location. Unfortunately, the focus given only on the characteristics that influence on the time performance. Very few research can be referred to in order to determine the characteristics of civil engineering project performance. Due to lack of evidence in literature on the characteristics of civil engineering that influence the project performance, this study will identify project characteristics variables gained from literature that are appropriate to be utilized to describe the characteristics of civil engineering project. Therefore, this study refines the project characteristics from the aforementioned previous studies and establish nine (16) project size; procurement method; type of Standard Form of Contract; contractual arrangement; type of design and supervision; project location; construction complexity; design completion before construction start; clear definition of project scope; tight project milestone; ease of site access; variety of stakeholders; environmental uncertainties; technological advances; multicultural team; resources availability.

Meanwhile, the contractual behavior of key participants is referring to the extent the responsibility stated in contract is implemented by the people who makes decision by the contract. In other words, it is related to the action as well as the commitment of the participants towards their task and responsibility as stipulated in the contract. [5] Observed that most of the problems in the construction project are not because of technical, financial, contract or environmental issues where most of them are due to the behavior of the participants. Therefore, the behavior of the participants in complying the responsibility as stipulated in Standard Form of Contract important to be investigated in the present study. This study would explore the behavioral of the participants towards the task and responsibility stipulated in Standard Form of Contract for instance in the aspects of giving instructions; obeying instructions; communication; making decision; making payment; variation; site possession.

3. Methodology and analysis

The study focused on civil engineering projects in Malaysia. The focus of research was limited to the roads projects in Malaysia including their associated structures such as bridges and tunnels. A specific area and locality is determined to ensure that the research would be covered within the limited time of study. The data for this study was obtained from G7 of CIDB's listed contractors and certified professional engineers registered under The Boards of Engineers Malaysia (BEM). In addition, to provide better analysis of the project characteristics that influence civil engineering project performance, the study would cover civil engineering projects with the contract value of more than RM10 million.

In order to achieve the objectives of the study, a mixed-method approach was adopted. The study started with an intensive literature review to identify the characteristics of civil engineering project and the common contractual behaviour of the key participants towards the project which have the possibility in affecting the project performance. Since there are very few studies can be found focusing on civil engineering projects as well as to incorporate the scenario of civil engineering projects problems in Malaysia, content analysis method was adopted. The National Audit Reports and construction report prepared by G7 contractors was used to investigate the common contractual behaviour of key participants in civil engineering projects. This is an appropriate solution to overcome the shortage of information available in literature.

Next, based on the variables gain from literature reviews and content analysis, preliminary questionnaire was developed and the distributed to the G7 contractors and professional engineers. The purpose is to ensure the validity of variables gained from literature reviews and report content analysis. Besides, since the study involves large number of respondents, the adoption of questionnaire survey is appropriate because it capable to approach respondents regardless of their location and answer the same set of questions effectively.

A total of 50 preliminary questionnaires were distributed. However, only 10 questionnaires were returned by the cut-off date, resulting in a 20% response rate. The views were therefore predominantly those of contractor. Majority of all respondents have more than 3 years of working experience in procuring construction projects with 50% of their companies' revenue comes from civil engineering projects.

3.1 Project characteristics variables affect the performance of civil engineering works

Civil engineering projects are complex and difficult to manage. Hence, this study attempts to investigate the level of influence of project characteristics towards performance. As indicated in Table 1, all respondents are very certain that construction complexity and environmental uncertainties are the main characteristics that give impacts on performance. This is followed by design completion before construction start and clear definition of project scope. Meanwhile procurement method, project location, and resources availability were rated as the characteristics that give low influence on performance. This result is in contrast with building project characteristics where project size is the most critical factors that influence project performance [12].

3.2 Contractual behaviour of key participants

Human factors are prevalent and always a major factor. In an industry where involvement from numerous parties is common, behavioural factors are always predominant. The following results examine the contractual behaviour among the parties. The data observed the most frequent behaviour practiced by the parties. The listed behaviours are alleged to be the factors of disputes in civil engineering works where eventually affect the performance.

Based on the result tabled in Table 2, most of the respondents rated very high frequency for the late of client in delivering interim payment. This is followed by the S.O give unauthorised instructions to the contractor and the contractor obey them. This is accord with the study by [7] and [13] who state that delay in paying progress payment to contractor is the most critical factors contributing to conflict among them and eventually can reduce the project performance.

IOP Conf. Series: Earth and Environmental Science **117** (2018) 012040 doi:10.1088/1755-1315/117/1/012040

Table 1. Civil engineering project characteristics

Item	Civil engineering project characteristics	Frequency: Not certain 1 No	%	Neutral 2 No	%	Very certain 3 No	%	TR	MR	Ranks	Freq
1	Project size	0	0%	3	30%	7	70%	10	2.70	5	Very certain
2	Procurement method	2	20%	6	60%	2	20%	10	2.00	16	Neutral
3	Type of Standard Form of Contract	0	0%	5	50%	5	50%	10	2.50	7	Very certain
4	Contractual arrangement	3	30%	1	10%	6	60%	10	2.30	12	Neutral
5	Type of design and supervision	3	30%	1	10%	6	60%	10	2.30	10	Neutral
6	Project location	6	35%	4	24%	7	41%	17	2.06	15	Neutral
7	Construction complexity	0	0%	0	0%	10	100%	10	3.00	1	Very certain
8	Design completion before construction start	0	0%	2	20%	8	80%	10	2.80	4	Very certain
9	Clear definition of Project scope	0	0%	1	10%	9	90%	10	2.90	2	Very certain
10	Tight project milestone	0	0%	5	50%	5	50%	10	2.50	7	Very certain
11	Ease of site access	0	0%	4	40%	6	60%	10	2.60	6	Very certain
12	Variety of stakeholders	0	0%	5	50%	5	50%	10	2.50	7	Very certain
13	Environmental uncertainties	0	0%	0	0%	10	100%	10	3.00	1	Very certain
14	Technological advances	0	0%	6	60%	4	40%	10	2.40	11	Very certain
15	Multicultural team	0	0%	5	50%	5	50%	10	2.50	7	Very certain
16	Resources availability (material, labour, plants)	2	17%	6	50%	4	33%	12	2.17	14	Neutral

IOP Conf. Series: Earth and Environmental Science **117** (2018) 012040 doi:10.1088/1755-1315/117/1/012040

Table 2. Contractual behaviour of key participants in civil engineering projects

Contractual behavior among the parties in civil engineering works	Frequency						TR	MR	Ranks	Freq
	Low		Neutral		Very high					
	1		2		3					
	No	%	No	%	No	%				
How frequent did the client give direct instruction (without going through the S.O)	2	20%	5	50%	3	30%	10	2.10	4	Neutral
How frequent did the S.O give unauthorised instruction to the contractor	2	20%	2	20%	6	60%	10	2.40	2	Very high frequent
How frequent did the contractor obeyed to the unauthorised instruction from the S.O	2	20%	4	40%	4	40%	10	2.20	3	Neutral
How frequent the client fail to pay the interim payment within the stipulated time	1	10%	3	30%	6	60%	10	2.50	1	Very high frequent

Table 3. Communication between contractor and S.O

Contractual behavior among the parties in civil engineering works	Agreement on effectiveness						TR	MR	Ranks	Freq
	Not agree		Neutral		Very agree					
	1		2		3					
	No	%	No	%	No	%				
How do you rate the effectiveness of communication between the S.O and the contractor		0%	6	60%	4	40%	10	2.40	1	Very agree

Concisely, communication between parties is rated from neutral to any positive scale, indicating that parties accept the communicational relationships within civil engineering works environment. Even though there is a preconceived notion that communicational problems are predominant in the industry, it is not observed here.

4. Conclusions

The performance of civil engineering projects influenced greatly by the ability to handle the unpredictable character of the civil engineering projects and adequate behavioural management of the key participants. Construction complexity and environmental uncertainties are the main characteristics of civil engineering projects that give big impact on performance. In Malaysia, the PWD203a Standard Forms of contract series are prevalent as most civil engineering projects are government initiated. The documents provided sufficient details to resolve conflicts, even though the contents are not generally favoured by the parties. An improvement to suit the characteristics of the civil engineering works would be advantageous. Most behavioural issues are induced by communication problems. There is a preconceived notion that communicational problems are predominant in the industry, but performance is observed to prove otherwise. This is probably due to the acceptance towards the current communicational practice. Evidently, performance is also influenced by high quality of workmanship, avoidance or well managed conflicts and satisfaction level. A balanced in the aforementioned factors will ensure performance in the civil engineering projects.

Acknowledgements
The authors would like to thank the Institute Pengurusan Penyelidikan dan Inovasi (IRMI) UiTM for supporting this research under Geran Penyelidikan Lestari 2015, 600-RMI/DANA 5/3/LESTARI (23/2015)

References

[1] F. Guo, Y. Chang-Richards, S. Wilkinson, and T. C. Li, "Effects of project governance structures on the management of risks in major infrastructure projects: A comparative analysis," *Int. J. Proj. Manag.*, vol. 32, no. 5, pp. 815–826, Jul. 2014.

[2] C. C. Cantarelli, B. van Wee, E. J. E. Molin, and B. Flyvbjerg, "Different cost performance: different determinants?," *Transp. Policy*, vol. 22, pp. 88–95, Jul. 2012.

[3] S. Verweij, I. van Meerkerk, and I. a. Korthagen, "Reasons for contract changes in implementing Dutch transportation infrastructure projects: An empirical exploration," *Transp. Policy*, vol. 37, pp. 195–202, Jan. 2015.

[4] P. Doan and K. Menyah, "Impact of Irreversibility and Uncertainty on the Timing of Infrastructure Projects," *J. Constr. Eng. Manag.*, vol. 139, no. MARCH, pp. 331–338, 2013.

[5] S. Toor and S. O. Ogunlana, "Construction professionals' perception of critical success factors for large-scale construction projects," *Constr. Innov.*, vol. 9, no. 2, pp. 149–167, Apr. 2009.

[6] F. Y. Y. Ling, Y. Ke, M. M. Kumaraswamy, M. Asce, and S. Wang, "Key Relational Contracting Practices Affecting Performance of Public Construction Projects in China," *J. Constr. Eng. Manag.*, vol. 142, no. March, pp. 1–12, 2013.

[7] N. Jaffar, a. H. A. Tharim, and M. N. Shuib, "Factors of Conflict in Construction Industry: A Literature Review," *Procedia Eng.*, vol. 20, pp. 193–202, 2011.

[8] W. Lu, L. Zhang, and L. Zhang, "Effect of Contract Completeness on Contractors' Opportunistic Behavior and the Moderating Role of Interdependence," *J. Constr. Eng. Manag.*, vol. 142, no. 6, p. 4016004, 2016.

[9] T. Y. Lo, I. W. H. Fung, and K. C. F. Tung, "Construction Delays in Hong Kong Civil Engineering Projects," *J. Constr. Eng. Manag.*, vol. 132, no. June, pp. 636–649, 2006.

[10] A. A. Othman, J. V. Torrance, and M. A. Hamid, "Factors influencing the construction time of civil engineering projects in Malaysia," *Eng. Constr. Archit. Manag.*, vol. 13, no. 5, pp. 481–501, 2006.

[11] B. Liu, T. Huo, Y. Liang, Y. Sun, and X. Hu, "Key Factors of Project Characteristics Affecting Project Delivery System Decision Making in the Chinese Construction Industry: Case Study Using Chinese Data Based on Rough Set Theory," *J. Prof. Issues Eng. Educ. Pract.*, vol. 137, no. January, pp. 206–210, 2012.

[12] S. Demirkesen and B. Ozorhon, "Assessing the Impact of Project Characteristics on Construction Project Success," *12th Int. Congr. Adv. Civ. Eng. ACE2016.*, pp. 1–8, 2016.

[13] W. Alaghbari, M. Razali A. Kadir, A. Salim, and Ernawati, "The significant factors causing delay of building construction projects in Malaysia," *Eng. Constr. Archit. Manag.*, vol. 14, no. 2, pp. 192–206, 2007.

IOP Conf. Series: Earth and Environmental Science 117 (2018) 012041 doi:10.1088/1755-1315/117/1/012041

Comparative study of cocoa black ants temporal population distribution utilizing geospatial analysis

N A Adnan[1], S Bakar[2], A H Mazlan[1], Z Mohd Yusoff[2], A R Abdul Rasam[1]

[1]Applied Remote Sensing and Geospatial Research Group, Centre of Studies Surveying Science & Geomatics, Faculty of Architecture, Planning and Surveying, Universiti Teknologi MARA Shah Alam, Selangor
[2]Malaysian Cocoa Board, 5 – 7th Floor, Wisma SEDCO, Lorong Plaza Wawasan,Off Coastal Highway, Locked Bag 211, 88999 Kota Kinabalu, Sabah, Malaysia

*nor_aizam@salam.uitm.edu.my

Abstract. Cocoa plantation also subjected to diseases and pests infestation. Some pests not only reduced the yield but also inhibit the growth of trees. Therefore, the Malaysia Cocoa Board (MCB) has explored Cocoa Black Ants (CBA) as one of their biological control mechanism to reduce the pest infestation of the Cocoa Pod Borer (CPB). CPB is capable to cause damage to cocoa beans, and later on will reduce the quality of dried cocoa beans. This study tries to integrate the use of geospatial analysis in understanding population distribution pattern of CBA to enhance its capability in controlling CPB infestation. Two objectives of the study are i) to generate temporal CBA distribution of cocoa plantation for two different blocks, and ii) to compare visually the CBA population distribution pattern with the aid of geospatial technique. This study managed to find the CBA population pattern which indicated spatially modest amount of low pattern distribution in February of 2007 until reaching the highest levels of ant populations in September 2007 and decreasing by the end of the year in 2009 for two different blocks (i.e 10B and 18A). Therefore, the usage of GIS is important to explain the CBA pattern population in the mature cocoa field. This finding might to be used as an indicator to examine the optimum distribution of CBA, which needed as a biological control agent against the CPB in the future.

1. Introduction

Cocoa crop is one of the national commodities and plays an important role for the economy in Malaysia. These plants can bear fruit throughout the year, so it can be a source of daily or weekly income to growers [1]. Cocoa like other plants also has problems with diseases and pests wherever it is grown. Some pests are not only reducing the yield but also inhibit the growth of trees. In the South East Asian region, the Cocoa Pod Borer (CPB), *Conopomorpha Cramerella* Snellen is well-known as the most devastating pest of cocoa. CPB infestation may produce low quality of dried cocoa beans, which resulted from clumped and unusable beans. The whole pod must be discarded when the infestation is severe [2]. Yield losses may up to 100% if the pest if left untreated [3]. As CPB persisted, smallholders have no choice either to adapt strategies in combating this pest or to suffer with a dramatic loss of cocoa income [4]. Various control approaches had been implemented including combination of recommended agronomic practices, chemical control, resistant planting materials, fertilizer programming and Integrated Pest Management [5]. Unfortunately, no unilateral technique was proven successful in managing CPB; even chemical control is the most preferable technique by the cocoa growers. As an alternative towards heavy reliance on chemical, the use of biological control in the management of CPB, particularly using ant species had been previously studied in Malaysia. Ants are insects which is abundance on the earth's surface, with more than 12,000 species. Among that, 7600 species from 250 genera have been named and described scientifically [6]. Few ant species were observed to have potential in controlling cocoa pests, especially the Cocoa black ants (CBA), *Dolichoderus thoracicus* Smith (Hymenoptera: Formicidae).

CBA is an endemic species and is feasible where cocoa is inter-planted under coconut [3]. CBA may thrive well due to the availability of nesting sites on the coconut crown. CBA is less hazardous due to their non-stinging behaviour, which given no harmful effects to cocoa growers. CBA have the ability to spread their population in the block with favourable condition. The population of the CBA can be enhanced by augmentation of artificial ant nests made from a plastic bag and stuffed with dry cocoa leaves. Nests were hung on the cocoa trees attended by CBA, and by increasing the numbers of artificial nests, the populations of CBA were enhanced [7]. CBA was able to reduce CPB infestation, in term of increasing the percentage of good cocoa pods and reducing the damage severity. CBA having potential as oviposition deterrence and may hinder CPB from laying their eggs on the pod surface. The use of CBA shows great potential as an alternative approach in controlling CPB, which produce almost similar results provided by insecticide application [8]. However, the CBA must be distributed evenly on every tree and can cover almost all pods for better protection against CPB. Hence, augmentation of CBA artificial nests should be encouraged for an alternative control of CPB.

In agriculture, GIS products are very useful for predicting the area and production of agricultural commodities, determination central agriculture, mapping of land resource potential, development of agro-industry and the prediction of the spread of pests and diseases of plants. In general, the map is a means to get some idea of scientific data contained on the surface of the earth by describing the various signs and descriptions, making it easy to read and understand. There are several interpolation method are often used in GIS, each method has its characteristics and advantages and disadvantages of each. Interpolation methods in GIS are Inverse Distance Weighted (IDW), Natural Neighbor, Kriging and Spline. Kriging interpolation method is the spatial interpolation method that utilizes spatial values at sample locations to produce value at another location that is not a sample. Stochastic estimation kriging method is similar to IDW [9]. Kriging interpolation normally estimates of the core surface by a weighted normal of the information, with weights declining with separation between the point when the surface is being evaluated and the areas of the information points [10]. In this study only Kriging interpolation method was utilised to produce CBA distribution map.

A current, most of the study of CBA population [3, 7, 8, 11] were analyzed using conventional methods such as utilizing the Microsoft Excel coupled with appropriate statistical analysis. However the data density of CBA distribution is not clearly seen and unable to reveal any distribution pattern of CBA distribution [3]. Therefore, with the advancement of technology such as Geographical Information System (GIS), the existing data can be mapped and analysed any pattern inherent in the data spatially. The aim of this project is to study the CBA population density formed within two different blocks of cocoa plantation. Hence the objectives of this study are to i) to generate temporal CBA distribution of cocoa plantation for two different blocks ii) to compare visually the CBA population distribution pattern with the aid of geospatial analysis.

2. Materials and Methods

The study was conducted at the Cocoa Research and Development Center (CRDC), Malaysian Cocoa Board Hilir Bagan Datuk, Malaysia (Longitude E.100 M, 52' 0', Latitude N3 53' 42). Two different blocks known as Block 18A high existing block of CBA population and 10B is categorized as medium existing block of CBA population [11]. Both blocks harbored almost similar conditions in term of tree number, tree age, shade trees as well as control and agronomic practices. Age of mature trees at these blocks ranging between 16 to 18 years during the data. CBA data were collected for 30 months starting in February 2007 through July 2009. All the data were recorded manually and saved in the tabular data format without information of x and y location for each tree sampling data.

Position of each sampling cocoa tree within these two blocks were also collected to make it compatible to be analyse in GIS environment. Global Positioning System (GPS) handheld receiver was used to obtain the position of each of the cocoa tree. GPS receivers use the World Geodetic System 1984. In this study, WGS84 projection was then transform to Malaysia Rectified Skew Orthomorphic (MRSO) to combine with other base map of study area GIS data. The coordinate of tree combine with related attribute such as number of ants counted for each trees was converted into shapefile in GIS system. The data divided into 4 different plots to become one block, known as Plot A, B, C and D in Block 18A and similar plotting were implemented at Block 10B. The total sample of cocoa tree is 50 for one plot with overall 200 sample cocoa tree for one block. However only Plot A for each block will be use as an analysis and further presented in the result section later due to data complexity and to present preliminary findings.

Then, GIS geospatial analysis such as interpolation technique is used. The interpolation technique was performed to the CBA population that gathered by the MCB research team for 30 months data (2007 to 2009). Interpolation analysis is a method for producing a prediction surface that is continuous from the group of data samples [12]. Kriging interpolation was used because it able to produce an estimate of the core surface by a weighted normal of the information, with weights declining with separation between the point when the surface is being evaluated and the areas of the information points [10]. Therefore Kriging interpolation will apply to the CBA population and other ant type species. CBA population ranges group as highest population when every trees have more than 500 ants and reach to 1000 ants and however when less than 500 cocoa black ants recorded, it is consider as low population [9].

Methods to identify high, moderate or low CBA distribution through observation visualization was based on quadrants estimation. Four quadrants which any quadrant has a CBA above 500 is considered high and less than 500 is low. When the ratio is ¾ or all quadrants covered by CBA above 500 considered high. While the moderate when the ratio of half or ½ and cocoa black ants are lower at a ratio of ¼ or quarter. The key plan for each of the block as drawn as shown in Figure 1.

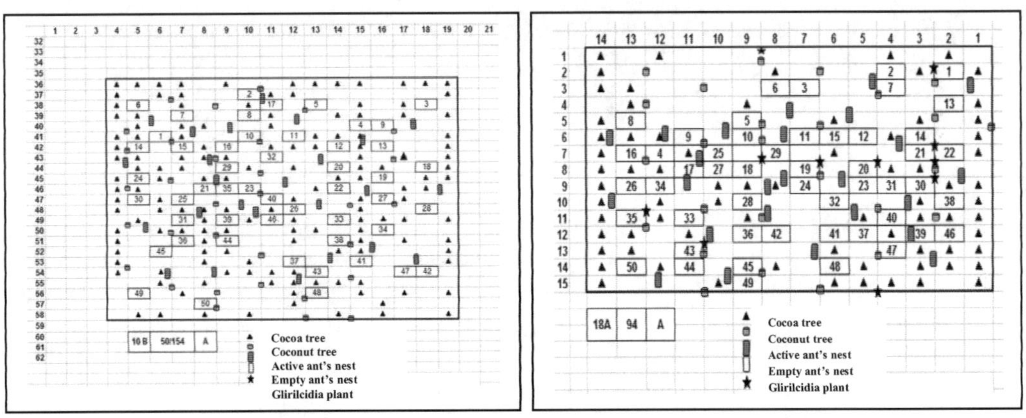

Figure 1. Key plan plot for Block 10B and 18A

3. Result and Discussion

3.1 Spatial CBA population distribution map according plot and year

Kriging Interpolation technique has been used to produce visible images for CBA population in the location plots. Each plot has 50 trees were selected as sample and have been on average the number of ants in the monthly value. Figure 2 and Figure 3 show population pattern of CBA for block 10B and Figures 4 and 5 show population pattern of CBA for block 18A. A total of 21 classes for the population range represented from red orange (zero number of ants) to dark blue (thousand number of ants). A cocoa plot can be grouped as highest population CBA distribution when every tree have more than 500 and reach to 1000 ants and any trees with less than 500 CBA is grouped as low population. Methods to

IOP Conf. Series: Earth and Environmental Science **117** (2018) 012041 doi:10.1088/1755-1315/117/1/012041

identify high, moderate or low classes of CBA distribution in the cocoa plot was based on quadrants subdivision analysis. Every plot subdivided into four quadrants as explained in Section 2 above.

The 10B plot is starting from February 2007 until July 2009, the populations of CBA began with a modest low class in February, 2007 until reaching the highest levels of ant populations in September, 2007 and decreasing by the end of the year of 2007 (Figure 2). Two months were found to have the highest rate of CBA in the months of April and October for the year 2008. While in 2009 there was a month that has the highest rate of CBA in June. When viewed in terms of population spots CBA there are deficiencies in plot A for ten months in 2007, eight months in 2008 and four months in 2009. For the three years, plot A has balance numbers of CBA on March and August 2008 also July 2008.

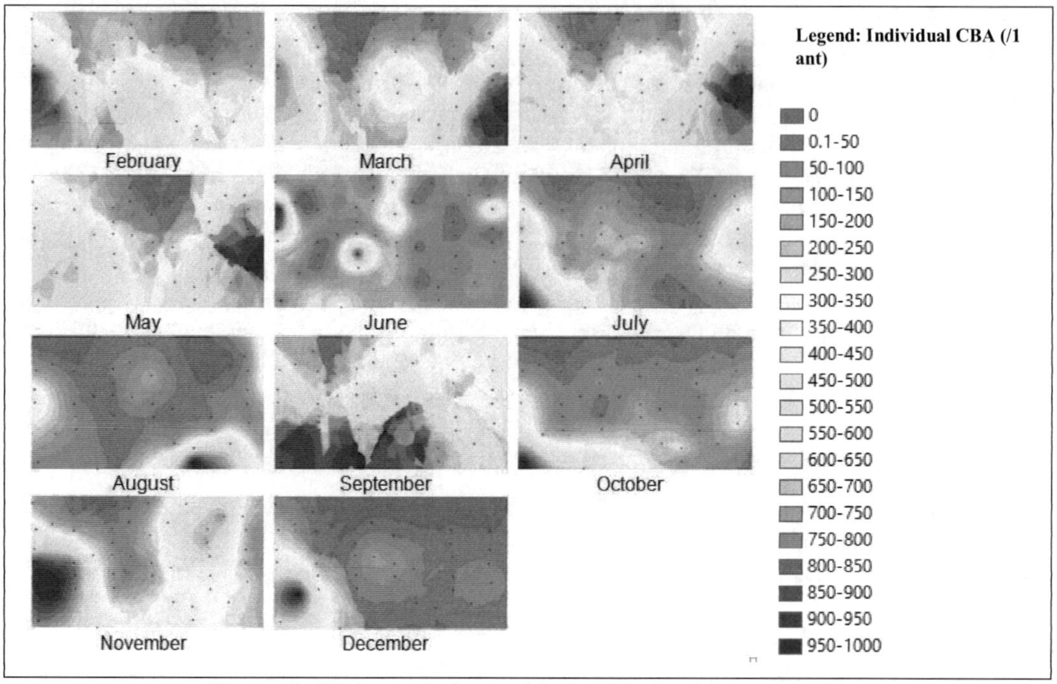

Figure 2. CBA populations at block 10B plot A in year 2007

The block 18A plot A (Figure 3), the dispersion of CBA increment quickly to it is top in the first year (2007), then stay at its low numbers in the second year (2008) and progressively increment in the third year (2009).

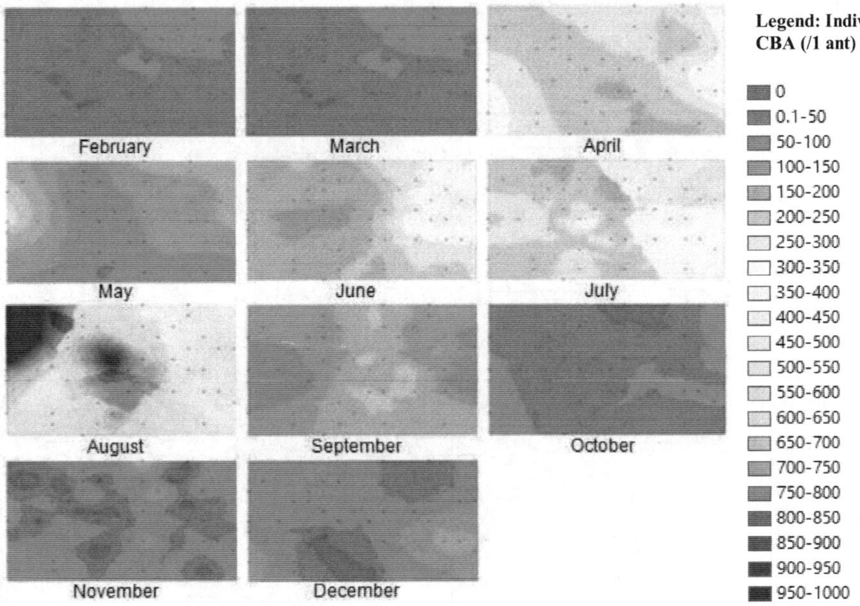

Figure 3. CBA populations at block 18A plot A in year 2007

3.2 Geospatial Comparison of CBA population distribution

The data that used in this study comprised of two different blocks known as Block 10B and 18A with the same data collection activities for the CBA's population. Therefore, to understand better the CBA's pattern according to month, analysis on the differences and similarity of CBA for Plot A was performed. The comparison was done by investigating each plot for the same year starting from 2007 to 2009 (i.e. 30 months). Figure 4 to Figure 6 illustrate the comparison of CBA distribution results from the Kriging interpolation for these plots for monthly data in 2007 to 2009.

Figure 4 shows that observation plot A 2007 on block 10B with high peak CBA on eighth (September) months and block 18A on seven (August) months from the starting observation. The first months on February for the two blocks are low populations which is less than 500 CBA.

Figure 5 shows that observation plot A 2008 on block 10B there was high peak CBA on seventh (April) and sixth (October) months from the starting observation. However does not have high peak from block 18A.

IOP Conf. Series: Earth and Environmental Science **117** (2018) 012041 doi:10.1088/1755-1315/117/1/012041

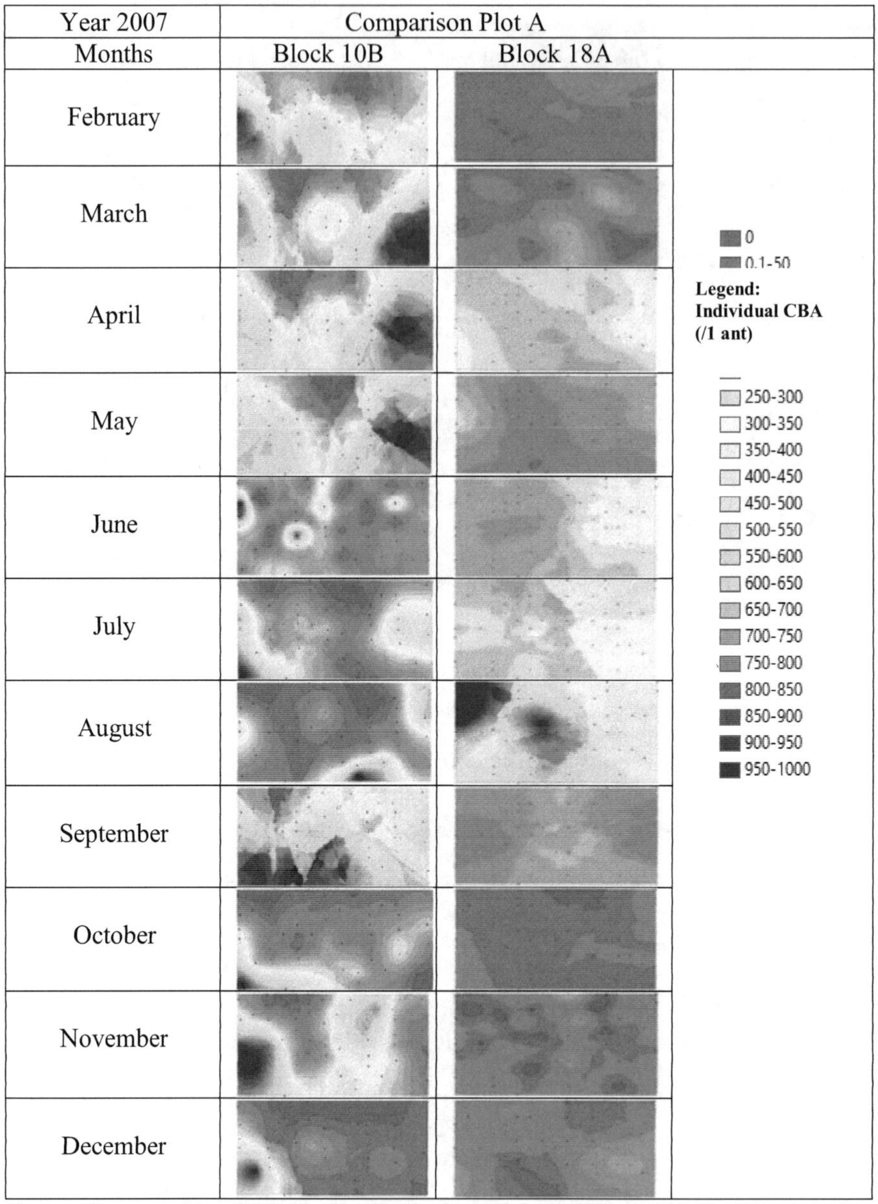

Figure 4. Comparison graphic of CBA's distribution result for 10B and 18A for Plot A in the year 2007

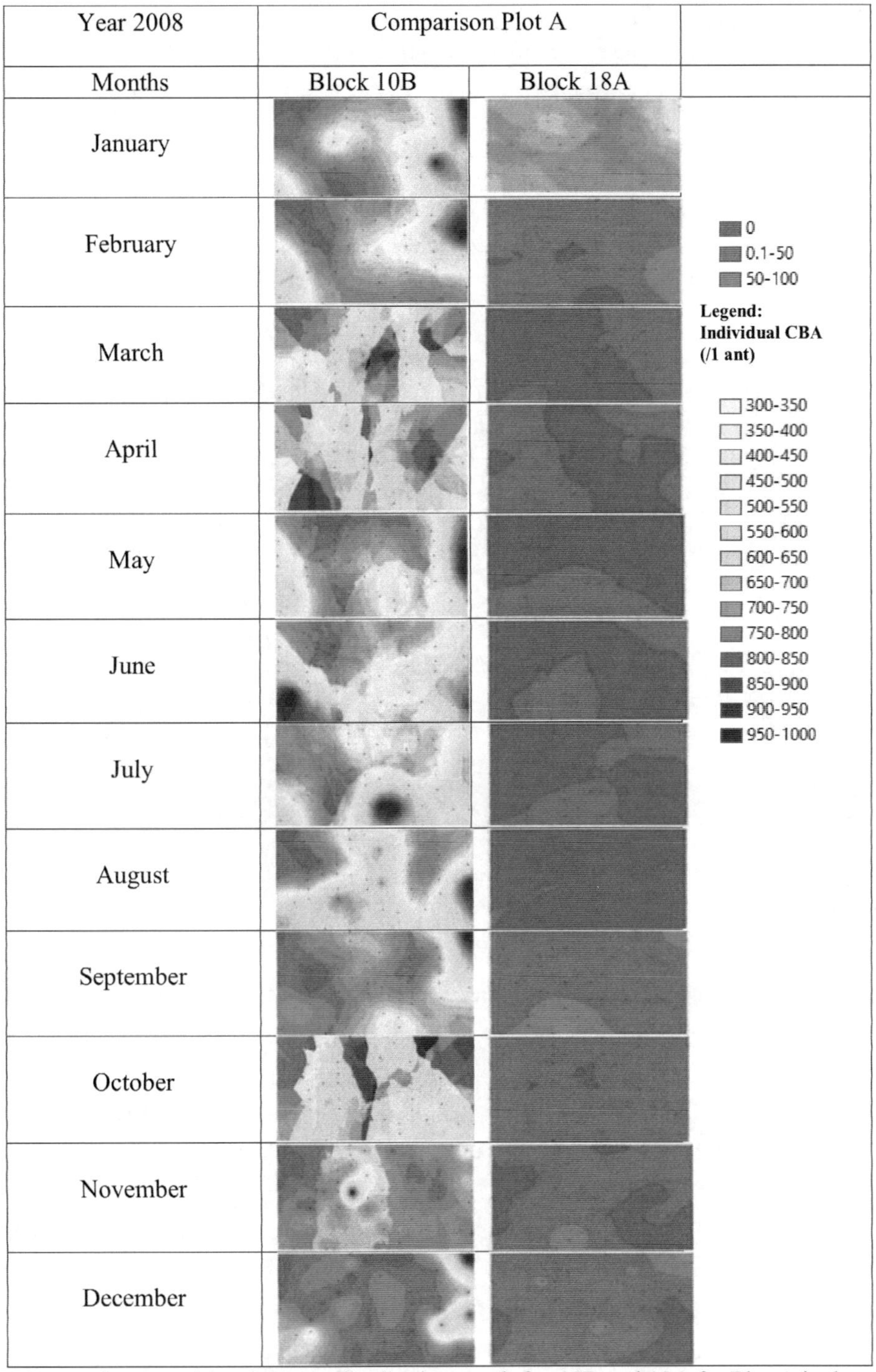

Figure 5. Comparison graphic of CBA's distribution result for 10B and 18A for Plot A in the year 2008

Figure 6 shows that observation of Plot A 2009 on block 10B represented high peak of CBA population on the 8th months (May) from the starting month of observation. However conversely, Block 18A not represented by any high peak of CBA populations. The differences obviously identified with 10B represented by the medium and high classed range the Block 18A dominated by the very low class of CBA population. Therefore, this type of analysis can serve as managing tool to enhance CBA population in any block with lower numbers of CBA population. This analysis able to indicated precision site specific biological input to any cocoa trees and a artificial nest to enhance CBA population can easily been identified.

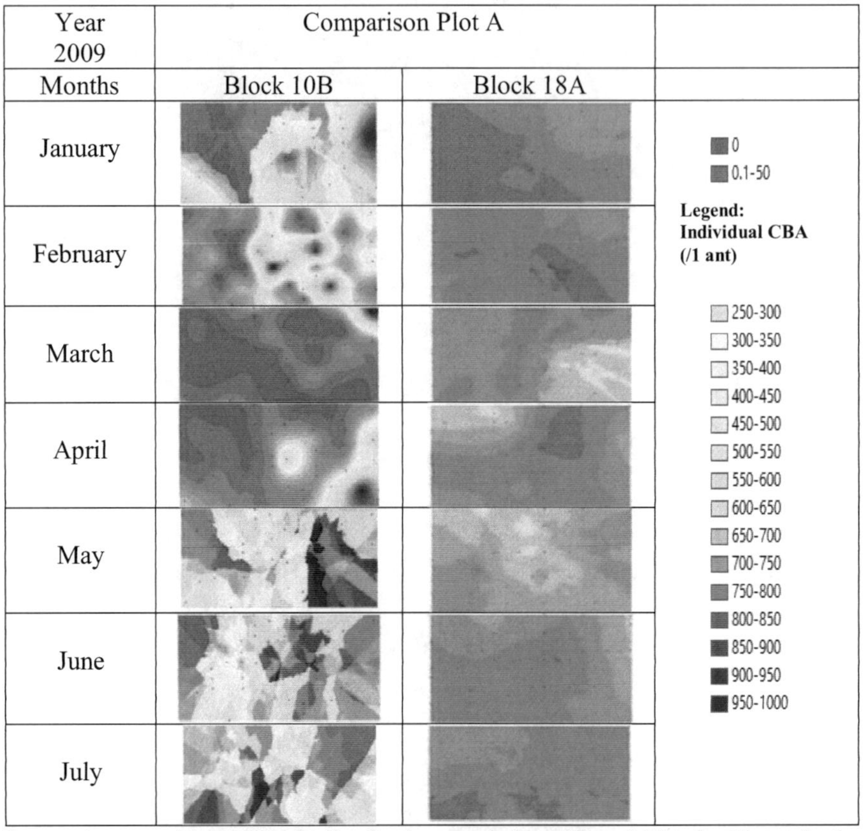

Figure 6. Comparison graphic of CBA's distribution result for 10B and 18A for Plot A in the year 2009

4. Conclusion

In conclusion the method for implementation of CBA as an effective biological control can be site specifically study for each cocoa tree. Hanging of artificial nest on the trees with appropriate amounts can be proposed to enhance CBA population and distributions. The study able to visually map the CBA population density formed within two different blocks in the cocoa plantation area. The CBA has its own unique pattern in which their population will peak in a particular month of the 7th and 8th months and declined in most of other months. Therefore, the usage of GIS able to explained the CBA pattern population and might able to be used to lessen the impact of pest infestation from CPB for further studies.

References

[1] Azhar, I. (1992). Role of Black Ants in Cocoa Pod Borer natural control. *MAPPS*, 36.
[2] McMahon, P., Husing, P., Lambert, S., Mulia, S., Nurlaila., Susilo, A. W., Sulistyowati, E., Sukamto, S., Israel, M., Saftar, A., Amir, A., Purwantara, A., Iswanto, A., Guest, D. & Keane, P. (2015). Testing local cocoa selections in three provinces in Sulawesi: (i) Productivity and resistance to cocoa pod borer and Phytophtora pod rot (black pod). Crop protection 70:28-39
[3] Saripah, B. (2012). Five years of using cocoa black ants, to control cocoa pod borer at farmer plot –

an epilogue. *Malaysian Cocoa Journal* 7: 8 - 14.

[4] Saripah, B. & Alias, A. (2016). Evaluation of best management practices for cocoa pod borer in cocoa ecosystem. *Malaysian Cocoa Journal* 9(1): 108-120.

[5] Azhar, I. (2007). The ways towards sustainability of cocoa industry in Malaysia. ICCO Round Table on a Sustainable World Cocoa Economy, 3-6th October 2007, Accra, Ghana.

[6] Babin, R. (2010). Spatial Relationships Between Dominant Ants And The Cocoa Mirid. *Agricultural and Forest Entomology*, 66-67.

[7] Saripah, B. (2014a). Sustaining Cocoa Black Ant *Dolichoderus thoracicus* (Smith) using artificial nest in the cocoa ecosystem. *Malaysian Cocoa Journal* 8: 23-30.

[8] Saripah, B. (2014b). Control of CPB using insecticides and Cocoa black ants. *Malaysian Cocoa Journal* 8: 14-22.

[9] Yulianto, F. (2014). Detection Settlements and Population Distribution Using GIS. *Emerging Technology and Advanced Engineering*, 21-33.

[10] Briggs, H. (2010). Spatial Autocorrelation The Single Most Important Concept in Geography and GIS. *Geosta*, 12-41.

[11] Saripah, B. & Azhar, I. (2010). Usaha menggalakkan pekebun kecil menggunakan teknik pembalutan buah dalam mengawal serangan Ulat Pengorek Buah Koko. *Confertech*, 343 -348.

[12] Junita, M. (2012). Perbandingan Teknik Interpolasi dengan Natural Neighbor. *Penginderaan Jauh*, 12-22.

IOP Conf. Series: Earth and Environmental Science **117** (2018) 012042 doi:10.1088/1755-1315/117/1/012042

Maintenance and Safety Practices of Escalator in Commercial Buildings

Nurul Afida Isnaini Janipha[1], Sharifah Nur Aina Syed Alwee[1], Raihan Mohd Ariff[1], Faridah Ismail[2]

[1]Centre of Studies for Quantity Surveying, Faculty of Architecture Planning and Surveying, Universiti Teknologi MARA, Shah Alam, Selangor, Malaysia.
[2]Centre of Studies for Postgraduate, Faculty of Architecture Planning and Surveying, Universiti Teknologi MARA, Shah Alam, Selangor, Malaysia.

*nurulafida@salam.uitm.edu.my

Abstract. The escalator is very crucial to transport a person from one place to another. Nevertheless, there are many cases recorded the accidents in relation to escalator. These may occur due to lack of maintenance which leads to systems breakdown, poor safety practices, wear and tear, users' negligence and others. Thus, proper maintenance systems need to be improvised to prevent and reduce escalator accident in future. This research was aimed to determine the escalator maintenance activities and safety practices in a commercial building. Three case studies were selected within Selangor area. Semi-structured interviews were conducted for collecting data from these three case studies. To achieve the aim of this research, the study was carried out on the maintenance activities, safety practices and cost related to escalator maintenance. As one of the important means of access in building, it is very crucial to increase effectiveness of escalator particularly in commercial building. It is expected that readers will get clear information on the maintenance activities and safety practices of escalator in commercial building.

1. Introduction

The Buildings are constructed to provide shelter for human beings to carry out daily activities and for having their own space for privacy. The main criteria of a good shelter, it must be sufficient to provide comforts besides other facilities must be able to function well [4]. For example, services such as plumbing system, air-conditioning system, ventilation system, elevators and escalators must be provided especially for commercial buildings. Escalators or moving stairs are one of vertical transportation under the mechanical equipment, which is commonly used to transport people from one place to another, not only in commercial but also in residential buildings. In shopping complexes, it plays a vital role in transporting passengers to above and below floor levels. However, the system must be able to perform well to avoid problems particularly during peak hours. The installation in a building needs to consider certain major aspects such as the design and location. This is important to ensure the system equipped is comfortable to the users, safe and problem such as sick building syndrome can be avoided. Thus, the maintenance aspects must be done regularly and diligently to make sure the escalator can serve well and fit for its purpose.

Maintenance management is another aspect that is crucial to be taken care by the building owner. He or she must think on how the costs for unnecessary maintenance can be avoided and wrong maintenance practices can be avoided. The lifespan of escalator system can usually sustain without the need for major maintenance or replacement of new equipment if the maintenance management is carried out sufficiently. Besides, pre-requisite requirement must be complied according to what has been stated in the standard operating procedure. Since there are many escalator accidents have been recorded, both maintenance and its safety practices are the significant elements that should not be ignored by all parties involved.

2. Maintenance Activities And Safety Practices Of Escalator

Maintenance activities can be described as keeping structural components or fixtures in a proper schedule, programme or routine [7]. Maintenance is generally carried out based on recurring basis and it is normally done to ensure the equipment can achieve its fully functional state [3]. Maintenance can also be defined as a process to uphold the system in a good condition. Some of the affects due to proper maintenance management can be such as improving safety issues, energy efficiency, cost optimisation, and improve integrity, conducive environment and many more [6]. Maintenance of escalator is important to assure it is safe when it is operated. By having regular maintenance, it can increase the lifespan of an escalator.

Nowadays, maintenance plays a central role in bearing all aspects of physical, financial and sociological needs in an organisation. There are a few types of maintenance works such as Run to Failure Maintenance (RTF), Preventive Maintenance (PM), Corrective Maintenance (CM), Improvement Maintenance (IM), and Predictive Maintenance (PDM). Maintenance activities could either be repair or replacement activities, which are necessary for the system to reach its highest degree of productivity. The activity should be reasonable to be carried out at the optimal cost. The aspects of maintaining escalator include but not limited to the followings; all safety signages and escalator's directions are clearly displayed, the edges of each step must be marked with yellow lines, safety devices are equipped to prevent passenger's feet from getting trap on or at the sides of step, switch for emergency breaks is well functioning and can stop when it is pressed and others.

There are a few safety practices that need to be precaution when using the escalator. These can be such as the safety features, traffic of escalator, location of escalator, and other physical considerations such as the surrounding nearby. Type of accident that is commonly happen when using the escalator is falling people or object. Therefore, safety features for escalator must be properly considered. The safety performance during normal operation depends on the number of safety features associated with the escalator. Safety features can be such as the speed on handrails and steps should be able to move concurrently to ensure the users are balance, width of the steps must be designed to prevent slipping, comb plates are placed at each of landing to avoid tripping when the escalator is moving and the balustrades are designed to prevent user's cloth from getting stuck. Besides, the automatic service break is also installed so that the escalator will stop smoothly in situation if the drive chain or step chain is broken or if an object stucks into the handrail's inlet. The emergency button must be functioning well in situation if a power failure happens [5].

Hence, safety is very important to be considered in the vertical transportation industry [1]. Even though escalator is very useful for transmitting users in the vertical way, lack of safety will cause dangerous to the users. Some people are reluctant to use vertical transportation due to those safety issues as been highlighted before. Without safety systems and reliable devices, building cannot be served at its supposed to be. Besides, the building owner might face some difficulties to obtain approval from the local authorities. Though the escalator can cause injurious to passenger in the state of dangerous, somehow it still has many advantages, which outweigh its flaws. Any system that deals with people must consider safety as priority.

The design criteria for an escalator include the traffic pattern when directing people to use the escalator [2]. Other than that, the maximum number of passengers per riding should be counted as well, for instance, a single width escalator travelling at about 450mm (1.5 feet) per second should be able to

IOP Conf. Series: Earth and Environmental Science **117** (2018) 012042 doi:10.1088/1755-1315/117/1/012042

move an approximate 170 persons for every five minutes ride. Certain escalators can travel up to 600mm (2 feet) per second, which can cater 450 passengers at a time. Hence, the capacity to carry the load has a direct relationship with the expected peak traffic demand. This is crucial at times where number of passenger increases during the prime hours. Escalator system must be safe and any breakdown or failure should be avoided.

Another crucial factor in determining the safety for escalator is the location to place the system. The most convenient place shall be located to areas where public users can see and move easily from one place to another. In shopping complex, it should be situated at the area where merchandise can display and attracted passengers to visit the stores [2]. The entrance and end step of escalator should not be placed at confined spaces because some safety issues may arise if it is located remote from the crowd. As a horizontal and vertical transporting system, it must be functioned at open spaces to maintain the line of traffic. This is different compared to lift system whereby it is normally situated at closed areas.

Safety on the floor openings can be done by installing the fire-resistant shutter as the exit door and installing fire sprinklers at the above area. Besides, hall that is constructed with fire-resistant walls must be provided. To avoid danger due to overheating at the floor opening, there must be an adequate ventilation system which allocating gears and motors. In compliance with the building by law, staircase must be adjacent to the escalator so that if any incident happens, the escalator can be functioned as the mean of escape to move passengers to the safer place. Another element that must be considered would be to allocate escalator for disable people i.e. for those who are using wheelchairs.

3. Research Methodology

The research process begins with the identification of research topic and follows the research methodology process to identify the aim of research. There are three objectives in this research which derived from the aim, to highlight the maintenance activities and safety practices for the escalator in a commercial building, to determine maintenance activities and safety practices for the escalator in Malaysia commercial building and to analyse the cost related to maintenance activities and safety practices for the escalator.

The data collected were based on the three objectives and they were gathered from primary and secondary sources. For the first objective, data collected from secondary sources like journals, proceeding papers and reliable articles. Secondary sources were information obtained from researchers who carried out similar areas but different aspects from this research. These sources aid to gain the clearer understanding of the topics covered. Data for the second and third objective were based on primary sources which were semi-structured interview and site visits.

Primary sources can be considered as the main data collection where it was composed from the interview session with the selected respondent and data which were gathered by attending site visits. The interview was based on semi-structured, which uses an interview schedule and flexibility in terms of the interviewee's responses. Site visits can be considered as very helpful to understand the escalator components and the functions that are equipped in the selected shopping complexes. All data obtained from the primary and secondary sources were compiled and analysed using Microsoft Office Excel Worksheet 2013 and presented in tabular format.

Semi structured interview were done based on three case studies. A representative for each case study was interviewed based on their experiences as technical staff specific in controlling and monitoring the escalators' operations and maintenance.

4. Analysis Of Findings

4.1. Maintenance Activities for Escalator

The maintenance activities vary depending on the company's policy. Based on the information gather from the interviewees, it can be noted that all escalators have both daily and monthly maintenance. There are certain types of escalator, which executing regular maintenance in daily or in monthly basis. However, this is subjected to the policy and regulation provided by the supplier of this system.

Basically, the daily and monthly maintenance is very well recognised as the preventive maintenance and improving maintenance, respectively in the literature review. Based on the review, the maintenance activities were identified based on the theory on maintaining escalator. While from the case study, differences from the theory were found. Practical ways for escalator maintenance activities will depend on the certain set of rules that must be facilitated properly by the building operator.

Hence, the efficiency of escalator can depend on the maintenance activities executed. Maintenance of escalator will keep the system work sufficiently. Besides, the lifespan of escalator system can be enhanced. Based on the study conducted on maintenance activities of the escalator in shopping malls, this has achieved the first objective, which is to determine the maintenance activities of the escalator in a commercial building.

4.2. Safety Practices of Escalator

From the analysis, it can be proved that all case studies were fully equipped with safety features. Each escalator applied different approach for it safety practices. It cannot be neglected that escalator suppliers still considered and maintained safety cultures in their businesses. Table 1 shows the differentiation of safety equipment installed at for different case studies.

Table 1. Safety equipment of escalator for different case studies.

Safety Equipment	Case Study 1	Case Study 2	Case Study 3
Emergency stop button & switch	√	√	√
Safety signage	√	√	√
Handrail inlet switch	√	√	√
Yellow line/ demarcation	√	√	√
Brush skirt guard	√	√	√
Audio message	√		
Auto start sensor	√		
Bollard		√	
Green step			√
Power saving features			√

There are some common features from all cases studies that were equipped by the installer, they are emergency stop button, safety signage, handrail inlet switch, yellow line, and brush skirt guard. As shown in Table 1, those safety features are the crucial components. In another situation, there are five safety equipment installed at each case study that makes it vary from one to another. The audio message, auto start sensor, bollard, green step and power saving features are the equipment that makes the escalator deviated in terms of its safety characteristics. Based on this finding, the added value on the escalator will depend on their safety equipment where it will be more efficient and protective to be used.

From the research, different safety features in each case study were due to building requirement, marketing purpose, assurance of the safety to uses, budget for providing escalator in building, improvements from previous experience or lesson learnt gathered from existing buildings.

4.3. Cost Related To Escalator Maintenance Activities

Table 2. Overall estimated cost for escalator maintenance.

Components	Case Study 1	Case Study 2	Case Study 3
Escalator	RM380 000.00	RM300 000.00	RM400 000.00
Set of motor and brake	RM200 000.00	RM200 000.00	RM200 000.00
Maintenance Charges	RM450.00	RM400.00	RM450.00

The table 2 shows the estimated costs for the escalator maintenance. The cost components in relation to the maintenance works is influenced by the lifetime of equipment, travelling usage by the users, equipment breakdown and others. The installation cost of escalator in case study 1is around RM 380 000.00 for one escalator. This case study has two number of the escalator which is for up and down. The technician had mentioned that if the motor break is broken, the motor should change the whole part of the motor. The cost for motor changes was quite expensive which is RM 200 000.00. The maintenance charge is RM450 for each escalator for every month. If any breakdown of the component will be no charge except for change of the component.

The installation cost of escalator in case study 2 is around RM 300 000.00 for one escalator. The maintenance cost depends on the type of maintenance activities. If it is for daily maintenance no charge required. If there is equipment breakdown, the charge depends on the equipment cost plus with a service charge. The maintenance charge estimated is RM400 for each escalator for every month. If any breakdown of the component will be no charge except for change of the component.

The overall cost of each escalator for case study 3 is around RM 400 000.00 for one escalator. There is a cost for servicing escalator, where the owner pays to the contractor. If there is any escalator breakdown, the owner will call a contractor in charge in servicing the escalator. Any replacement for the component will also be a charge. But, for daily inspection, it will not be charge any amount as the inspector is working under building supervision. The maintenance charge is RM450 for each escalator for every month. If any breakdown of the component will be no charge except for change of the component.

5. Discussion And Conclusion

There are various types of maintenance activities such as run to failure maintenance, preventive maintenance, corrective maintenance, improvement maintenance, and predictive maintenance. Safety practices for escalator are also important. Generally, safety practices can be achieved when these factors are considered in the system such as safety features, traffic, location and physical of escalator. All these factors contribute to conveniences and comfortable of the escalator to the users. For example, there are safety features installed in each escalator such as emergency stop button, yellow demarcation, brush skirt guard, auto start sensor and other safety component. All these safety features help to reduce accident during riding of escalator.

The maintenance activities for escalator are also important. From the three case studies, works are carried out as per scheduled. It can either be a daily and monthly maintenance activity. It can be concluded that all maintenance works based on the case studies undergo similar process. This shows that the maintenance activities are crucial, not only to escalator, but to other equipment. From the case studies, it can be said that the design criteria take into consideration the safety features. The safety aspects of these escalators are also sufficient on the safety of the users. The system is equipped complete with the basic and addition safety components to ensure the effectiveness of the escalator during riding. Usually, the safety sign attaches to the escalator as a guideline to the customer as the safety precaution.

It is important to include cost breakdown for the maintenance work and it should be planned from the beginning stage to avoid problems in the later stage. It is useful to prevent from delayed the maintenance activities of escalator. Based on the case studies, the average cost of escalator maintenance activities is around RM430/day. There might be some other additional costs, such as cost to replace escalator equipment. Nevertheless, due to difficulty to obtain information, the exact data for that particular cost could not be analyzed.

There are a few recommendations that can be highlighted for improvement in future, for instance, a quality standard for escalator need to be standardized, further research on the maintenance activities and safety practices in Malaysia not only for shopping malls are needed to be expended. Road tour on the public awareness about the uses of escalator should also be conducted. Based on the case studies, there are difference maintenance activities of each case study. It is suggested that a Standard Maintenance Guideline should be proposed and suppliers need to follow and comply with the requirements. The escalator maintenance activities observed from the case studies are some of the good examples since the inspection is carried out daily and monthly basis, to make sure the escalator is in good condition and safe.

Since this research only focuses on the area of Selangor, it would be better if the research is expanded all over Malaysia. Therefore, we can know the broader aspects of maintenance activities and safety practices of escalator in Malaysia. The information for the escalator maintenance activities and safety practices will be more precise. All escalator system from the case studies taking the safety of the passenger as the main priority. Even though safety features are provided, as a passenger, we should aware and obey with Dos and Don'ts for our own protection.

Acknowledgement
The authors wish to thank the Universiti Teknologi MARA via LESTARI grant, 600-RMI/DANA 5/3/LESTARI (82/2015), for funding this research paper. A highly appreciation all people that involved directly and indirectly in the success of this research. Special thanks also go to the reviewers for their valuable comments on this paper.

References

[1] Andon, A. (2015). Escalator Safety. [online] Elevator World Unplugged. Available at: http://www.elevatorworld.com/blogs/escalator-safety/ [Accessed 26 Oct. 2016]. reference

[2] Barney, G. (1986). Elevator technology. 1st ed. Chichester: Published for the International Association of Elevator Engineers by Ellis Horwood.

[3] Bfs.ucsb.edu. (2011). What is the difference between Maintenance and Construction? | Business & Financial Services - UCSB. [online] Available at: http://www.bfs.ucsb.edu/faqs/what-difference-between-maintenance-and-construction [Accessed 10 Apr. 2017]. references

[4] Chandler, I. (1988). Building technology. London: Mitchell in association with the Chartered Institute of Building.

[5] McGuinness, W., Stein, B. and Gay, C. (1971). Mechanical and electrical equipment for buildings. 5th ed. New York: Wiley, pp.1459-1473.

[6] Moubray, J. (2004). Maintenance Management-A New Paradigm. [online] Available at: http://www.relogica.com/main/getfilesp/3 [Accessed 26 Oct. 2016].

[7] Osha.gov. (2003). 08/11/1994 - Construction vs. Maintenance. | Occupational Safety and Health Administration.[online] Available at: https://www.osha.gov/pls/oshaweb/owadisp.show_document?p_table=interpretations&p_id=21569 [Accessed 10 Apr. 2017].

IOP Conf. Series: Earth and Environmental Science **117** (2018) 012043 doi:10.1088/1755-1315/117/1/012043

ANFIS multi criteria decision making for overseas construction projects: a methodology

W P Utama[1*2], A P C Chan[2], Zulherman[3], H Zahoor[3], R Gao[1] and D Y Jumas[2]

[1] Department of Building and Real Estate, The Hong Kong Polytechnic University
[2]Department of Quantity Surveying, Universitas Bung Hatta, Indonesia.
[3]Department of Construction Engineering and Management, National University of Science and Technology, Risalpur Campus, Pakistan.

*wahyudi.utama@connect.polyu.hk

Abstract. A critical part when a company targeting a foreign market is how to make a better decision in connection with potential project selection. Since different attributes of information are often incomplete, imprecise and ill-defined in overseas projects selection, the process of decision making by relying on the experiences and intuition is a risky attitude. This paper aims to demonstrate a decision support method in deciding overseas construction projects (OCPs). An Adaptive Neuro-Fuzzy Inference System (ANFIS), the amalgamation of Neural Network and Fuzzy Theory, was used as decision support tool to decide to *go* or *not go* on OCPs. Root mean square error (RMSE) and coefficient of correlation (R) were employed to identify the ANFIS system indicating an optimum and efficient result. The optimum result was obtained from ANFIS network with two input membership functions, Gaussian membership function (*gaussmf*) and hybrid optimization method. The result shows that ANFIS may help the decision-making process for go/not go decision in OCPs.

1. Introduction

In the topic of internationalization of construction enterprises, the application of Multi-criteria Decision Making (MCDM) methods can be addressed to assess the capability of enterprises to operate outside their market origin and evaluate the potential projects to be considered for go or for bid [1]. In such decision, intuition and experience of decision makers in judging the capability of the firm may be more dominant. Several approaches such as Analytical Hierarchy Process [2], Cross Impact Analysis [3] and Analytical Network Process [4] have been adopted to solve the decision making problems in OCPs. Indeed, the intuition of decision makers plays important role in strategic decision making [5].

This study aims to demonstrate the integration of neuro-fuzzy system to support management decision in choosing OCPs. Adaptive Neuro-Fuzzy Inference System (ANFIS), a type of amalgamation of neural network (NN) and fuzzy logic (FL), was utilized to develop the decision model. NN can be used for dealing with a given problem if a set of data sample are available. Learning and computational abilities are the power of the NN [6]. Hence, a mathematical formulation and the previous knowledges are not necessary. The problem is that the solution deriving from the learning process in the networks cannot be interpreted because the character of the neural network seems a black box. On the other hand, the FL can be adopted to settle the problem if there is knowledge about the solution and the system can be created by using the linguistic *if-then* rules. This system does not need training data and a mathematical model of the problem of interest [7]. The integration of the two systems will bring advantages to existing ones.

2. Concept of ANFIS

The basic concept of ANFIS is to create the stipulated input-output pairs through assembling a set of fuzzy if-then rules with suitable membership functions through implanting the Fuzzy Inference Rule (FIS) into the structure of adaptive networks [8]. The ANFIS model structure is constructed by both ANN and fuzzy logic which allow the model to work with uncertain and imprecise information [9]. It utilizes the NN training process to tune the membership function (MF) and the related parameter approaching the desired data sets [10]. Structurally, ANFIS consists of three devices namely, a rule base, a database, and a reasoning mechanism. The rule base constitutes fuzzy if-then rules following a linear function as described by Takagi and Sugeno [11].

Rule 1: IF X_1 is A_1 and X_2 is A_1 THEN $Y_1 = p_1x_1 + q_1x_2 + r_1$

Rule 2: IF X_1 is A_2 and X_2 is A_2 THEN $Y_2 = p_2x_1 + q_2x_2 + r_2$

Where X_1, X_2 and Y_1 and Y_2 are numerical inputs and outputs respectively, A and B are numerical variables, and *p, q* and *r* are the parameters determining the relation between input and output. ANFIS algorithm is composed of five layers as follows:

Layer 1: This layer shows the number of numerical inputs belonging to different fuzzy sets. Every node *i* in this layer is represented by square node with the output function as shown in Equation 1.

$$\left\{ \begin{array}{l} {}_i = \mu\ {}_i(x\ ) \\ {}_i = \mu\ {}_i(x\ ) \end{array} \right. \quad (1)$$

Where $\mu\ {}_i(x\ )$ and $\mu\ {}_i(x\ )$ are membership functions of A and B fuzzy sets.

Layer 2: In this layer, all incoming signals are multiplied to obtain an output, ω by which operator AND or OR are used, known as firing strength. The output is calculated by Equation 2.

$$\omega_i = \mu\ {}_i(x\ ) \times \mu\ {}_i(x\ ) \quad (2)$$

Layer 3: Every node N in this layer calculates the average ratio of previous outputs to produce a new output $\bar{\omega}$It is obtained by Equation 3.

$$\bar{\omega} = \frac{\omega_i}{\sum \omega} \quad (3)$$

Layer 4: Square node in this layer produces an output $\bar{\omega}f_i$ based on Equation 4.

$$Y_i = \bar{\omega}f_i = \bar{\omega}(\ {}_ix\ +\ {}_ix\ +\ {}_i) \quad (4)$$

Layer 5: This is an output layer in which the node calculates all outputs from Layer 4 by Equation 5.

$$Y = \sum_i \bar{\omega}f_i = \frac{\sum_i \omega_i f_i}{\sum_i \omega_i} \quad (5)$$

Learning process in a neural network aims to create a stable structure. In ANFIS, the learning process of network combines the least square estimate (LSE) and the gradient descent method. This hybrid learning procedure is composed of a forward step in which the input signal passes forward until Layer 4, where the output parameters are adjusted using the LSE of the error between the estimated output and the actual output. Then, on the backward step, the error rates propagate back through the system, and MFs in Layer 1 are updated by the gradient descent method [12]. The process of these forward and backward propagations is called as epoch. The hybrid learning algorithm trains the MF parameters to mimic the training data set.

3. Methodology

The hierarchy structure of the model development for OCP selection is illustrated in Figure 1. It consists of four main steps; determination of attributes, data collection, ANFIS operation, and recommendation. The process begins with a determination of criteria for OCP assessment. These criteria were obtained from data analysis of Delphi survey. Based on the criteria, an evaluation form of OCP was designed as an instrument to build cases database.

IOP Conf. Series: Earth and Environmental Science **117** (2018) 012043 doi:10.1088/1755-1315/117/1/012043

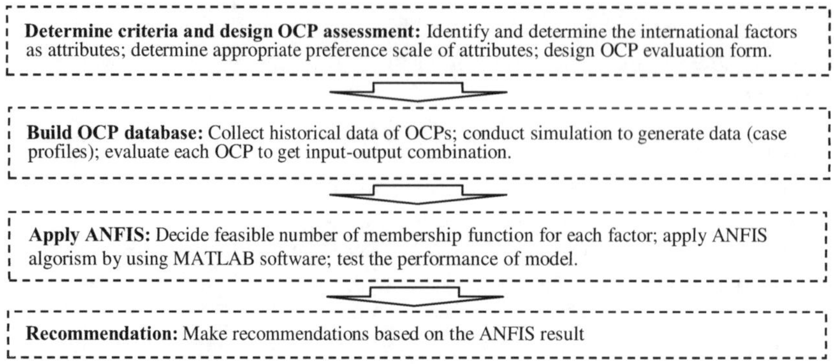

Figure 1. Hierarchical structure of model development

Two methods, convergence and generalization proposed by Refenes [13] were adapted to validate and verify the applicability and performance of the model. Convergence views the learning mechanism implemented for training data. It indicates the optimum performance of the model and the accuracy. The indicator to measure the performance is Root Mean Square Error (RMSE). Regarding its efficiency, the model is indicated by the correlation coefficient (R) formulated as:

$$\boldsymbol{R} = \frac{\sum_{i=} (\ _i - \bar{\ })(F_i - \bar{F}}{\sqrt{\sum_{i=} (\ _i - \bar{\ }) \times \sum_{\ i=} (F\ - \bar{F}}} \qquad (6)$$

Where A_i, F_i and N are actual scores, calculated score produced by model and number of data respectively. The ideal characteristics of the model have RMSE score of 0 and R closed to 1 or 100%.

4. ANFIS based 'Go/Not go' decision model

ANFIS system needs a set of input-output pair data. In preparation of such data, the OCP cases were then tabulated in a table presenting five input attributes (project, contract, owner, host country and business) and one output (Go/Not Go). The grades given by the experts were then normalized using equation $x' = \frac{x-\ i}{ax-\ i}$ where x' is normalized score, x is initial score, and *min* and *max* are the smallest and the largest scores. This equation casts the initial scores in one unified range [0 to 1]. The different ranges between input value and output value need the data to be normalized.

On the output side, the categorical scales on decision (Y) should be transformed into crisp number in which the program can recognize the input attribute. Linguistic variables of output were coded in a binary digit representing 1 for Go and 0 for Not Go. However, Lam et al. [14] advice that the use of 1 and 0 approach in imitative based learning algorithms result in very slow learning speed. They suggest assigning 0.95 and 0.05 instead of 1 and 0 to avoid expected slow convergence. The cases are then divided into training (70%), testing (20%) and checking (10%) data sets. The training data set is used for generating an initial ANFIS model, whereas the testing and checking data sets are used for validation and generalization of the model respectively.

Before starting the fuzzy inference system (FIS) training, an initial FIS model structure needs to be specified by choosing *grid partition* that generates a single-output Sugeno-type FIS by using grid partitioning on the data. While generating FIS, the number of membership functions (MFs) at INPUT, were set for [2 2 2 2 2] which indicate that each input has two MFs (Low – High), whereas their type of MFs were set at "trimf" indicating triangular MF, and at OUTPUT, "linear" was chosen as the type of MF. The ANFIS editor provides eight types of MFs. This setting creates an ANFIS model structure in which to obtain the optimum ANFIS model indicated by the training minimum error, the number and type of input MFs are tuned arbitrarily.

Given the above parameters, the system was trained with *hybrid* learning algorithm. Another set parameter in ANFIS editor is *Error tolerance,* which determines a stopping criterion of training related to the size of error. The training error is the variance between the output value of training data set and the output of the FIS for the same input value of training data set. The training error records RMSE of the training and checking data sets at each epoch.

Each Train-FIS process basically creates an ANFIS model which is a trained system. As stated above, the average training error records the RMSE, and to find the RMSE of each data set, the trained systems were then tested against training, testing and checking data set. Test-FIS in ANFIS editor is used to test the data set. The average train errors shown for each option constitute the RMSE for the given data as tabulated in Table 1.

Table 1. Result of Test-FIS of training, testing and checking data

No	Type of MF	Average training error (RMSE)					
		2 MFs of input			3 MFs of input		
		Train	Test	Check	Train	Test	Check
1	Trimf	1.14×10^{-5}	0.304	0.215	2.97×10^{-7}	0.277	0.295
2	Trapmf	6.52×10^{-6}	0.256	0.148	1.44×10^{-7}	0.227	0.323
3	Gbellmf	5.71×10^{-6}	0.347	0.153	3.17×10^{-7}	0.119	0.160
4	Gaussmf	5.31×10^{-6}	0.130	0.075	1.90×10^{-7}	0.220	0.287
5	Gauss2mf	7.89×10^{-6}	0.305	0.172	1.33×10^{-7}	0.203	0.396
6	Pimf	5.82×10^{-6}	0.290	0.139	1.23×10^{-7}	0.208	0.405
7	Psigmf	4.30×10^{-6}	0.277	0.109	1.66×10^{-7}	0.196	0.370
8	Dsignmf	4.30×10^{-6}	0.277	0.109	1.66×10^{-7}	0.196	0.370

Overall, the RMSE of each data for various numbers and types of MFs show very small value (close to zero). Generally, these values can be said that the network works well under all parameters. The minimum training errors of training data for two and three number MFs of input were obtained from *gaussmf* (5.31×10^{-6}) and *trapmf* (1.44×10^{-7}) respectively. The average training errors for testing the data of both parameters were 0.130 and 0.227, and for checking the data were 0.075 and 0.323. Based on the results, the ANFIS model for GO/Not Go decision on OCP was developed using three parameters: two input membership functions, Gaussian membership function (*gaussmf*), and hybrid optimization method.

Of the trained system, the ANFIS rule which is the ANFIS model for GO/Not GO decision making in OCP was obtained as illustrated in Figure 2. This IF-THEN rule displays the all records of FIS and enables management to make a quick choice of OCP by subtituting the input scores based on an analysis of a particular project. Changes made on score of each input generate a new output value. Thus, the decision makers can further determine a threshold output score in deciding Go or Not Go for OCPs under evaluation.

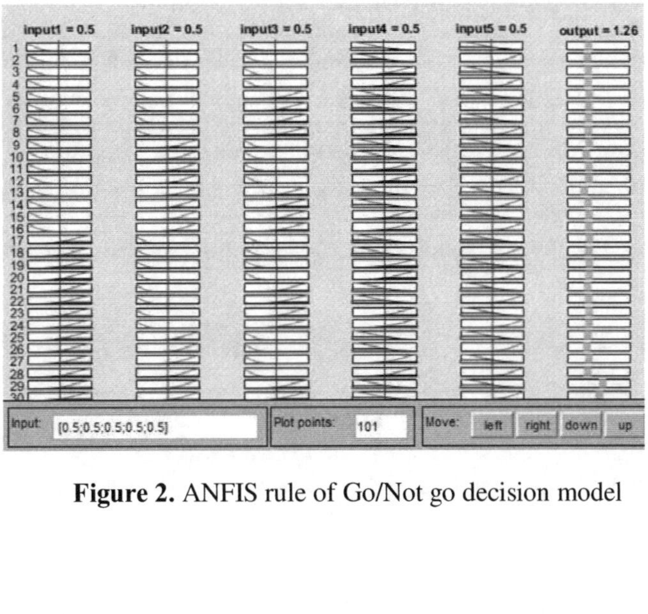

Figure 2. ANFIS rule of Go/Not go decision model

The correlation coefficient (R) between testing/checking data output and output result of trained ANFIS signifies the efficiency of the model. To calculate R value, first, the desired and predicted outputs were collected. Using ANFIS rule (Figure 2), the input scores of 22 testing and 11 checking data were set, while the generated outputs were then recorded. The score of R for testing and checking data set was then calculated using the given equation. The results are 0.995 and 0.976 showing a strong correlation and an indication of the fitness of the designed ANFIS model. Of both RMSE and R, values can be summarized that the performance of designed ANFIS model for Go/Not Go decision making in OCPs was found to be satisfactory.

The generalization ability of the ANFIS model is then examined further. Nine real cases of OCPs performed by Indonesian firms were used for this purpose. Before assessing the decision of the projects using ANFIS model for Go/Not Go decision making, the input data was normalized. Each normalized project data was then entered into trained system in which the ANFIS rule (Figure 2) processes the data and generates an output score. With nine real-life cases of OCPs executed by Indonesian large contractors, the ANFIS model could predict the desirable decision with 11.11% fault. The result of real cases suggests a very good generalization ability of the proposed ANFIS model for Go/Not go decision making in OCPs.

5. Conclusion

This paper demonstrates the ability of ANFIS for supporting the decision-making in OCPs selection. The developed model using this approach gives several advantages combining fuzzy reasoning (e.g. ability in handling uncertainty regarding imperfect information) and neural network (i.e. ability in learning and generalizing from a set of prior knowledges). The developed model was based on the simulation cases of OCPs which evaluates the Indonesian contractors. The evaluation results may be different from other contractors, so the model may also diver.

The use of simulation cases and small number of data sets are the limitation of this study. However, there exist several advantages of using simulation cases, such as a freedom to select attributes contributing to the complexity of the structure of data, and the possibility of tuning different scenarios tailored to meet various conditions which are not available in real data sets. In future, the use of an appropriate numbers of real data sets is encouraged. A cross-validation method, comparing results with other approaches, such as statistical techniques and case based reasoning, can be carried out to view the accuracy of each tool.

References

[1] Utama, W.P., Chan, A.P.C., Gao, R. and Zahoor, H. (2017). 'Making international expansion decision for construction enterprises with multiple criteria: a literature review approach', *International Journal of Construction Management*, doi: 10.1080/15623599.2017.1315527.

[2] Hastak, M. and Shaked, A. (2000). 'ICRAM-1: Model for International Construction Risk Assessment', *Journal of Management in Engineering*, Vol. 16 (1), pp. 59-69.

[3] Han, S.H., Diekmann, J.E. and Ock, J.H. (2005). 'Contractor's risk attitudes in the selection of international construction projects', *Journal of Construction Engineering and Management*, Vol. 131 (3), pp. 283-292.

[4] Bu-Qammaz, A.S., Dikmen, I. and Birgonul, M.T. (2009). 'Risk assessment of international construction projects using the analytic network process', *Canadian Journal of Civil Engineering*, Vol. 36 (7), pp. 1170-1181.

[5] Khatri, N. and Ng, H.A. (2000). 'The role of intuition in strategic decision making', *Human relations*, Vol. 53 (1), pp. 57-86.

[6] Canuto, A. M. D. P. (2001). *Combining neural networks and fuzzy logic for applications in character recognition*. Doctoral dissertation, University of Kent at Canterbury.

[7] Nauck, D., Klawonn, F. and Kruse, R. (1997). Foundation of neural neuro-fuzzy systems. Sussex, UK: John Wiley & Sons Ltd.

[8] Jang, J.S. (1993). 'ANFIS: adaptive-network-based fuzzy inference system', *IEEE transactions on systems, man, and cybernetics*, Vol. 23 (3), pp. 665-685.

[9] Liu, M. and Ling, F. Y. Y. (2003). Using fuzzy neural network approach to estimate contractors' markup. *Building and Environment*, 38(11), 1303-1308.

[10] Wu, J. D., Hsu, C. C. and Chen, H. C. (2009). An expert system of price forecasting for used cars using adaptive neuro-fuzzy inference. *Expert Systems with Applications*, 36(4), 7809-7817.

[11] Takagi, T. and Sugeno, M. (1985). 'Fuzzy identification of system and its application to modelling and control', *IEEE Transactions on systems, man and cybernetics*, Vol. 15, pp. 116-132.

[12] Opeyemi, O. and Justice, E.O. (2012). 'Development of neuro-fuzzy system for early prediction of heart attack', *International Journal of Information Technology and Computer Science (IJITCS)*, Vol. 4 (9), p. 22.

[13] Refenes, A.N. (1995). *Neural network design considerations.* In Refenes, A. N. (ed). Neural Networks in the capital markets. pp. 67-76, John Wiley, NY.

[14] Lam, K., Hu, T., Thomas Ng, S., Skitmore, M. and Cheung, S. (2001). 'A fuzzy neural network approach for contractor prequalification', *Construction Management & Economics,* Vol. 19 (2), pp.175-188.

The problems and abuse of performance bond in the construction Industry

A A Hassan[1], H Adnan[1]

[1]Faculty of Architecture, Planning & Surveying, 40450 Universiti Teknologi MARA, Shah Alam, Selangor MALAYSIA

*arzleehassan@gmail.com

Abstract. Performance bond is a bank-origin system adopted in the construction industry to act as a tool to ensure project delivery. The bond is taken up by an obliging party with endorsement by a third party; normally a banking institution, which would warrants the release of an insured amount should the obliging party fails to complete the contracted work on time or as agreed. The nature of the instrument is said to be perplexed and abusive. The research aims to investigate this claim. Applied legal research using both doctrinal and non-doctrinal legal research is employed. With the application of deductive reasoning, the data is processed and summarized. It is found that the term bond and guarantee, which are used intermittently but equivocally by the parties, should be distinguished accordingly to its obligation values, or otherwise would trigger major disputes. Interpretation, drafting and liabilities issues caused by archaic language and dated recycled wordings in the contracts contributed nothing to the convenience and application of this instrument. The loopholes in the contract made it easy for suspected fraudulent calls to be made by the beneficiary, whom are normally, the client. With several trusted professional bodies discouraged the use of this instrument, the legal system is compelled to provide direction to ensure continuous utilization of this project delivery assurance mechanism. Standardized system and structure is vital in order to obtain uniformity in its application.

1. Introduction

Completion of a contracted work is of pinnacle importance in any construction contract and employers' compromises against them are deemed conditional. Contractors are expected to plan for contingencies if encountered with anything preventing them from delivering the contracted work. A structured back-up plan needed to be contemplated. Thus, the concept of performance bond, guarantee and any insurance-like scheme has been adapted into the construction contract as a financial mechanism to ensure project delivery.

As the owner demand assurance for the completion of a project, this bank-origin system was initially adopted to ensure that the completion is guaranteed without any extra expenses to the owner [1]. However, researches revealed that problems and plausible abuse of this subsidiary tool exist. In the UK, few owners as the beneficiary are drawn to the idea of obtaining the sum through false calls.

1.1 Aim and Objectives

The aim of this paper is to examine the complication and plausible abuse in the application of performance bond in the construction industry. The paper seeks to unravel the components and key principles in its application from the law perspectives and obtain potential improvement. The objectives are as follows:

- To untangle the ambiguities within the related terminologies and their consistencies in the usage
- To observe professional attitudes towards the usage instrument
- To investigate the law governing the implementation of the instrument
- To examine potential solutions and preventive steps to be taken to ensure impartiality in its implementation

1.2 Problem Statement

Performance bond are observed to be 'archaic and are frequently expressed in outmoded language and create obligations whose legal interpretation may be unclear' [2]. Terms such as 'on-demand' and 'unconditional' in performance bonds are deemed to be susceptible to various abuse [3] and unfair from the perspective of the contractors. The proneness of abusive calls has led FIDIC, ICC and IBA to take the initiative of reducing the risk by proposing the introduction to the term 'performance security'.

Professional approaches towards performance bond and law governing the implementation are alarming. There are multiple views observed in the policies, the court and international professional bodies towards the practice. This mainly involves the usage of term and interpretations used by each party from various disciplines. The issue of overpayment and fraudulent or unreasonable calls (illegitimate calls made not out of genuine cause) also requires investigations.

2. Literature Review

Performance bond is an promissory for the contractor's (the obliging party) to delivery the contracted works to the employer (the offering party) through an endorsement by a third party (normally financial institution or insurer) on the contractor's capability of carrying them out. As an assurance should contractors fail to deliver completion, the owner relies on the financial capacity of the third party to provide compensation [4]. Performance bond is viewed commonly as a contract between *the contractor* (the obliging party, or called *principal* in the bond) and *the third party* (a guarantor, bondsman or *surety*), which is signed and will be void if the related contract is carried out as agreed but shall warrant a release in favour to *the owner* (the offering party, or *beneficiary*) if the contractor fails to meet the conditions.

As revealed in law cases, the courts differentiated bonds and guarantee as different subjects, while some members of the industry understand them as similar means, with very few and limited exceptions [5]. However, there are fundamental distinctions between the on-demand performance bond which contain a certain condition or *guarantee,* and the completely rigid and unconditional on-demand bond or *bond.* Thus, the thin line of distinction in terms should be measured according to their degree of obligations and never be underestimated [5].

It is pertinent that the term bond and guarantee are distinguished by comparing the weightage of the obligations, which can also be observed by classifying them into two types of obligation; primary obligation and secondary obligation. The general use of the term 'performance bond' in reference to all types is confusing, as the terms can be mistakenly used to refer both types of obligation [6]. Bankers would commonly describe them as primary obligation, while contractors understand them to be of secondary obligations. Thomas (2009) in *Dictionary of Banking Terms* deduced each term as follows:

Guarantee – *three-party agreements; including a promise by one party (guarantor) to fulfil the obligation of a person owing a debt, if that person fails to perform.*

Bond/Performance Bond – *bond that calls for specific monetary payment to a beneficiary if the purchaser or maker fails to do something or acts in violation of a contract. It may be a* *'surety bond' purchase from an insurance company or cash held in an escrow account by a bank or a third party.*

In summary, the term bond and guarantee can be distinguished by their amount of obligation values, the parties bound by the contract and the outcome whenever the principal contract terms are dishonoured. Whilst primary obligation refers to the contract between the insurer (or surety) and the employer, the secondary obligation is a genuine tripartite contract. A guarantee is a promissory of the obliging party to fulfil the duties in the principal contract, in which failure would amount to the release of the bond. Only a primary obligation contracts could trigger an on-demand calls under whatever circumstances as agreed in the contract. Therefore, most performance bonds should be characterized as the secondary obligations contracts. The trigger for release includes non-performance, failure to honour obligations under the contract and any eventualities that could amount to breach of contract terms.

2.1 Rights of Parties

Contractors as the obliging party would naturally favour guarantees over bonds, as guarantees provide more security and leniency in conditions of the contracts. The employers prefer the more autonomous bonds. The bonds are made easy for the beneficiary, or employer as demands can be called without being required to adduce proof of substantive default [7] and as effortless even via a simple phone call. [7] further expressed that on-demand bond exerts commercial pressure but contractors still reluctantly submits in order to compete; making it susceptible and abusive from the contractors' point of view.

Rules were provided to demand the employer to comply with procedures on calling for the bond, but were ignored [8]. The bank or insurance company has the prime obligation to honour the terms of the bond [8] on demand as long as all the requirements were satisfied [1]. Processing the call and claims are straightforward as it is not obligatory for them to question whether the call is bona fide or not. Hence, FIDIC discretion against the use of on-demand bonds due to their one-sided and confrontational nature [9] despite the sureties' preference towards bond contracts over guarantees.

2.2 Problems and Abuse

The ambiguity observed in the documents and the terms revealed confusion in deciphering whether the said instrument refers to the tripartite guarantee with lenient conditions *or* the autonomous unconditional and rigid on-demand bond. These confusion and potential trigger to dispute are caused by the lack of sufficient understanding of the exact nature of these instruments and the use of medieval language in their wording, compounded further by a miscellany of confusing terminology used virtually interchangeably [10]. The courts in the US and Canada had also criticized the practice, where it was discovered that many of the banking institutions and insurance companies continuously drafting contracts and bonds in a form that is archaic, obscure and full of traps for the uninitiated [11]. It was also observed that there are lacks in uniformity in the drafting process that led to difficulties in implementing the instrument and complication in monitoring the use of these tools. It is observed that the way the guarantee/bond documents are drafted to mainly benefit the employer. Sir Michael Latham urged the use of modern language with clear obligations and end date in guarantee/bond contracts to allow the beneficiary to claim without the need for litigation and this should be enforceable on default rather than demand [6]. Consequently, a clear, unambiguous and rigorous draft for guarantee *or* bond [7] is highly crucial.

[7] revealed that a bond could be easily called without the need to refer the claim through litigation or arbitration. It is also agreed that the bonds act as protection from non-completion where insolvency may be regarded as one of the factor. [12] revealed that the contractor's financial failure is the major cause for a bond to be called for. However, as discovered through law cases, insolvency alone is not the trigger for the release of a bond. It has become a thinly disguised solvency guarantee and a strong weapon in the employer's hands to ensure prompt completion [13].

3. Methodology

The suitability of doctrinal legal research and non-doctrinal research conducted for this research is due to the applied legal research nature [14]. Legal doctrinal research uses principles and law collected from known basic principles and established case law as basis of research. It is a study on the related law and its impact on a particular community; in this case the construction community. Relevant cases law related to performance bond were collected and examined to demonstrate the operation and efficacy of a particular law [15]. Therefore, all relevant cases enlisted are defined, categorized and analysed. To analyse the data, deductive reasoning is used. These include defining the principle, identifying a factual event and lastly, making a relation between the two to determine whether the principle applies or otherwise [14]. Governing rules from forms of contract and procedures supplied by the financial institution provide a basis and reference of comparison. Non-doctrinal research will focus on the application and study about the law. Inductive reasoning was also applied to identify whether a general accepted principles within a specific current event are complied or otherwise. Data were collected and sorted before a general content analysis was carried out. Finally, a comparative analysis was applied in the pursuit to better understanding the general pattern of the sets of data. Since the research mainly focuses on the black letter law, the implementation and their effects, the findings of the research were limited to that perspective excluding all insights and opinion from the very community in which the research wishes to address.

4. Data Analysis

4.1 The Use and Abuse

4.1.1 Disentangling Ambiguity

Observation gathered from case of *Workington Harbour Dock Board v. Trade Indemnity Co Ltd (No. 2)* [1938] 2 All ER 101where the court decided that the bond is in fact, a guarantee and the contractor is held liable base on the fact that '*the rights of the parties should be regulated on that footing'* referring to the true meaning and intention of the contract. An interesting commentary by the court on the contract and its confusing nature is that *'it is difficult to understand'*. The parties' persistence in *'entering upon considerable obligations in old-fashioned forms of contract which do not adequately express the true transaction'* was also criticized. As suggested, it can be observed that uniformity does not exist in the implementation of this instrument. Monitoring is difficult as no specific scheme can be used as a model.

After the decision in the Workington case, it was again disclosed in the Trafalgar House case where a typical example of inaccuracy is the use of the term 'bond' in a contract, initially intended to be regarded as a 'guarantee' by parties in a contract. This consistency was the long awaited benchmark in the effort of setting the needed guideline. In another classic leading case of *Edward Owen Engineering Ltd v. Barclays Bank International* [1978] 1 QB 159 where the differentiation between bonds and guarantees was highlighted. The court described the similarity in nature within the implementation between bond and the letter of credit. The court also established the importance in proving of fraudulent calls before any challenge and injunctive relief on a demand can be made.

4.1.2 Drafting Issues

The critics against the usage of 'recycled' guarantee/bond documents and promoting the use of modern clearer and non-confrontational documents is becoming prominent. In the case of *Oval (717) Ltd v Aegon Insurance Co (UK) Ltd* [1997] 85 B.L.R. 97, the court stated that a need to pay attention to the careful construction of the contracts and commitment towards them should never be underestimated. The case established that it is important to make sure all necessary terms are in the contract, and if such duties are intended they are deemed to be included. This was firmly supported by the decision held in the case of *Marubeni Hong Kong (MHK) & South China Ltd v. Mongolia* [2005] where the Lord Justice Carnwath elaborated that *'if MHK had wanted the additional security of a demand bond, one would have expected them to have insisted on appropriate language to describe it, in both the instrument itself, and in the Legal Opinion'.*

4.1.3 Fraudulent Calls and Threat Issues

A claim for fraudulent call would need to be supported by solid evidence. This would mean that the accuser would also impliedly suggest that the surety is an accomplice and knows that the fraudulence exists. This would be quite difficult to prove if the dispute is between the contractor and the surety. Such occurrence was revealed in *HLC Engenharia E Gestao de Projectos S.A. v. ABN Amro Bank NV*

& Others [2005] *Banque Saudi Fransi (A Company Incorporated in the Kingdom of Saudi Arabia) v. Lear Siegler Services Inc* [2007] where there were clear evidence *'that the beneficiary of the performance bond had made a fraudulent demand of which the bank was aware'*. Unfortunately, evidence is hardly generous in many other occurrences.

The issue of threatening the contractor using on demand call is also alarming. Threats are considered as abusive call and are equally devious as a fraudulent call. Therefore, providing evidence is similarly challenging. In the case of *TTI Team Telecom International Ltd v. Hutchison 3G UK Ltd* [2003], the beneficiary threatened to call on the advance payment bond and it was held to be a threat to an anticipatory breach of the contract.

In a situation where the contractor was expected to fail due to their bankruptcy, the court in *Perar BV v. General Surety & Guarantee Co Ltd* [1994] 66 B.L.R. 72 CA, *Balfour Beatty Civil Engineering & ANR, (Together T/A The Balfour, Beatty/Costain (Cardiff Bay, Barrage) Joint Venture) v. Technical & General Guarantee Company Ltd* [2000] held that insolvency was not a default and would not easily trigger a call for a bond.

4.2 Injunctive Relief

Injunctive relief can be made by the principal (contractor) to avoid the release of guarantee/bond through fraudulent calls. However, the ground for what validates an injunction relief application is seen to be too strict. Evidently, the court would only grant injunctive relief under three circumstances: 1) a clear and proven case of fraud, as established in in the *Edward Owen* case and *Consolidated Oil* case, 2) a potential abusive calls such as a threat to breach a contract by any member of the contracting parties such as in the case of *TTI Telecom v. H3G* case, and 3) the ambiguous and controversial *balance of convenience* such as in the *HLC–Amor* case where the court held that if in any case the obligor is actually entitled for leave to defend to avoid a *potentially irredeemable injustice*. A strict canon is pertinent in giving injunctive relief. Regardless of the risk in having the appealing party misuse the injunction to escape a due obligation, fair opportunity to defend against unfair calls is due in the effort to ensure justice to the deserving party.

4.3 Governing Law

The need for performance bond from the employer's perspective is very important. However, there are need to investigate the criticism from professional bodies' such as FIDIC who claims that the use of 'on-demand' bonds as not adding to the likelihood of high quality performance [9]. Problems needed to be identified and improvement is highly essential. Modern construction contracts such as the JCT Standard Building Contract 2005 Edition [16], governing clauses and forms to properly support the implementation are absent. Furthermore, The FIDIC Condition of Contracts series and [16] agreed with the decision held by the court in the *Perar BV* case and do not consider insolvency as breach of contract. Since the initial objective behind the performance bond was to mainly protect against non-completion (and insolvency may as well be a considerable reason), a well-drafted form of the bond/guarantee contract is vital. The Association of British Insurers (ABI) had continuously supplied the industry with their model of guarantee/bond contracts and forms in an attempt to govern the use of this instrument. Unfair Contract Terms Act 1977 is also available for parties who wishes to draft and construct a bespoke guarantee/bond forms. Unfortunately, they are bound to return to the archaic and biased contents.

5. Conclusion

Performance bond purpose itself as an immediate fund to recover from incomplete work, by reasons of financial or technical disruption on the contractors' side. The construction of contract using archaic wording in the contract has done more damage to the implementation of this tool. The contract exhibits confusing interpretations of terminologies, unclear liabilities and obligations, potential misuse of authority and questionable rights in the contract; contributed to eventual disputes within the tripartite contract. The parties failed to observed the fundamental principles and objective of the instrument. The one-sided interpretations found in the contracts are the cause of lacks in uniformity and inconsistency in the contents of guarantee/bond forms. Moreover, the parties neglected the ideal surface guideline and definition provided by the banking system.

The courts remained firm in dealing with the subject but contribute to minor inconsistency in their decisions, presumably due to the existing inconsistency and lack of uniformity in the contract itself. The court's conservative approach observed in dealing with this bank-origin system has managed to provide a fair guideline to the users in addressing the contents of the contract through its true meaning and the intended purposes. By imposing uniformed definitions and interpretations, establishing transparent definitions within the industry and standardizing scheme through collaboration between ABI and the professional bodies such as JCT, ICE and FIDIC to create an ideal model form could ensure fairness in the implementation of this instrument.

Table of Cases

[1] *American Cyanamid Co v. Ethicon Ltd* [1975] AC 396, HL
[2] *Balfour Beatty Civil Engineering & ANR, (Together T/A The Balfour, Beatty/Costain (Cardiff Bay, Barrage) Joint Venture) v Technical & General Guarantee Company Ltd* [2000] C.L.C. 252; 68 Con. L.R. 180
[3] *Banque Saudi Fransi (A Company Incorporated in the Kingdom of Saudi Arabia) v Lear Siegler Services Inc* [2006] EWCA Civ 1130; [2007] 1 All E.R. (Comm) 67; [2007] 2 Lloyd's Rep. 47;
[4] *Edward Owen Engineering Ltd v. Barclays Bank International* [1978] 1 QB 159
[5] *Workington Harbour Dock Board v. Trade Indemnity Co Ltd (No. 2)* [1938] 2 All ER
[6] *HLC Engenharia E Gestao de Projectos S.A. v ABN Amro Bank NV & Others* [2005] EWHC 2074 (QB)
[7] *Marubeni Hong Kong & South China Ltd v Mongolia* [2005] EWCA Civ 395; [2005] 1 W.L.R. 2497; [2005] 2 All E.R.
[8] *Oval (717) Ltd v Aegon Insurance Co (UK) Ltd* [1997] 85 B.L.R. 97; 54 Con. L.R. 74;
[9] *Perar B.V v. General Surety and Guarantee* [1994] 66 BLR 72
[10] *TTI Team Telecom International Ltd v Hutchison 3G UK Ltd* [2003] EWHC 762 (TCC); [2003] 1 All E.R. (Comm) 914
[11] *Spiersbridge Property Developments Ltd v. Muir Construction Ltd [2008] CSOH 44*

References

[1] Dingman, MS. (2000) *How the Owner Derives Benefit From A Performance Bond* [online].[London,England]; *Reed Smith: Publications*
[2] Furst, S & Ramsey V. (2006) *Keating on Construction Contracts*, 8th Ed. London: Sweet & Maxwell
[3] Duncan Wallace I.N. (1998) *Perar Revisited: When Will They Ever Pay? When Will They Ever Learn?* [online]. [London]; *Construction Law Journal*. Const. L.J. 1998, 14(5), 322-328
[4] Warren, L. (2008) *Performance Bond* [online]. [Qatar]; *Clyde & Co* [cited 16 October 2008] http://www.clydeco.com/knowledge/articles/performance-bonds.cfm.
[5] Lurie, J. (2008), On Demand Performance Bond: Is Fraud The Only Grounds for Restraining Unfair Calls? – *The International Construction Law Review* Vol.25 Part 4 Oct. 2008, Informa Law, London: MPG
[6] Davis, R. (1999) Construction Insolvency 2nd Ed., Ch.14, Bembridge: Palladian Law Publishing
[7] Dunham, P. (2008), The Use and Abuse of First Demand Guarantees In International Constrcution Projects – *The International Construction Law Review* Vol.25 Part 3 Jul. 2008, Informa Law, London: MPG
[8] Atkinson, D. (1998) *Secure The Right Bond* [online].[London, England]; *Atkinson Law* [cited 16 October 2008] http://www.atkinsonlaw.com/library/ article.php?id=118
[9] Anon (2001) *About FIDIC- Policies: Guarantees, bonds and retention relating to professional services* [online] Switzerland]; *FIDIC Federation*
[10] Ndekugri, I. (1999) Performance Bond and Guarantees: Construction Owners and Professionals Beware – *Journal of Construction Engineering and Management* Nov/Dec 1999 Vol. 125 No. 6,American Society of Civil Engineers Construction Division

IOP Conf. Series: Earth and Environmental Science **117** (2018) 012044 doi:10.1088/1755-1315/117/1/012044

[11] Hughes, W., Hillebrandt, P. and Murdoch, J. (1998) Financial Protection in the UK Building Industry, Reading Construction Forum, London: E& FN Spon.
[12] Levine, M. and J. Wood (1991) Construction Insurance and UK Construction Contracts, London: Lloyd's of London Press Ltd
[13] Chappel, D., Marshall, D., Powell-Smith, V. & Cavender, S. (2001) *Bulding Contract Dictionary* 3rd ed, Oxford: Blackwell Science Ltd.
[14] Chynoweth, P. (2008), *Legal Research* in Knight, A. & Ruddock, L (2008) ed.,*Advance Research Method in the Building Environment*, Oxford; Wiley & Blackwell Corbin, J. & Strauss, A. (2008) *Basic of Qualitative Research* 3rd Edition, California: Sage Publication
[15] McConville, M & Wing, HC (2007), *Research Method for Law*: Edinburgh University Press
[16] JCT (2005) *JCT Standard Building Contracts With Quantities SBC/Q,* The Joint Contract Tribunal Limited

IOP Conf. Series: Earth and Environmental Science **117** (2018) 012045 doi:10.1088/1755-1315/117/1/012045

The relationship between cost estimates reliability and BIM adoption: SEM analysis

N A A Ismail[1], N H Idris[1], H Ramli[1], R R Raja Muhammad Rooshdi[1], S R Sahamir[1]

[1]Faculty of Architecture, Planning And Survey ,Universiti Teknologi MARA, Shah Alam, Malaysia

*noorakmaladillah@gmail.com

Abstract. This paper presents the usage of Structural Equation Modelling (SEM) approach in analysing the effects of Building Information Modelling (BIM) technology adoption in improving the reliability of cost estimates. Based on the questionnaire survey results, SEM analysis using SPSS-AMOS application examined the relationships between BIM-improved information and cost estimates reliability factors, leading to BIM technology adoption. Six hypotheses were established prior to SEM analysis employing two types of SEM models, namely the Confirmatory Factor Analysis (CFA) model and full structural model. The SEM models were then validated through the assessment on their uni-dimensionality, validity, reliability, and fitness index, in line with the hypotheses tested. The final SEM model fit measures are: P-value=0.000, RMSEA=0.079<0.08, GFI=0.824, CFI=0.962>0.90, TLI=0.956>0.90, NFI=0.935>0.90 and ChiSq/df=2.259; indicating that the overall index values achieved the required level of model fitness. The model supports all the hypotheses evaluated, confirming that all relationship exists amongst the constructs are positive and significant. Ultimately, the analysis verified that most of the respondents foresee better understanding of project input information through BIM visualization, its reliable database and coordinated data, in developing more reliable cost estimates. They also perceive to accelerate their cost estimating task through BIM adoption.

1. Introduction

The implementation of Building Information Modelling (BIM) technology in improving the construction industry has been widely reported across the globe [1]–[5]. There are several studies that addressed many benefits of adopting BIM by the practitioners and organizations within the industry; such as cost and time saving, reduced human resource, quality and performance improvement, clash detection, improved accuracy, increased profitability, enhanced collaboration and communication, better presentation and documentation process, improved planning and design, better visualization, and improved information [6]–[12].

BIM usage also gives impact on the reliability of construction cost estimates, typically part of vital tasks of the Quantity Surveyors for any construction projects. The BIM technology is claimed to potentially enhance the traditional method of cost estimating that are constantly facing errors, inaccuracies, omissions and ethical flaws [13]. Especially towards improving project input information supplied to the project, the application of BIM is capable of assisting the cost estimators to attain better understanding in the preparation of more accurate and reliable cost estimates. With regards to improving cost estimates reliability, BIM visualization through its 3D models, allow the demonstration of the construction sequences through simulation by project stages [14]–[17]. Additionally, the provision of digital database via BIM platform reinstates the limitations of using conventional 2D drawings; allows access to all disciplines in the team to obtain integrated project information for cost estimating purpose [18][19]. Furthermore, coordinated data generated by BIM applications permits multiple views of building models, thus helps in examining any clashes between the elements in the building system [19][20].

As interpreted by Lu et al. (2012), the diffusion of the BIM technology could be characterized by the perceived benefits of the practitioners in the industry when adopt BIM [21]. Hence, it could be assumed that the merits of BIM innovation in facilitating cost estimates preparation might influence the Quantity Surveyors as cost estimators to adopt BIM in their practice. Therefore, by utilizing SEM method, this paper aims to examine the relationship between cost estimates reliability and BIM adoption amongst the Quantity Surveyors as cost estimators.

2. Research methods

For this study, the data were collected through a survey questionnaire distributed to the registered Quantity Surveyors with the Royal Institution of Surveyors Malaysia (RISM). There are 1140 members listed, in which 294 were selected as samples from the overall population [22], upon systematic random sampling technique [23]. It resulted from the survey, the satisfactory response rate of 68.5% [24][25], where only 202 responses were usable for further analysis.

The questionnaire employed Likert scale rating in requiring the respondents to imply their point of views towards the effects of BIM technology on the reliability of cost estimates leading to adoption of the technology. Four main variables namely *BIM perceived benefits, BIM improved information, cost estimates reliability* and *BIM adoption* were evaluated through the questionnaire. Prior to the survey instrument development, six hypotheses were inferred from the relationships that might exist between the variables. The hypotheses are: (H1) Improved Information has a positive and significant effect on Cost Estimates Reliability; (H2) Improved Information has a positive and significant effect on Perceived Benefits; (H3) Perceived Benefits has a positive and significant effect on BIM Adoption; (H4) Cost Estimates Reliability has a positive and significant effect on BIM Adoption; (H5) Improved Information has a positive and significant effect on BIM Adoption; and (H6) Cost Estimates Reliability has a positive and significant effect on Perceived Benefits. The questionnaire was pre-tested amongst the experts beforehand to identify and rectify potential issues on the questionnaire setting [26]–[28].

Primarily, the inferential analysis via SEM method engaging SPSS-AMOS was used to analyse the studied variables and their relationships. Prior to the main analysis, the raw data obtained from the survey were coded, screened and cleaned in compromising with missing values in the dataset [29].

3. Structural Equation Modelling (SEM) approach

SEM approach was deployed to model relationships amongst the variables involved in this study [30]. As described by Kaplan (2009), the SEM method incorporates the processes of: model specification to fit the theory; gathering samples and measures of variables; estimation of the model to obtain parameters; modification to fit the assessment done; and interpretation and discussion of reported results [31]. Two types of SEM models which are measurement model (also known as Confirmatory Factor Analysis (CFA) model) and full structural model were developed throughout the analysis. The Confirmatory Factor Analysis (CFA) assessing uni-dimensionality, validity and reliability, and fitness index were carried out for all measurement models representing each variable respectively. According to Awang (2015), uni-dimensionality in the measurement models is achieved with the factor loading value of ≥ 0.60 [32]. Whereas, the validity evaluated by Average Variance Extracted (AVE) must attain ≥ 0.50 and reliability assessed by Composite Reliability (CR) must range ≥ 0.60 in term of values. Table 1 shows the AVE and CR formula, and Table 2 tabulates the values of Fitness Index based on categories that need to be achieved for model fit. Eventually, the finalized measurement models that have been evaluated accordingly, were combined and further analysed as SEM full structural model until it reached the model fit requirement.

Table 1. AVE and CR formula [32]

AVE $= \frac{\Sigma K^2}{n}$	**CR** $= \frac{(\Sigma K^2)}{\lvert(\Sigma K^2)+(\Sigma 1-K^2)\rvert}$
K = factor loading of every item n = number of items in the model	

Table 2. Categories of Fitness Indexes and Level of Acceptance [32]

Name of Category	Name of Index	Index Full Name	Level of Acceptance
Absolute fit	**Chi-Square**	Discrepancy Chi Square	P-value > 0.05 *(not applicable for large sample size more than 200)*
	RMSEA	Root Mean Square of Error Approximation	RMSEA < 0.08
	GFI	Goodness of Fit Index	GFI > 0.90
Incremental fit	**AGFI**	Adjusted Goodness of Fit	AGFI > 0.90
	CFI	Comparative Fit Index	CFI > 0.90
	TLI	Tucker-Lewis Index	TLI > 0.90
	NFI	Normed Fit Index	NFI > 0.90
Parsimonious fit	**Chisq/df**	Chi Square/Degrees of Freedom	Chi-Square/df < 3.0

4. Results and discussion

Four measurement models were developed for each variable involved in this study. Items in the questionnaire were exploited as indicators to measure the variables accordingly. Initially, those measurement models were assessed and all undervalued items that failed the related tests were eliminated from the model. Analysis was continuously done with items for every variable were tested until all measurement models achieved model fit. The final results of CFA validation for each measurement model are shown in Table 3. Finally, Figure 1 portrays the SEM full structural model combining all evaluated measurement models for all variables.

Table 3. CFA validation final results

Measurement models (variables)	AVE (≥ 0.5)	CR (≥ 0.6)	P-value > 0.05	RMSEA < 0.08	GFI > 0.90	AGFI > 0.90	CFI > 0.90	TLI > 0.90	NFI > 0.90	ChiSq/df < 3.0
Improved Information	0.877	0.955	0.052	0.062	0.974	0.935	0.995	0.990	0.988	1.777
	0.893	0.944								
	0.785	0.879								
Perceived Benefits	0.851	0.966	0.527	0.000	0.992	0.975	1.000	1.001	0.997	0.832
Cost Estimates Reliability	0.941	0.970	0.098	0.062	0.983	0.941	0.997	0.993	0.994	1.781
	0.931	0.964								
	0.912	0.954								
BIM Adoption	0.884	0.968	0.589	0.000	0.997	0.987	1.000	1.002	0.999	0.530

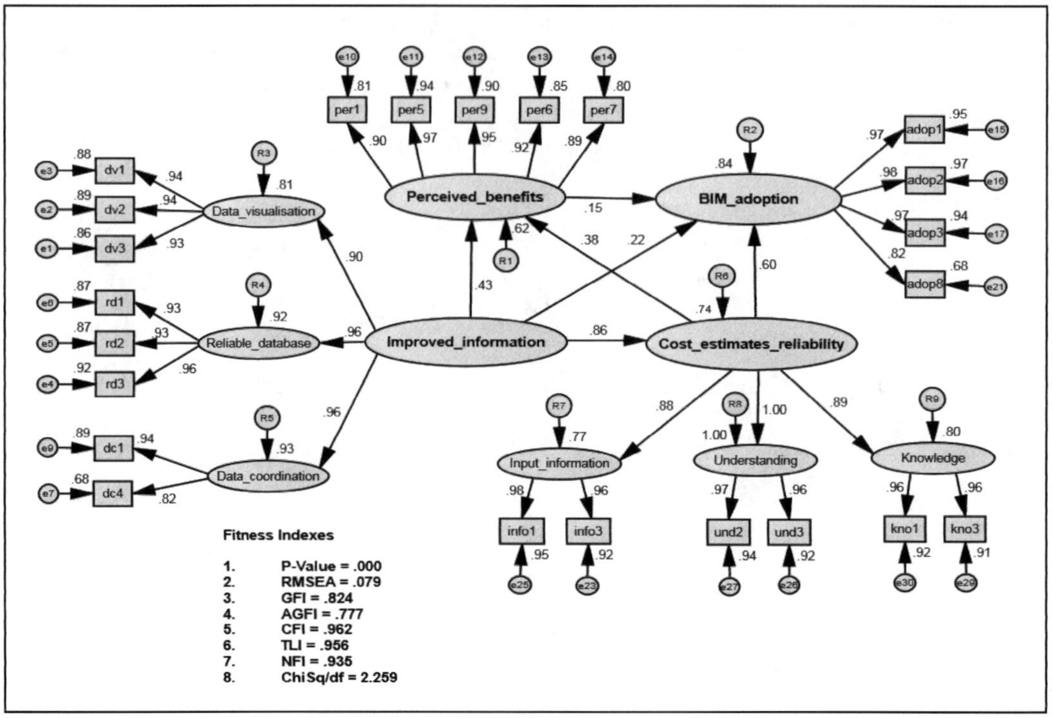

Figure 1. The SEM full structural model

Table 4. The results of hypotheses testing for variables

Hypotheses Statements	Estimate	P-value	Results on Hypotheses
H1: Improved Information has a positive and significant effect on Cost Estimate Reliability	0.862	<0.001	Supported
H2: Improved Information has a positive and significant effect on Perceived Benefits	0.434	<0.001	Supported
H3: Perceived Benefits has a positive and significant effect on BIM Adoption	0.149	0.005	Supported
H4: Cost Estimate Reliability has a positive and significant effect on BIM Adoption	0.601	<0.001	Supported
H5: Improved Information has a positive and significant effect on BIM Adoption	0.219	0.004	Supported
H6: Cost Estimate Reliability has a positive and significant effect on Perceived Benefits	0.382	<0.001	Supported

The final results as shown in Figure 1 indicate the overall fitness measure for the final SEM structural model: P-value=0.000, RMSEA=0.079, GFI=0.824, AGFI=0.777, CFI=0.962, TLI=0.956, NFI=0.935 and ChiSq/df= 2.259. All values demonstrated in the model have achieved the required level of Fitness Index except for GFI =0.824<0.90 and AGFI =0.777<0.90 which might be due to sensitivity of the values towards sample size [33][34]. With the majority of the values reached the level of acceptance for Fitness Index, the final SEM structural model was considered fit for the study. Following the full structural model, the results of hypotheses also yielded the significance p-value (<0.05) for all variables tested, hence all hypotheses are supported (refer Table 4).

The results further consummated that there are significant relationships exist between cost estimates reliability and BIM adoption. Information is improved via BIM platform through its data visualization, reliable database, and coordinated data throughout the construction projects. Evidently by the BIM procedures establishing those aspects, it accommodates better understanding of construction project information amongst the cost estimators. It will then lead towards knowledge improvement in the cost estimates preparation. The mechanism subsequently serves the perceived benefits towards cost decision making incorporating BIM technology by the Quantity Surveyors as cost estimators. For the BIM innovation to be useful within the estimator's roles, it is anticipated that the BIM tools could accelerate the cost estimating tasks, in concurrent with increasing the accuracy of the cost estimates. With the BIM operation manifesting its capabilities of data visualization, reliable database and data coordination, that could improve their cost expertise, it ultimately motivates the cost estimators to adopt the BIM technology.

5. Conclusion

Conclusively, Structural Equation Modelling (SEM) method was statistically used to model relationships and estimate its values, which exist between variables in the study dataset. Hypotheses were formulated that preceded the development of the SEM models. There were two parts of SEM models, namely measurement model and structural model. All measurement models were established individually representing every variable that were further validated through CFA validation assessing uni-dimensionality, validity, reliability and fitness index; to be next combined as a structural model. Subsequently, the full structural model was established following the confirmation on its model fit. Some key points were highlighted upon the verification of the final model for this study. It was acknowledged that the BIM-improved information through its data visualization, reliable database and coordinated data, could provide accurate data to better understand and interpret project data, leading towards improving knowledge in estimating process. These factors become the drivers that motivate the respondents to adopt BIM in their practice, in anticipating that it would assist in performing cost estimating tasks more quickly. The conducted SEM analysis in establishing SEM full structural model in this study could become a guideline to further develop a framework in strategizing the cost estimating practice incorporating BIM technology.

References

[1] McGraw Hill Construction, "SmartMarket report: The business value of BIM in Europe," 2010.
[2] Masterspec, "New Zealand National BIM Survey 2012," 2012.
[3] Masterspec, "New Zealand National BIM Survey 2013," 2013.
[4] NBS, "NBS International BIM Report 2013," 2013.
[5] NBS, "International BIM Report 2016," 2016.
[6] H. Yan and P. Damian, "Benefits and barriers of Building Information Modelling," in *12th International Conference on Computing in Civil and Building Engineering 2008*, 2008.
[7] G. Aranda-mena, A. Chevez, J. Crawford, R. Wakefield, T. Froese, J. Frazer, S. Gard, and D. Nielsen, "Business Drivers for Building Information Modelling," Brisbane, Australia, 2008.
[8] B. Becerik-Gerber and S. Rice, "The perceived value of Building Information Modeling in the U.S. building industry," *J. Inf. Technol. Constr.*, vol. 15, pp. 185–201, 2010.
[9] R. Liu, R. R. A. Issa, and S. Olbina, "Factors influencing the adoption of Building Information Modeling in the AEC Industry," in *Proceedings of the International Conference on Computing in Civil and Building Engineering (ICCBE) 2010*, 2010.
[10] D. Zhang and Z. Gao, "Project time and cost control using Building Information Modeling," in *ICCREM 2013: Construction and Operation in the Context of Sustainability*, 2013, pp. 545–554.
[11] A. AlabdulQader, K. Panuwatwanich, and J. Doh, "Current use of Building Information Modelling within Australian AEC industry," in *Hokkaido University Collection of Scholarly and Academic Papers : HUSCAP*, 2013.
[12] L. Zhang, G. Wang, T. Chen, and G. He, "Survey of BIM application status and characteristics in China," in *Proceedings of the 17th International Symposium on Advancement of*

Construction Management and Real Estate, 2014, pp. 969–979.

[13] O. A. Olatunji and W. D. Sher, "A comparative analysis of 2D computer-aided Estimating (CAE) and BIM estimating procedures," in *Handbook of Research on Building Information Modeling and Construction Informatics: Concepts and Technologies*, IGI Global, 2010, pp. 170–189.

[14] N. Thurairajah and D. Goucher, "Advantages and challenges of using BIM: A cost consultant's perspective," in *49th ASC Annual International Conference Proceedings, California Polytechnic State University (Cal Poly)*, 2013.

[15] A. Z. Sheth and S. M. Malsane, "Building Information Modelling, a tool for green built environment," in *All India Seminar on Innovation in Green Building Technology, Green Build 2014*, 2014.

[16] R. Eadie, H. Odeyinka, M. Browne, C. Mckeown, and M. Yohanis, "An analysis of the drivers for adopting Building Information Modelling," *J. Inf. Technol. Constr.*, vol. 18, pp. 338–352, 2013.

[17] C. Furneaux and R. Kivvits, "BIM – Implications for Government," CRC for Construction Innovation, Brisbane, 2008.

[18] InfoComm International, "Building Information Modeling (BIM) Guide," 2011.

[19] R. P. Kumanayake and R. M. P. S. Bandara, "Building Information Modelling (BIM); How it improves building performance," in *International Symposium on Ensuring National Security Through Reconciliation & Sustainable Development*, 2012, pp. 357–365.

[20] D. A. Campbell, "Building Information Modeling: The web 3D application for AEC," in *Web3D 2007*, 2007, pp. 173–177.

[21] W. Lu, Y. Peng, Q. Shen, and H. Li, "A generic model for measuring benefits of BIM as a learning tool in construction tasks," *J. Constr. Eng. Manag.*, no. 139, pp. 195–203, 2012.

[22] R. V Krejcie and D. Morgan, "Determining sample size for research activities," *Educ. Psychol. Meas.*, vol. 30, pp. 607–610, 1970.

[23] J. W. Creswell, *Education research: Planning, conducting and evaluating quantitative and qualitative research*, 4th ed. Boston: Pearson, 2012.

[24] Y. Baruch, "Response rate in academic studies - A comparative analysis," *Hum. Relations*, vol. 52, no. 4, pp. 421–438, 1999.

[25] T. P. Johnson and J. S. Wislar, "Response rates and nonresponse errors in surveys," *JAMA J. Am. Med. Assoc.*, vol. 307, no. 17, pp. 1805–1806, 2012.

[26] L. Frazer and M. Lawley, *Questionnaire design & administration*. Milton, Qld: John Wiley & Sons Australia, 2000.

[27] D. R. Cooper and P. S. Schindler, *Business research methods*, 8th ed. New York: McGraw-Hill, 2003.

[28] W. G. Zikmund, S. Ward, B. Lowe, and H. Winzar, *Marketing research*. Victoria, Australia: CENGAGE Learning, 2007.

[29] J. Pallant, *SPSS survival manual: a step by step guide to data analysis using IBM SPSS*, 5th ed. Crows Nest, N.S.W: Allen & Unwin, 2013.

[30] R. H. Hoyle, *Handbook of structural equation modeling*. New York: Guilford Press, 2012.

[31] D. Kaplan, *Structural equation modeling: foundations and extensions*, 2nd ed. Sage Publications, 2009.

[32] Z. Awang, *SEM made simple: A gentle approach to learning Structural Equation Modelling*. Selangor, Malaysia: MPWS Rich Publication Sdn. Bhd., 2015.

[33] D. Hooper, J. Coughlan, and M. R. Mullen, "Structural equation modelling: Guidelines for determining model fit," *Electron. J. Bus. Res. Methods*, vol. 6, no. 1, pp. 53–60, 2008.

[34] S. Sharma, S. Mukherjee, A. Kumar, and W. R. Dillon, "A simulation study to investigate the use of cutoff values for assessing model fit in covariance structure models," *J. Bus. Res.*, vol. 58, no. 7, pp. 935–943, 2005.

Commitment to sustainability: A content analysis of website for university organisations

M S Hasim[1], A E Hashim[1], N R M Ariff[1], Z Sapeciay[1], A S Abdullah[1]
[1]Faculty of Architecture, Planning and Surveying, Universiti Teknologi MARA, Shah Alam, Selangor, Malaysia.

*moham315@salam.uitm.edu.my

Abstract. This research aim on investigating the commitments of organisations towards sustainability. For this research context, 'commitment' refers to the extent of information provided by universities in their website which demonstrated initiatives towards achieving the sustainability goal. The objective of this study was to identify sustainability initiatives highlighted within university websites using Australia as a case study. Thirty-nine (39) websites were reviewed and web content analysis was performed to publicly available data including any relevant accessible PDF documents attached to the universities website. Specific websites information was reviewed to detect sustainability themes in the broad university management and operations (i.e., in general policies, corporate mission statements, research activities, positions available and strategies). The commitment of Australian universities was significant and well established with a set of twenty (20) related themes were identified. The findings have some limitations because the established themes only emerged from the websites' content without human validation which possibly weakens the correlations between website information and organisations actual practice. This possibility is recognised and for this reason, further assessment may be advantageous to provide verification of the findings. Therefore, further studies using other techniques are suggested such as interviews or observations for validation of data and reinforce the entire conclusions. An interesting aspect of this study is the validity of reviewing organisational websites for gauging actual practice and a number of researchers supporting this approached as indicated in methodology section of this paper.

1. Introduction

The concept of sustainability refers to efforts to adopt environmental, economic and social dimensions, as well as practices [1, 2]. Generally, organisations are adopting the sustainability principle into practice when they undertake initiatives such as using sustainable, green or socially-responsible materials, involve using environmentally-safe cleaning products; renewable energy for lighting and transportation; recycling or reuse of all waste; applying leadership to influence the industry, that is, buying green energy from green suppliers to drive the industry; and contributing to the community, that is, providing programs to help solve a social problem [1]. Moreover, for the progress organisation, sustainability values must penetrate into all aspects of the organisation [1] and must embed the sustainability agenda within their organisational structure and governance (i.e., by establishing a vision and mission policy incorporating sustainability).

There are various reasons why university organisations are important with regard to sustainability issues. For example, universities can set an example to the community by influencing the way forward by enhancing the sustainability agenda through education, research and their operation. As cited in Amaratunga and Baldry [3], sustainability practices are important to universities due to the following:

1) Universities have a range of diverse building types which need more different operational tasks than other organisations;
2) Universities are trying to improve their efficiency while facing the challenge of rising operating costs and increasing user expectations;
3) Massive expansion has forced universities to achieve more economic use of their facilities.

Therefore, this study is significant to universities as they need new knowledge to accelerate their effort to improve sustainable practice and expand their organisational sustainability outcomes.

This paper highlighted the commitments of organisations towards sustainability by reviewing organisational websites for evaluating actual implementation. An interesting aspect of this paper is the validity of reviewing organisational websites for measuring actual practice. The focus is on university organisation as these institutions are in a unique position to influence the attitudes of the community, government and the private sector towards sustainability. The findings may encourage global discussion to influence the sustainable practice. The paper is divided into five parts. Starting from this introduction, the literature on university commitment towards sustainability, the methodology, followed by results and analysis. Finally, the conclusions are drawn related to the overall discussion.

2. University commitment towards sustainability

Universities organisations have great potential to play a fundamental role in achieving the sustainability goal. The importance of universities with regard to sustainable development is reflected in various international agreements such as the Copernicus University Charter for Sustainable Development issued by the International Association of Universities and the Kyoto Declaration on Sustainable Development issued by the UNESCO World Declaration on Higher Education [4]. These agreements promote universities as key institutions in developing sustainable societies. For example, in the Kyoto Declaration, universities are encouraged to adopt a sustainable development approach in their physical operation and by embedding the environment in education. According to the Association of University Leaders for a Sustainable Future (ULSF 1999) cited in Van Weenen [5], 'universities can be involved in sustainable development such as in management, planning, development, education, research, operations, community service, purchasing, transportation, design, new construction, renovation and retrofit'(p. 21). Thus, the findings from these organisations can be used as a benchmark in measuring the commitments of other major building owners and occupiers in adopting sustainability. Moreover, the university is one of the significant institutions in the public sector and is a key player in encouraging new approaches. It plays a crucial role in initiating change [6, 7] and is a significant component in the implementation of government policies.

At present, the discussion on sustainable universities has been carried out in the areas of research, education and administration including campus operation [8], and numerous universities around the world are taking steps to improve their management practices to achieve sustainability [9]. For example, 350 universities representing 40 countries have signed a voluntary environmental agreement, the 'Talloires Declaration', for institutions of higher learning to show their willingness and readiness to address the sustainability challenges [9]. However, Hitchcock and Willard [1] argued that a single organisation such as a university on its own cannot be responsible for making all of society sustainable. For the successful of this agenda, every organisation should examine its inputs, outputs, processes and effects on the larger system in which it operates. This means that commitments and participation are needed from all parties, both private and public organisations to achieve sustainability goal.

3. Methodology

The facts about the adoption of sustainability in the university organisations are not well explored. Moreover, there are varying commitments to sustainable practices between organisations. Thus, to address this knowledge gap, there are needs to conduct a research on the extent of sustainability adoption in university and an investigation of their commitment toward sustainability.

This study performed the website analysis to gather information concerning sustainability initiatives in university organisations. Traditionally, information has been communicated to stakeholders via hard-copy medium such as published annual reports, brochures and newsletters. However, with technological changes and common acceptance of the internet as an 'organisational' tool for communication [10], the website has been created to post relevant information [11, 12]. These changes have led to the richness of information on websites and there is evidence of the rising use of websites by corporations as a medium to communicate organisational responsibilities. This is in line with findings by Nejati, Shafaei [13], based on their studies on university mission and vision statements in developed countries, where they found that the majority of universities were highlighting sustainability issues through their websites. In addition, according to Zuo, Zillante [14], in a study of 50 international contractors, they found that almost 80% of the companies revealed their sustainability policies through their websites. Moreover, websites are used to publicise information relating to organisational matters [15] and have been chosen as a data source by many researchers for measuring sustainability practices [16]; identifying corporate social responsibility among the top 10 universities [13]; exploring e-commerce in developing countries [17]; investigating market orientation in local government [18]; investigating hotel information quality [19]; investigating tourism [15, 20, 21]; and examining government sustainability policy [22]. Therefore, this study ascertains that web content analysis is a legitimate research tool.

To establish the themes, content analysis was used to analyse the content of the websites. Content analysis analyses and explores the correlation between texts and possible themes or concepts [23] and it is most appropriate and well accepted in social and environmental research [13]. This method has been used by many researchers such as Zuo, Zillante [14], Denskus and Esser [24], Kim and Weaver [25], Hine [26], Zhang [27], Braun and Clarke [28], Capriotti and Moreno [29], Lambert and O'Halloran [30], and Gill, Dickinson [31].

As for this study, the website analysis involved thirty-nine (39) Australian universities and each website was reviewed manually. This stage reviewed sustainability initiatives in broad university management and operations: concerning general policies, corporate mission statement, goals, research activities, positions available, and strategies. This technique was carried out by utilising publicly available data including any additional accessible relevant documents attached to the university websites. The systematic review of the websites considered the contents posted within the specific timeframe of 1 August to 15 September 2010. The review was conducted by typing specific words (Sustainability, Environment, Green practice, Facilities Management Office) for each university website to find target themes and relevant sustainability information.

The rationale for selecting and using these words was that studying the websites showed that these were the commonly used descriptors. These words were the most relevant for identifying themes and were also consistent with the research objectives. By thoroughly reading the possible sources from the websites, it was then possible to become familiar with the sustainability agenda setting within the universities. Figure 1 below outlines the flow chart for the content analysis of the selected websites.

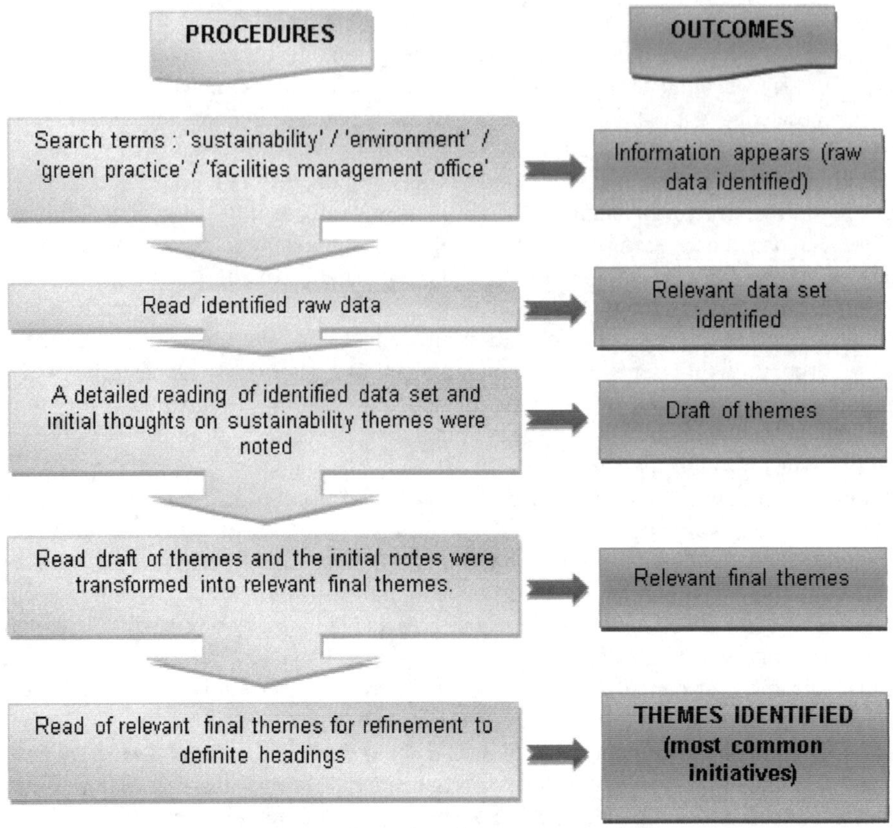

Figure 1. Flow chart for content analysis

The detailed procedure was as follows:

1) The information was read numerous times and relevant data sets were identified. These data sets were copied and pasted to the tables in a document with the specific name for each university. A detailed reading was carried out and initial thoughts on sustainability themes were noted in the specific column. These notes related to the sustainability concepts and practices.

2) The relevant data were read again and the initial notes were transformed into final themes that were significant and relevant to the research aim and objectives. Any unsuitable and unclear initial notes were excluded.

3) The data were read again for refinement of the themes into definite headings based on common practices and articulated within university operations. For the purpose of this research, only the most common themes relevant to the research objectives were identified and chosen. Based on this study, initial thoughts concerning sustainability indicators were established.

4. Result and analysis

This section reports on the results of the website analysis which collected data on sustainability initiatives undertaken by universities. A set of twenty (20) relevant themes were identified to represent initiatives undertaken by universities as their commitment towards sustainability. Then, the themes were compiled based on the universities' descriptions of their initiatives on their websites. Table 2 depicted the twenty (20) themes emerged and an examples of initiatives in thirty-nine (39) Australian universities websites.

Table 1. Sustainability themes emerged in websites of Australian universities

No	Themes Emerged	Examples of Sustainability Initiatives
1.	Statement of future vision	The Australian Catholic University [32] 'is committed to making a substantial contribution to environmental sustainability through the development of its physical infrastructure, recycling of materials and efficient use of all resources'. Macquarie University [33] stated their sustainability vision as: 'Macquarie University is ecologically sound, socially just and economically viable in all of its activities'.
2.	Policy and plan	Australian National University (ANU) [34] created their Environmental Management Plan to fulfil the ANU's environmental policy and other objectives. For the Australian Catholic University [32], their Environmental Management Plan 'ensures all campus operation processes and procedures for environmentally sustainable developments comply with university policy across all campuses'.
3.	Special position (officer/ manager)	Various positions such as Project Sustainability Engineer and Environmental Services Manager at the University of Wollongong, Director of Sustainability and Manager of Operational Sustainability at Macquarie University and Recycling and Sustainable Systems Manager and Pollution Prevention Officer at ANU.
4.	Committee/ team	The Environmental Management Planning Committee at ANU advises on the performance of environmental management whereas this function is undertaken by the Sustainability Working Group at Macquarie University and the Environmental Advisory Committee at the University of Wollongong.
5.	Department/office/ unit	The Green Office at Charles Sturt University [35] coordinates and monitors sustainability performance and the Sustainability Office (ANUgreen) at ANU was established under the ANU environment policy to implement the university's environmental management plan.
6.	Reporting	The Southern Cross University [36] reports on their sustainability initiatives in education and operational activities. RMIT [37] reports its energy and utilities' consumption to both state and federal government.
7.	Energy efficiency	For the University of South Australia [38], various energy efficient projects are featured such as intelligent energy management systems, power factor correction and on-demand lighting. These have contributed to a 700,000 kWh saving per year. In Deakin University [39], a number of energy efficiency programs have been implemented across campus together with energy and emissions audits.
8.	Water efficiency	The University of Tasmania [40] implemented the consumption of reuse water by the installation of rainwater tanks. A program of 'Water Conservation and Management' to reduce water consumption and minimize the pollution of water discharge was introduced in the University of Western Sydney [41].
9.	Waste management (reduction/reuse/ recycling)	La Trobe University [42] is establishing strategies for the recycling of organic waste and e-waste, toner cartridges and mobile phones. In Monash University [43], priority was given to a waste recycling program in the area of materials, products and packaging selection.
10.	Biodiversity/ ecosystem preservation	The University of Queensland [44] promotes the preservation of birds and possums. ANU's themes on sustainable biodiversity involved drafting a biodiversity management plan and partnership with other parties for a 'Life in the Suburbs' project.
11.	Pollution prevention/ emission control	Curtin University of Technology [45] conducted an emissions audit of greenhouse gases (GHG) as part of achieving a sustainable operation. The Comparative Environmental Risk Assessment Method (CERAM) was established at ANU to manage pollution risks to water, air and land.
12.	Green transportation/ travelling	Flinders University [46] established the Green Transport Office to enhance student and staff awareness about sustainable transport with the aim being to minimize the number of vehicles with solo occupants travelling to the university. Edith Cowan University [47] introduced 'Travelsmart@ECU' as a medium for communication to guide students and staff in sustainable transport application.
13.	Green building and infrastructure	Bond University [48] has developed the Mirvac School of Sustainable Development which was the first educational building in Australia awarded a 6 Star 'Green Star – Education PILOT Certified' rating for design. The Australian Catholic University

		constructed the 'National Centre for Health and Wellbeing' which was rated the highest Green Star performance for environmental sustainability.
14.	Green procurement/ purchasing	La Trobe University approached sustainable purchasing via supporting energy efficient transport and encouraging sustainable life cycle materials and services. Griffith University [49] stated that 'employees have a commitment to procuring and purchasing environmentally, socially and economically sustainable products.
15.	Green IT, office and laboratory	The Green Office Program was introduced at La Trobe University to promote sustainability practices and, in addition, 'EnviroSmart' was launched to improve awareness in recycling program practices. The University of Adelaide [50] has embarked on sustainable IT projects such as the replacement of the out-dated data centre, screen saver timing and automated lights.
16.	Education (academic programs) and research	The University of Southern Queensland [51] offers several sustainability programs such as the Bachelor of Science (Environmental and Sustainability). Murdoch University [52] has established the Centre for Research into Energy for Sustainable Transport (CREST).
17.	Community engagement	The University of New South Wales [53] has developed a website 'Teach Sustainability' to support local schoolteachers in promoting sustainability to the younger generation. The University of Technology, Sydney [54] organised a 'Sustainable Business Forum' in the form of dialogue and consultation focusing on corporate social responsibility.
18.	Knowledge enhancement (campaigns and training programs)	The University of Technology, Sydney [54] has created the 'UTS Alumni Sustainability Network' to enhance awareness and interest among staff and students. The University of New South Wales [53] has introduced a sustainability support group, 'Enviro Collective', involving students conducting activities and campaigning for a greener world.
19.	Alliances with other higher education association	Macquarie University [33] has a commitment with ACTS, the body concerned with sustainability in the Australian and New Zealand tertiary sectors; AASHE, an association responsible for creating a sustainable future for colleges and universities in the USA and Canada.
20.	Collaborations	The University of Ballarat [55] has collaborated with the National Centre for Sustainability (NCS) to support the development of sustainable practices. ANU [34] staff are collaborating with colleagues from other universities in Australia such as via the Intergovernmental Panel on Climate Change (IPCC).

5. Conclusion

Generally, the Australian universities appeared to be progressive in their commitment towards sustainability. Twenty (20) emerged sustainability themes were detected using web content analysis of the selected university's websites and varied across thirty-nine (39) universities. They demonstrates their commitment to sustainability by upholding numerous sustainable practices throughout the organisation. The commitments refer to the extent of information provided by universities in their website that demonstrated initiatives towards achieving the sustainability goal. Indeed, this study has shown that most universities in Australia were committed in adopting sustainability in their organisational agenda.

The data from this study provided understandings for the research aim of determining the level of sustainability commitments in organisations focusing on Australian universities. However, the findings have some limitations because the established themes only emerged from the websites' content without human validation and, therefore, the results could be influenced by other factors such as technological issues and the level of transparency of the information that could lead to an inadequate judgment. Therefore, further assessment may be advantageous to provide confirmation of the findings and complementary studies using other techniques are suggested such as interviews or observations to provide validation of data and reinforce the overall conclusions.

Acknowledgements
I would like to extend the appreciation to Associate Professor Dr Stephen Pullen and Dr Alpana Sivam from University of South Australia (UniSA), Adelaide, Australia for their advice and guidance for this research.

References
[1] Hitchcock, D.E. and M.L. Willard, *The business guide to sustainability: practical strategies and tools for organizations*, ed. M.L. Willard. 2006, London, UK: Earthscan Ltd.
[2] Petrini, M. and M. Pozzebon, *Integrating Sustainability into Business Practices: Learning from Brazilian Firms.* Brazilian Administration Review, 2010. **7**: p. 362-378.
[3] Amaratunga, D. and D. Baldry, *Assessment of facilities management performance in higher education properties.* Facilities, 2000. **18**(7/8): p. 293-301.
[4] Sabir, M., *The role of universities in sustainable development with special focus on Pakistan.* Universiti Malaysia Perlis (UNIMAP) Digital Library Journal, 2008: p. 60-68.
[5] Van Weenen, H., *Towards a vision of a sustainable university.* International Journal of Sustainability in Higher Education, 2000. **1**(1): p. 20-34.[[
[6] Miranda, L. and L. Marulanda, *Sustainable Construction in Developing Countries. A Peruvian Perspective*, in *Agenda 21 for Sustainable Construction in Developing Countries*. 2002.
[7] Osman, M.M., S.N.S.A. Rashid, and N. Ahmad, *Local Agenda 21 in Malaysia : Issues and problem faced by the stakeholder in the participation process*, in *Ecocity World Summit 2008*. 2008: San Francisco, Calfornia, USA.
[8] McMillin, J. and R. Dyball, *Developing a Whole-of-University Approach to Educating for Sustainability: Linking Curriculum, Research and Sustainable Operations.* Journal of Education for Sustainable Development, 2009. **3**(1): p. 55-64.
[9] Bajracharya, B. and L. Too, *Building a sustainable campus: A case study of Bond University*, in *State of Australian Cities national conference*. 2009: Perth, WA, Australia. p. 15.
[10] Stuart, H. and C. Jones, *Corporate Branding in Marketspace.* Corporate Reputation Review, 2004. **7**(1): p. 84-93.
[11] Adams, C.A. and G.R. Frost, *Integrating sustainability reporting into management practices.* Accounting Forum, 2008. **32**(4): p. 288-302.
[12] Adams, C.A. and G.R. Frost, *Accessibility and functionality of the corporate web site: implications for sustainability reporting.* Business Strategy and the Environment, 2006. **15**(4): p. 275-287.
[13] Nejati, M., et al., *Corporate social responsibility and universities: A study of top 10 world universities' websites.* African Journal of Business Management, 2011. **5**(2): p. 440-447.
[14] Zuo, J., et al., *Sustainability policy of construction contractors: A review.* Renewable and Sustainable Energy Reviews, 2012. **16**(6): p. 3910-3916.
[15] Baggio, R., *A websites analysis of European tourism organizations.* Anatolia, 2003. **14**(2): p. 93-106.
[16] Sharma, S. and I. Henriques, *Stakeholder influences on sustainability practices in the Canadian forest products industry.* Strategic Management Journal, 2005.
[17] Sambhanthan, A. and A. Good, *Implications for Improving Accessibility to E-Commerce Websites in Developing Countries-A Study of Hotel Websites.* arXiv preprint arXiv:1302.5491, 2013.
[18] Saez Vegas, L. and I. Periáñez Cañadillas, *Market Orientation in Local Government Through the Analysis of Municipal Website Content: A Framework for its Measurement.* Global Journal of Business Research, 2013. **7**(2): p. 47-58.
[19] Chung, T. and R. Law, *Developing a performance indicator for hotel websites.* International Journal of Hospitality Management, 2003. **22**(1): p. 119-125.
[20] Govers, R. and F.M. Go, *Projected destination image online: Website content analysis of pictures and text.* Information Technology & Tourism, 2005. **7**(2): p. 73-89.

[21] Stepchenkova, S. and A.M. Morrison, *The destination image of Russia: From the online induced perspective.* Tourism Management, 2006. **27**(5): p. 943-956.
[22] Carter, A.J. and J. Pisaniello, *Moving from rhetoric to effective implementation for Australian governments' sustainability policies.* Journal of the Asia Pacific Centre for Environmental Accountability, 2012. **18**(1): p. 25-55.
[23] Jose, A. and S. Lee, *Environmental reporting of global corporations: A content analysis based on website disclosures.* Journal of Business Ethics, 2007. **72**(4): p. 307-321.
[24] Denskus, T. and D.E. Esser, *Social Media and Global Development Rituals: a content analysis of blogs and tweets on the 2010 mdg Summit.* Third World Quarterly, 2013. **34**(3): p. 405-422.
[25] Kim, S.T. and D. Weaver, *Communication Research about the Internet: a Thematic Meta-analysis.*
New Media & Society, 2002. **4**(4): p. 518-538.
[26] Hine, C., *Social Research Methods and the Internet: A Thematic Review.* Sociol. Res. Online, 2004. **9**(2).
[27] Zhang, J., *Content Analysis of Web Sites from 2000 to 2004: A Thematic Meta-Analysis*, in *Science and Technology Journalism*. 2005, Texas A&M University.
[28] Braun, V. and V. Clarke, *Using thematic analysis in psychology.* Qualitative Research in Psychology, 2006. **3**(2): p. 77-101.
[29] Capriotti, P. and A. Moreno, *Corporate citizenship and public relations: The importance and interactivity of social responsibility issues on corporate websites.* Public Relations Review, 2007. **33**(1): p. 84-91.
[30] Lambert, S. and E. O'Halloran, *Deductive thematic analysis of a female paedophilia website.* Psychiatry, Psychology & Law, 2008. **15**(2): p. 284-300.
[31] Gill, D.L., S.J. Dickinson, and A. Scharl, *Communicating sustainability: A web content analysis of North American, Asian and European firms.* Journal of Communication Management, 2008. **12**: p. 243-262.

IOP Conf. Series: Earth and Environmental Science **117** (2018) 012047 doi:10.1088/1755-1315/117/1/012047

Automatic classification techniques for type of sediment map from multibeam sonar data

R Zakariya[1], M A Abdullah[1], R Che Hasan[2] and I Khalil[1]
[1]School of Marine and Environmental Sciences, University Malaysia Terengganu, Terengganu, Malaysia.
[2]UTM Razak School of Engineering and Advanced Technology, University Teknologi Malaysia (UTM), Jalan Semarak 54100 Kuala Lumpur, Malaysia.

*Azhafiz88@gmail.com

Abstract. Sediment map can be important information for various applications such as oil drilling, environmental and pollution study. A study on sediment mapping was conducted at a natural reef (rock) in Pulau Payar using Sound Navigation and Ranging (SONAR) technology which is Multibeam Echosounder R2-Sonic. This study aims to determine sediment type by obtaining backscatter and bathymetry data from multibeam echosounder. Ground truth data were used to verify the classification produced. The method used to analyze ground truth samples consists of particle size analysis (PSA) and dry sieving methods. Different analysis being carried out due to different sizes of sediment sample obtained. The smaller size was analyzed using PSA with the brand CILAS while bigger size sediment was analyzed using sieve. For multibeam, data acquisition includes backscatter strength and bathymetry data were processed using QINSy, Qimera, and ArcGIS. This study shows the capability of multibeam data to differentiate the four types of sediments which are i) very coarse sand, ii) coarse sand, iii) very coarse silt and coarse silt. The accuracy was reported as 92.31% overall accuracy and 0.88 kappa coefficient.

1. Introduction

Fauna communities harbored different space that suits their lifestyle in order to survive. In the aquatic or marine world, benthic communities are the one that has to meet this condition in order to survive. Benthic communities are organisms that spend their lives on, in or near the ocean floor or sediment [1]. Different benthic organisms have different preferences on the habitat that they lived on which specifically based on the sediment type which is called zonation [1]. Example of this zonation can be seen at coastal area from the intertidal zone to sub tidal zone where sediment types are different in each zone.

Since sediments have different size and texture [2], it would provide all sorts of information and data for scientists. Besides, physical data, the distribution of sediment on certain area could also provide biological data on species that live there. This means that area of different sediment distribution provides different habitat based on the organisms adaptation or specificity.

Multibeam echosounder is a device with the ability to map the seafloor in large scale by emitting active sonar. Multibeam echosounder can sounding with broad scope, high accuracy and can be used to determine characteristics of seabed type [4]. One of the methods is to use backscatter or intensity returns from underwater acoustic sonar such as multibeam echosounder to remotely gather seafloor sediment types [5]. Backscatter can be used to identify hard and soft seafloor surfaces which can later be used to predict sediment classes. An analysis of the amplitude of the return wave sound allows for information on the structure and hardness of the base waters, which is used to classify sediment Strong types that return to the hard surface (rock, gravel) and weak signals returning to a finer surface (silt, clay). This is because the greater the impedance of a medium the larger the reflection coefficient.

There are many benefits that could be obtained by using multibeam echo-sounder in marine science applications. First, it helps to decrease the expenses and required less time for fieldwork. The data obtained which specifically on sediment mapping can be used for future studies. Stephens and Diesing (2014) stated that a sediment map of high resolution is in demand for better environmental assessment and management. On top of that, it can also be used by a third party like Department of Fisheries so that they can conduct optimum yet effective research on locality of fish and invertebrate in the area. Department of Environment also will benefit from sediment mapping as the information can be used to scrutinize the sediment distribution, somehow related to pollution study. The aim of this study to determine types of sediment map using Multibeam sonar data

2. Materials and methods

2.1 Study area

Pulau Payar is situated far apart from the coast of Kedah state, between Pulau Langkawi and Pulau Pinang. The study area covers part of the north of Pulau Payar. The coverage of this survey area is approximately 239385 m². The area is selected as the study area due to the availability of bathymetry data and backscatter data to classify the sediment map characteristics. The height range of this area is 17 meter up to 25.76 meter. Figure 1 shows the location of the study area.

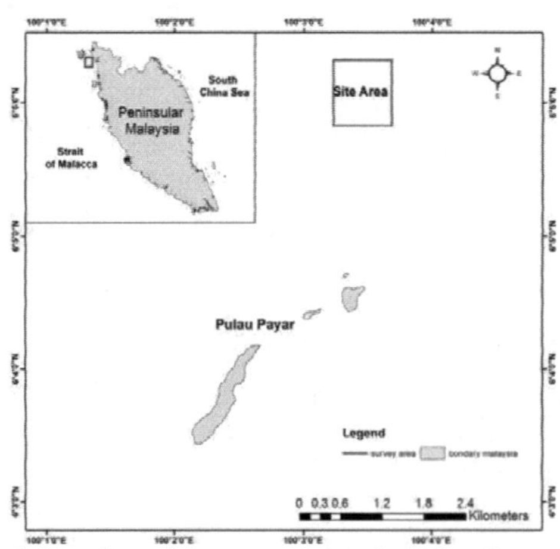

Figure 1. location of the survey area in Pulau Payar, Kedah

2.2 Acoustic data

Acoustic data were acquired in Pulau Payar on 13th to 17th January in 2017. Multibeam echo sounder system (MBES) R2Sonic 2020 was used in this study for collecting bathymetry image and backscatter image. The average speed for survey vessel is 5 knots. This instrument produces 256 beams equiangular arrayed over an arc of 130° and used frequency at 400 kHz. Differential Global Position System (DGPS)

Trimble 461 was used for system navigation and heading (gyro compass). The vessel's motion was quantified using DMS10 motion sensor with 0.05° accuracy for heave, roll and pitch. Real time speed of sound for MBES transducer used Valeport Mini sound velocity sensor (SVS). Software QINSy v8.0 was used during a survey for data logging, real time quality control, display and navigation. The sound velocity profile (SVP) collects from cast away-CTD used for measuring sound velocity during data acquisition to correct the effect of refraction.

Software Qimera version 1.4 was used for data processing on bathymetry raw data. The bathymetry raw data processing follows step correction for sound velocity variation though the water column, correction for tide in water depths, correction of motion sensor information for data bathymetry and remove for spike or noise [6]. FMGeocoder Toolbox or FMGT version 7.4 was used to process the raw MBES backscatter and obtain backscatter mosaic.

2.3 Ground truth data

At the sampling site (figure 2a), ponar grab was used to collect the sediment sample because it was suitable and handy. The grab was tied to a rope prior to sampling. At each sampling point, the grab was deployed until it hits the bottom, collect the sediment and finally was retrieved. The sediment was scoped into a plastic bag, labelled and sealed. The sediment samples were brought back to the laboratory for further analysis.

The sediment sample analysis was carried out using dry and wet sieving techniques to determine the grain size of the sediment. Using dry sieving techniques, it is suitable for coarse fraction and particles with diameter greater than 63 μm. Approximately; a 100g of dried sample was weighed. The sample was then placed on top of the sieve which was stacked with other sieves below it with different mesh size. A series of sieves was shaken using sieve shaker for 15 and 20 minutes. Post to the shaking process, each sieve containing the trapped samples was weighted. For wet sieving the laser diffraction method (using a Particle Size Analyzer (PSA) CILAS 1180) was employed to obtain the statistical value.

The type of the sediment classes was based on sediment grain size [7-9]. In sedimentological calculation, mm unit is commonly used as a metric unit, using to several geological terms i.e. mean size. Mean size (X_Φ) is an indicator of sediment size were measured based on its weight. The decreasing of a mean size value indicates an increase in sediment grain size and vice versa. The value of the mean was obtained either from the calculation of the gravity centre under the grain size curve or the gravity centre of the cumulative distribution curve. The classification systems of sediment grain size typically as Table 1.

Table 1. Grain size class

Grain diameter (mm)	Descriptive terminology
4 – 2	Granule
2 – 1	Very coarse sand
1 – 0.5	Coarse sand
0.5– 0.25	Medium sand
0.25– 0.125	Fine sand
0.125– 0.0625	Very fine sand
0.0625 – 0.031	Very coarse silt
0.031– 0.0156	Coarse silt
0.0156 – 0.0078	Medium silt
0.0078 – 0.0039	Fine silt
0.0039 – 0.0020	Very fine silt
> 0.0020	Clay

2.4 Classification

According to [2], both techniques of supervised and unsupervised have been extensively proposed in benthic studies. These include the easy-to-implement maximum likelihood and ISO cluster classifiers found in most GIS packages [10 – 11]. In this research, supervised technique has been chosen to classify the data. The selected data have been classified by using maximum likelihood supervised.

Supervised classification requires class-describing information that must be as accurate, representative and complete as possible. Such information on the class characteristics might be acquired by training classifiers (e.g. maximum likelihood, k-nearest neighbour, classification tree, random forest, support vector machine and Bayesian network) with ground samples. Maximum likelihood classification method [11 – 12] was used in the present study. This method was chosen as it assigns class to the pixels. Data of sediment type from the dry and wet sieving techniques will be used as guideline to classify certain pixels in order to know the distribution of the sediment based on the pixels characteristic which correlated with different type of sediment.

The backscatter variable is the use of automatic classification were backscatter, mean, mode, median. The statistic data derived from the analysis of variance in backscatter in FMGT version 7.4 software. All information collected will be compared with each other based on the backscatter data to know how reliable the data, based on the error matrix. Plus, kappa coefficient will also be used to determine the accuracy to evaluate the reliability of the data. Kappa coefficient compares the expected accuracy with the observed accuracy of the ground truth data.

A kappa value of 1 implies complete agreement, while a value less than 0 implies more disagreement than a random assignment of classes to one of the maps [13]. Kappa considers all the cells of an error matrix and thus incorporate more information [14 – 17]. Besides that, Kappa is suitable for comparison between different error matrices because it removes chance agreement [18 – 19]. Accuracy of more than 80% marked the reliability of the classified data and concludes it as accurate.

3. Result and Discussion

3.1 Acoustic analysis

The results of the bathymetry map processing show in figure 2 (a) the depth value of natural reef (rock) ranged from 17.626 to 25.760 meters with composition in depth less than 20 meters indicated on natural reef (rock). The average depth in the dominated depth ranges from 20 to 25 meters.

Backscatter map of natural reef (rock) mapped using MBES is shown in Figure 2(b). The backscatter strength, in decibels (dB) ranges from -38 to -80 dB presented as a greyscale image. High backscatter strength can be seen on white part of the map while weak or low backscatter strength focused on darker region on the map. High decibel indicates coarse particles as rough surface reflects sound wave acts on it. Low decibel on the other hand indicate soft surface as some of the sound waves are being absorbed.

3.2 Type of sediment analysis

The analysis of particle size of the total number of samples for each sedimentary class was: very coarse sand is 5 samples, coarse sand is 3 samples, very coarse silt is 4 samples and coarse silt is 1. The mean value in phi indicates the size of particle diameter where higher values exhibit smaller diameter or fine particles and vice versa. Overall all phi values can be classified into 4 types of sediment, i.e. very coarse sand, coarse sand, very coarse silt and coarse silt. The high sediment grain diameter size is 1.57 mm that was very coarse sand for station 13 and the low sediment grain diameter size is 0.02 that was coarse silt. Therefore, it can be stated that all situations it is a particle that is different from the size of the smooth and rough. Table 2 shows the results of classification type of sediment sample wit multibeam data.

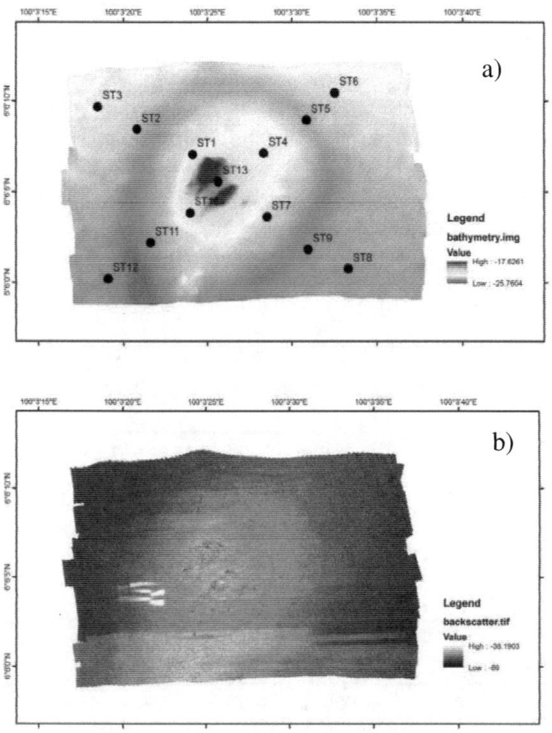

Figure 2. Acoustic data of the study area: (a) multibeam bathymetry with the positions of the sediment stations (b) relative backscatter intensity image at natural reef (rock) in Pulau Payar.

Table 2. Result of classification type of sediment sample with multibeam data.

Station	**Depth (m)**	**Intensity (dB)**	**Grain diameter (mm)**	**Classification**
1	23.738	-59.450	1.25	Very coarse sand
2	24.081	-63.351	0.63	Coarse sand
3	23.629	-64.771	0.02	Coarse silt
4	23.330	-62.291	0.91	Coarse sand
5	24.308	-61.231	1.42	Very coarse sand
6	23.815	-66.541	0.04	Very coarse silt
7	23.876	-63.710	1.01	Very coarse sand
8	24.568	-65.831	0.04	Very coarse silt
9	24.738	-65.121	0.06	Very coarse silt
10	22.689	-59.100	1.00	Very coarse sand
11	24.703	-64.411	0.88	Coarse sand
12	24.772	-65.121	0.04	Very coarse silt
13	20.761	-58.040	1.57	Very coarse sand

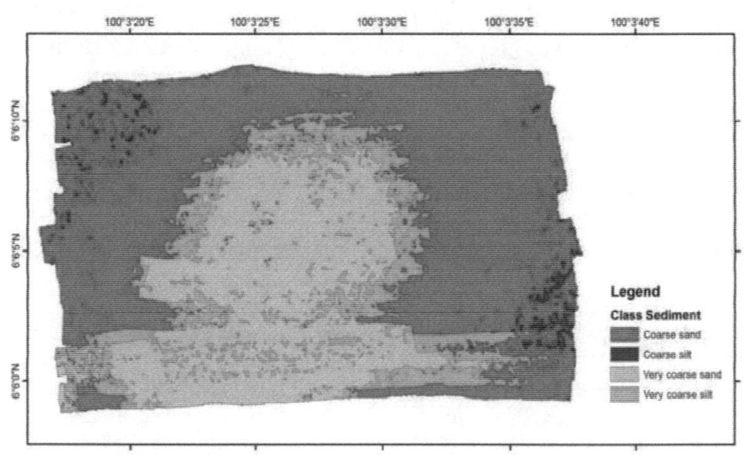

Figure 3. Classified sediment map based on integration between backscatter image, statistic backscatter image and ground truth data using maximum likelihood classification.

3.3 Sediment map accuracy

Figure 3 displays sediment map made using integration of data from both ground truth (sediment type) and multibeam (decibel of backscatter) data. Comparing with map in Figure 2 (b) the accepted map which represents as the final map for the present study is the second map (Figure 3) as it is more accurate. Five sampling stations were chosen for the accuracy assessment consists of ST 3, 6, 9, 11 and 13.

The error matrix of user accuracy for the classification method used is shown in Table 3. Commission error is possibly made when classifying the map based on the selected pixel which did not match the ground truth data. From the table below, very coarse sand was misinterpreted as coarse sand and very coarse silt. Table 4 shows the accuracy assessment for the classification method used. The overall accuracy is 92.31%, while the Kappa coefficient is 0.88. The greater confidence in the classification with the value closer to 1 for k range between 0 and 1.

Table 3. Error matrix of user accuracy for maximum likelihood classification method

Class	Very coarse sand	Coarse sand	Very coarse silt	Coarse silt	Total
Very coarse sand	5	0	0	0	5
Coarse sand	0	3	0	0	3
Very coarse silt	0	0	4	0	4
Coarse silt	0	1	0	0	1
Total	5	4	4	0	13
User accuracy (%)	100	100	100	0	

Table 4. Accuracy assessment of maximum Likelihood classification method

	Overall Accuracy	Kappa Coefficient
Maximum Likelihood	92.31	0.88

Overall, from a total of 23.9383 km², the area is mostly covered by coarse sand is 13.457 km²; secondly, the area is mostly covered by very coarse sand is 6.7478. Meanwhile, very coarse silt and coarse silt comprise only about 3.7142 km² and 0.0193 km² respectively. Besides understanding the hydrodynamic effect of the sediment distribution, mud (silt loam) area also serves to understand the health status of the sea. The fine size of the silt loam makes it vulnerable for pollutants to be absorbed. Through the knowledge of sediment transport, researcher can assess the rate of dispersion of the pollutants. Besides, different substrate composition might harbor different benthic habitat. Table 5 shows the estimated total coverage for sediment.

Table 5. Estimated total coverage for sediment area (km²)

Sediment Type	Area (km²)
Very coarse sand	6.7478
Coarse sand	13.457
Very coarse silt	3.7142
Coarse silt	0.0193
Total	23.9383

4. Conclusion

Analysis of size sediment of the sediment sample reveals that natural reef (rock) contains at least 4 types of sediment. The sediment was classified based on its texture from triangle plot, thus classifying the sediment as very coarse sand, coarse sand, very coarse silt and coarse silt. Furthermore, multibeam echosounder analysis helps to establish a backscatter image with values measured in dB that reflects its strength. Integration of both data creates the sediment map using maximum likelihood classification. In a nutshell, the map helps to understand the distribution of sediment along with the understanding of the physical parameter where hydrodynamic influence like tides, water velocity and depth stratified the sediments. The coarse sand is also has the highest coverage area of at natural reef (rock), followed by very coarse silt, very coarse sand and coarse silt. The overall accuracy and kappa coefficient obtained which are 92.31% and 0.88 respectively shows promising results. Future works on another site are suggested, especially in different sediment bottom type.

Acknowledgements

The authors gratefully to the Ministry of Education (MyBrain 15), University Malaysia Terengganu (UMT) School of Marine and Environmental Sciences and Universiti Teknologi Malaysia (UTM) Razak School of Engineering and Advanced Technology, for research facilities and financial assistance provided for this project.

References

[1] Castro, P. and Huber, M.E. (2010). Marine Biology (8th Edition) . New York: Mc Graw Hill.
[2] Prothero, D.R. & Schwab, F. (1996). Sedimentary Geology: An introduction to sedimentary rocks and stratigraphy. W.H. Freeman and Company, New York, pp. 1-25.
[3]Brown, C.J., Todd, B., Kostylev, J., Pickrill, R.A., V.E., 2011. Image-based classification of multibeam sonar backscatter data for objective surficial sediment mapping of Georges Bank, Canada. Continental Shelf Research 31,S1 10- S1 19.
[4] Brown, C.J., Smith, S.J., Lawton, P. & Anderson, J.T. 2011. Benthic habitat mapping: A review of progress towards improved understanding of the spatial ecology of the seafloor using acoustic techniques. Est. Coast. Shelf Sci. 92(3):502-520. doi:10.1016/j.ecss.2011.02. 007.
[5] Lecours, V., Dolan, M. F. J., Micallef, A., & Lucieer, V. L. (2016). Characterising the ocean frontier: A review of marine and coastal geomorphometry. Hydrol. Earth Syst. Sci. Discuss., 2016(March), 1–66. http://doi.org/10.5194/hess-2016-73
[6]Huang, Z., Siwabessy, J., Nichol, S.L., Brooke, B.P., 2014. Predictive mapping of seabed substrata using high-resolution multibeam sonar data: a case study from a shelf with complex geomorphology. Mar. Geol. 357, 37e52.
[7]Blott, S. J., Pye, K. (2001) - GRADISTAT: a grain size distribution and statistics package for the analysis of unconsolidated sediments. Earth Surface Processes and Landforms, 26, 1237-1248.
[8]Udden, J. A. (1914) - Mechanical composition of clastic sediments. Bulletin Geological Society of America, 25, 655-744.
[9]Wentworth, C. K. (1922) - A scale of grade and class terms for clastic sediments. Journal of Geology, 30, 377-392
[10]Brown CJ, Collier JS (2008) Mapping benthic habitat in regions of gradational substrata: An automated approach utilising geophysical, geological, and biological relationships. Estuarine, Coastal and Shelf Science 78: 203–214.
[11]Ierodiaconou, D., Monk, J., Rattray, A., Laurenson, L., Versace, V.L., 2011. Comparison of automated classification techniques for predicting benthic biological communities using hydroacoustics and video observations. Cont. Shelf Res. 31, S28eS38

[12]Che Hasan, R.C., Ierodiaconou, D., Rattray, A., Monk, J. & Laurenson, L. 2011, Applications of multibeam echosounder data and video observations for biological monitoring on the south east Australian continental shelf. ISG & ISPRS. Shah Alam, Malaysia

[13]Rositter, D.G., 2014. Statistical methods for accuracy assessment of classified thematic maps. Department of Earth System Analysis[14]Rosenfield, G. H., and K. Fitzpatricklins. 1986. A coefficient of agreement as a measure of thematic classification accuracy. Photogrammetric Engineering and Remote Sensing 52 (2):223-227.

[15]Fung, T., & E. Ledrew. 1988. The determination of optimal threshold levels for change detection using various accuracy indices. Photogrammetric Engineering and Remote Sensing 54 (10):1449-1454.

[16]Dicks, S. E., and T. H. C. Lo. 1990. Evaluation of thematic map accuracy in a land-use and land-cover mapping program. Photogrammetric Engineering and Remote Sensing 56 (9):1247-1252.

[17]Jansen, L. L. F., & F. J. M. van der Wel. 1994. Accuracy assessment of satellite derived land-cover data: a review. Photogrammetric Engineering and Remote Sensing 60 (4):419-426.

[18]Congalton, R.G., Oderwald, R.G & Mead, R.A., 1983. Assessing Landsat classification accuracy using discrete multivariate analysis statistical techniques. Photogrammetric Engineering and Remote Sensing 49 (12):1671-1678.

[19]Congalton, R.G., R.G. Oderwald, & Mead, R.A., 1991. A review of assessing the accuracy of classifications of remotely sensed data. Remote Sensing of. Environment 37 (1):35-46.

Application of Delphi expert panel in joint venture projects

H Adnan[1], M R Rosman[2], Z Z Ahmad Rashid[3], N Mohamad Yusuwan[1] and N A Bakhary[4]

[1]Faculty Architecture, Planning and Surveying, Universiti Teknologi MARA, Shah Alam, Selangor
[2]Faculty Architecture, Planning and Surveying, Universiti Teknologi MARA, Seri Iskandar, Perak
[3]School of Housing Building and Planning, Universiti Sains Malaysia, Gelugor, Penang
[4]Faculty of Civil Engineering, Universiti Teknologi MARA, Shah Alam, Selangor

*mimad856@gmail.com

Abstract. This study was conducted with the aim to identify the application of the Delphi Technique in validating findings obtained from questionnaire surveys and interviews done in-depth on the subject of joint venture projects in Malaysia. The Delphi technique aims to achieve a consensus of opinion amongst expert panellist that were selected on the primary factors in JV projects. To achieve research objectives, a progressive series of questions was designed where a selected panel of expert to confirm and validate the final findings. The rationale, benefits, limitations and recommendations for the use of Delphi were given in this study. From the literature review done, twenty-one factors were identified as critical factors to the making any joint venture project successful. Detail information from contractors were obtained by using the questionnaire survey method and forty-three in-depth interviews were carried out. Trust between partners, mutual understanding, partner selection criteria, agreement of contract, objective compatibility, conflict, and commitment were confirmed by the Delphi panel to be the critical success factors besides another fourteen factors which were found to be the Failure Reduction Criteria. Delphi techniques has proven to successfully assist in recognising the main factors and would be beneficial in supplementing the success of joint venture arrangements application for construction projects in Malaysia.

1. Introduction

Construction JVs in Malaysia have become more well-known in multinational construction and local government to achieve their individual objectives. The contractors need to know the factors, which were considered critical if they want their JVs to be successful.

Since the 1990's, there have been contractors from foreign countries involving in construction projects in Malaysia and the number has been increasing. Bumiputera Contractors have started to accept the culture of joint-venture for the reason of political stability, economic growth, and relatively low cost of labour and others resources. Foreign firms see this is a way into the Malaysian market through joint venture for projects with Malaysian and Bumiputera Contractors. Participation in the regional and global market of the local contractors were also encouraged by the Malaysian government based on the contractor's knowledge and skills on building, infrastructure, highways, power generations, ports, and airport coasts construction or development projects.

The liberalization of policies along with foreign participation in several sectors was announced in March 1988 by the Malaysian government. The foreign participation of Malaysian sectors include privatized infrastructure projects. Apart from the completion of Petronas Twin Towers, Second Link to Singapore, and Kuala Lumpur International Airport in Sepang, there are many other projects associated with infrastructure of Malaysia that are needed to be implemented. The companies running in Malaysia have ventured with countries like China, Bosnia Herzegovina, India, and some parts of South America and Africa. These projects are likely to provide excellent opportunities to the companies in Malaysia for teaming up with the foreign contractors to explore efficient business opportunities.

2. Problem statements

The Malaysian market is reducing in size due to the growing number of construction companies. Efforts to make necessary changes in these companies by means of creating new strategies, consolidating the departments, and be prepared for exit or replacement have to be thought through by these companies. Normally, the method of taking part in large infrastructure projects in Malaysia by the foreign companies is through JV with local partners. However, a minimum of 30% indigenous Malay or more commonly known as Bumiputera, is required in the said JV bids.

The welcoming of foreign participation by the Malaysian government is evident through the amount payments made for numerous contracts and professional services to them. Identification of any difficulties or inconveniences is necessary as they may be a reason for instability and prone to trouble during the execution of the project with the respective foreign venture. Establishing trust may also be a problem together with the uncertainty of partnering success factor and the existence of competitiveness strategy in international construction.

Apart from that. language, ethnics' culture and local standards were also proven to be hindrances in these joint venture projects. Cultural shock which may have impact on the organizational, working and coordination may occur when these cultures cross. Technological, financial, legal, equity control, profit margin, size compatibility, socio-cultural and political are also additional joint venture projects' other problems to be looked into. It is necessary for a planned approached by both sides of the JV agreement throughout all these aspects in order to ensure successful implementation of the project. Careful analysis of political, cultural, economic, and social environment is required for assembling the feasibility and desirability of a joint venture that needs to be managed and implemented.

3. Literature on Delphi background

The name Delphi was taken from the Greek Oracle's skills of interpretation and foresight. The Delphi was developed in the 1950's by the Rand Corporation for the US Air Force concerning the use of expert opinion [1]. The procedure was designed to obtain the most reliable consensus of opinion of a group of experts by a series of intensive questionnaires interspersed with controlled opinion feedback, with the results of each round being fed into the next round [2]. It involves the selection of procedures for suitable experts, development of appropriate questions to be put to them and analysis of the answers given by them [3]. The intended outcome is that by the final round the experts will have reached a consensus of opinion on the issues put before them.

An example of iterative forecasting procedure, in this case the Delphi method, is characterized by three distinct features [4]: anonymity; iteration with controlled feedback; and statistical response. Panel members remain unknown to one another respond to a series of questionnaires. The iterative nature of the procedure generates them to modify their assessments and projects them beyond their own subjective opinions. It can represent the best forecast available from the consensus of experts [5]. The process is continued until a consensus is reached on the various issues under consideration, or until it becomes evident that no further consensus can be developed. Generally, Delphi runs to two to seven rounds of questioning, at more. The major difficulties of Delphi, however, lie in maintaining the high level of response and in reaching and implementing a consensus [1].

The aim of this paper is to report how the Delphi Technique was used to validate and confirmed a list of factors that was considered critical from the literature review, questionnaire survey and interviews on the critical success factors on construction JV projects in Malaysia. It is also aim to derive a consensus on the most critical success factors from expert panellists for the final findings. The effectiveness of the Delphi method will be evaluated and the difficulties in conducting Delphi survey will be discussed.

4. Methodology and discussion

Following a thorough literature research, 21 factors critical to the success of construction JV projects was identified. These factors were then assembled into questionnaire survey that was distributed to 1630 local and 70 foreign contractors in Malaysia. A response rate of twenty (20%) 341 was obtained. The purpose of the questionnaire survey was to discover which companies had experience of JV projects, the extent of that experience and their views of key factors associated with joint ventures.

The 2nd part of the research involved in-depth interviews conducted in Malaysia which involved 43 participants from various foreign and local companies. It was carried out with the Chief Executive Officers, General Managers to provide detailed information in their company structure, management and experience of any JV projects undertaken in Malaysia. Results from interviews confirmed the postal survey results of having at least 90 per cent the same 12 ranking CSF.

To confirm that the CSFs discovered by this process are reliable, a Delphi Technique which involved expert panels in Malaysia was conducted to validate/confirm the final findings. A triangulation approach had been adopted for this study which enables to integrate the quantitative results (questionnaire surveys) with the results from the qualitative method (interviews). This will prove useful; as the research process will be more robust enable a much higher quality reliable data to be gathered and results to be achieved.

5. Implementation of Delphi study

In this study, 22 experts were selected based from the recommendation from the contractors involved in the previous in-depth interviews conducted in Malaysia. Experts in Malaysia were contacted via email with the intention to get their approval to participate in the Delphi group. The 19 expert panels were Chief Executive Officers, Executive Directors, and Senior General Managers with vast experience and numbers of past and present JV local and overseas works and 3 academicians from the Malaysian universities whom involved in JV projects.

5.1 Delphi Round 1

The questionnaire consisted of 21 factors critical to success of construction JV projects were sent out followed by email phone calls to encourage participation. There was 86% response with 19 experts (out of 22) returned the questions. The experts were asked to rate each statement on a 4 point Likert-type rating scale in terms of their criticality. They were encouraged to enhance the result by making additional comments on the topic after the questionnaire finishes. [6] suggested that "an expert may be defined as someone with special skills or knowledge evidence by leadership in professional organizations, holding office in professional organization, presenter at national conventions, published in recognized journals.

5.2 Delphi Round 2

On the second round of this study, the expert panel provided 21 statements of contents with some amendments received from the experts' comments during the first round. They were given the numbers of response of each factor based on the scale of criticality again. To achieve consensus in the statements added by the panel during the first round, the experts were directed to review their rating again in terms of their criticality. The second questionnaire received a 77% responses with 17 experts returned the questionnaire. A majority of the experts reviewed their score and made necessary alterations.

5.3 Results and Analysis

The responses that were obtained were compiled and analysed using the procedure and formula mentioned as above. The questions were analysed by providing numerical scores on each of the scale of criticality with 100 points on Extremely Critical (4), 50 points on Critical (3), 10 points on (Less Critical) and 0 points on (Not Critical). These scores were then translated into importance indices to establish their relative ranking of the attributes. A Relative Importance Index (RIX) was then developed to convert the E-Scores into a decimal figure using the following formula taken from [7]:

$$\frac{\sum W}{A \times N} \qquad (1)$$

Where

W = weighting given to each factor in the evidence scale,
A = the highest weight applied,
N = total number in sample.

Consensus was achieved on 7 CSFs, which include inter-partner trust, criteria for partner selection, mutual understanding, agreement of contract, compatibility of objective, and management control. [8] points out that consensus can simply mean a 'group opinion, general agreement or group solidarity in sentiment and belief. The panel selected these CSFs as being high importance to the implementation of construction JV work in Malaysia.

The factors with scores of 0.80 or more on the RIX scale were considered extremely critical success factors and it shows that where the factors were included, there may be an increase of success chance but if left out, it will increase the chance of failure [9]. The factors that score less than 0.80 on the RIX scales were not considered as critical. Fourteen factors were designated as Failure Reduction Criteria (FRC) which if included in the project, might reduce the chance of failure but would not increase the chance of success [9].

In addition, to the RIX scale, the percentage of respondents scoring 4 (extremely critical) with 80% and more was used to rank the most critical success factors. If the RIX were equal, the distribution of the percentage of each scale will be taken into account. These were confirmed too by the scores of 50% to 74% from results in Round One.

There were increased on convergence of opinion among the responses agreed in the factors during the second round. Although a three round sequences seem to be the most popular and recommended method, there were two cases where two rounds were found to be satisfactory [10]. It was judged from the response of the experts to be the second round that the marginal gain from the third round may be small compared to the effort required to perform it, because of the declining participation and it was agreeable to stop the Delphi study at the second round.

This confirmed with [11] that answers are most accurate on Round Two and became less accurate on subsequent rounds. The Delphi final results confirmed the questionnaire surveys and in-depth interviews, which produced the similar results where 85% of the respondents agreed on the seven CSFs and the fourteen FRCs.

Table 1. Perceptions of the Finalized Project Success Factors from Round 1 and Round 2

Rank	Round 1	RIX	Round 2	RIX
1	Criteria for Partner Selection	0.87	Inter- Partner Trust	1.00
2	Mutual Understanding	0.77	Mutual Understanding	0.96
3	Inter- Partner Trust	0.74	Criteria for Partner Selection	0.94
4	Agreement of Contract	0.75	Agreement of Contract	0.91
5	Compatibility of Objectives		Compatibility of Objectives	0.88
6	Conflict		Commitment	0.85
7	Commitment		Conflict	0.82

8	Partner Experience	Communications	0.61
9	Cultural Understanding	Cultural Understanding	0.55
10	Coordination	Profit	0.54
11	Profit	Partner Experience	0.54
12	Communications	Financial Stability	0.53
13	Management Control	Cooperation	0.53
14	Cooperation	Equity Control	0.53
15	Equity Control	Management Control	0.48
16	Organizational Structure	Coordination	0.46
17	Financial Stability	Effective HRM	0.43
18	Effective HRM	Motivation	0.30
19	Motivation	Organizational Structure	0.24
20	Knowledge Transfer	Knowledge Transfer	0.17
21	Size Compatibility of Partner Firms	Size Compatibility of Partner Firms	0.12

6. Difficulties in conducting the Delphi techniques

It is important that panel members treat the work seriously, and devote the time necessary to provide thoughtful and reasoned responses to the questions. Secondly, the wording of the questions and the presentation format of the survey were extremely important [1]. It is very important to keep the whole panel responding to each round of Delphi and finally, the successful rounds of Delphi technique were time consuming. This study was taken with relative success in that a response rate of 77% was achieved and considered relatively high and acceptable for the purposes of this research.

7. Conclusion

The Delphi method was preferred to validate the final findings as it involved panellists whom are knowledgeable in their area of expertise to provide the reliable data on JV projects in Malaysia. This is because they were experts whom could confirm the results of this study. The Delphi Expert Panel resulted in the identification of seven critical success factors as being extremely high importance to the construction JV in Malaysia which require a significant level of application in order to facilitate additional success. Fourteen factors were identified as FRCs and without these elements a project will risk the significant chance of failure. The results of this study are expected to provide useful guidelines for forming and operating effective and efficient JV both in Malaysia and in other similar economies.

8. References

[1] Robinson, J. B. L 1991 *"Delphi Technology for economic impact assessment"*, Journal of Transportation Engineering, Vol 117, No. 3.
[2] Linstone, H. and Turoff, M. 1975 The Delphi Method: Techniques and Application, Addison, Wesley, Reading, pp 3-12.
[3] Outhred, G. P. 2001 *"The Delphi Method: A Demonstration of its Use for specific Research Types"*, Paper read at the RICS Foundation, Construction & Building Research Conference, Melbourne, Australia, Vol1, September.
[4] Dickey, J. and Watts, T. 1978 Analytic Techniques in Urban and Regional Planning, McGraw–Hill, New York.
[5] Corotis, R., Fox, R and Harris 1981 *"Delphi Methods: theory and design load application"*, Journal of the Structural Division, Vol 107, No. 6, pp 1095-105.
[6] Cabanis, K. 2002 *"Computer-Related Technology Use by Counsellors in the New Millenium – A Delphi Study"*, [online], Journal of Technology in Counselling, Vol 2-2, http://jtc.cplstate.edu/Vol2.2/cabanis/cabanis.htm.
[7] Kometa, S.T., Olomolaiye, P.O., Harris, F.C., 1994 *Attributes of UK construction clients influencing project consultants' performance*. Constr. Manage. Econ. Vol 12 Issue 5, pp 433–443.

[8] Mitchell, T. 1991 *The Limits of the State: Beyond Statist Approaches and Their Critics,* The American Political Science Review, Vol. 85, No. 1. pp. 77-96.
[9] Morledge, R. and Owen, K. 1999 *"Developing a Methodological Approach to the Identification of Factors Critical to the Success in Privatised Infrastructure Projects in the UK"*, Paper read at Harmony and Profit, ChiangMai, Thailand.
[10] Olsen, S. A. (1982) Group planning and problem solving methods in engineering management, John Wiley and Sons, New York, Chapter 4.
[11] Dalkey, N. (1969) *"Analysis from a Group Opinion Study"*, Futures, Vol. 1, No. 6, pp 541-555.

Research Frontiers and Way Forward

H Bahron[1]

[1]Institute of Research Management and Innovation, Universiti Teknologi MARA, 40450 Shah Alam, Selangor, Malaysia

*hbahron@gmail.com

Abstract. Researchers and academics do research to address the current and future needs of the nation. They have to be alert and sensitive to demands of their surrounding that can be largely be classed into the community, industry and government to complete the quadruple helix that intertwines the three into the fourth segment i.e. the academia. The current buzzwords that ought to catch the attention of academia are Translational Research, Niche Area, Sustainable Development Growth (SDG), Internet of Things (IoT), 4th Industrial Revolution (IR 4.0), Big Data, Transformasi Nasional 2050 (TN50) and Bottom 40% (B40). UiTM's niche areas that have been declared to KPT are (a) Sustainable Technology and Economy, and (b) Social Engineering and Creative Media. These are very much in line with 17 Sustainable Development Goals declared by the United Nation in 2012 with the goal of integrating planetary stability with the target of fighting poverty and securing human wellbeing. Internet of Things that swarms the globe brings about the seamless integration of physical objects with information network through internet connectivity. This flourishing development hails the advent of 4th Industrial Revolution (IR 4.0) that is inevitable where industrial automation and data exchange in the manufacturing technology rapidly includes cyber-physical system through IoT, cloud computing and cognitive computing. IR 4.0 should catapult productivity into hyper-drive to meet the needs of global consumers. The tremendous growth of cyber activities that encompass all aspects of life presents the inevitable progression into the dimension of Big Data. It is a new data ecosystem that is very large and complex, rendering the traditional data processing software inadequate and obsolete. A new breed of ICT experts are required to handle the enormousness of the task of capturing, storing, analyzing, searching, sharing, transferring, visualizing, updating and most importantly, information privacy. Closer to home, Malaysia is rapidly approaching the end of Wawasan 2020 and a new set of targets under Transformasi Nasional 2050 (TN50) have been set to guide the nation's journey into the year 2050. The goal is to be within the top 20 of world's nation in economic development, social progress and innovation. There are seven thrusts in TN50 that encompass all aspects of a mature and well-endowed nation. The welfare of the bottom 40% of the people of Malaysia is given a special attention by a dedicated effort the B40 agenda in elevating their socio-economic status. UiTM has designed an approach to develop a lush research ecosystem based on a synergic and concerted move of bringing its workforce in the faculties on a transdisciplinary platform of Communities of Research (CoRe). The effect of CoRe is very much apparent in research spotlights of UiTM competency map that shows a definite healthy growth towards multi- and interdisciplinary research activities.

IOP Conf. Series: Earth and Environmental Science **117** (2018) 012049 doi:10.1088/1755-1315/117/1/012049

1. Introduction

Research is a systematic activity or inquest carried out to find answers to questions, solutions to problems or simply progress to current situations. Research is essential to the survival and growth of a nation, giving an edge to catalyze its prosperity and boost the tangible and non-tangible wealth and gains.

Researchers are often driven by internal and external factors. Their inquisitive and curious nature when combined with natural or cultivated talent for investigation, keen observation and power of deduction could bear fruition to ground breaking innovations and generation of new knowledge. Most often, demand driven research would lead to relatively quicker return of investment than fundamental curiosity driven research, but in terms of knowledge creation, the latter has proven to be more valuable.

To remain relevant, it is essential for researchers to always be alert and sensitive to the needs of the community they serve. They must be aware of the challenges, possess far-sightedness and high creativity to come up with and deliver the best answers, solutions and innovations required. Researchers need to be at the forefront and carry out meaningful activities at the frontier of knowledge in their particular field, putting the best interest of their stakeholders on top priority.

2. Frontiers of Research

All research typically deals with the unknown, and any finding has potential to contribute to new knowledge. Along that argument, all research may be classified as frontier. However, to distinguish a genuine frontier research and normal science, a set of criteria may be used as a guide. These characteristics include, but not limited to the following [1]:

- The issues being addressed are shrouded with considerable controversy within the scientific community in the related field.
- The questions are hard to answer, particularly when only applying the normal methodological tactics.
- It requires the application of atypical methodologies and concepts.
- Addressing the issues are commonly initiated through tackling unexpected findings that defy the current dominant paradigm.
- It focuses on resolving issues that would confirm or refute the currently accepted paradigm
- The issues are huge challenges that the likelihood of finding successful answers or solutions has a high degree of uncertainty.

In the Malaysian context, researchers are urged to apply the quadruple helix approach in setting their research goals and objectives to include and intertwine the 4 important stakeholders of the nation, namely the academia themselves, government, industry and community. There are a few vogue, or in-trend, terms that researchers have to be mindful and be aware of:

- Translational research
- Niche areas
- Sustainable Development Goals (SDG)
- Internet of Things (IoT)
- Fourth Industrial Revolution (IR 4.0)
- Big data
- Transformasi Nasional 2050 (TN50)
- Bottom 40% (B40)

This paper briefly touches on each of the terms to put them in context with research direction and the way forward for researchers to concertedly pool their expertise and effort. This is pertinent to ensure that the interest of the nation, and human sustainability is addressed, at large.

2.1 Translational Research

Translational research is activities carried out by researchers that generates high impact outcomes and solutions that positively affect and directly benefit the government, industry and community. It is one of the ten initiatives in the Malaysian Education Development Plan 2015-2025 (Higher Education) gazette under Shift 7: Innovation Ecosystem. It was introduced during the 2017 New Year Address of the Minister of Higher Education [2] in response to the realization that RM5.58 billion has been invested in Malaysia into research in the period of 2007-2015. The returns of research investment (RORI) in the form of tangible and intangible gains were valued at RM7.17 billion (28.5%), as diverse sources of income from fundamental and commercialized research products.

Wealth and value creation across the quadruple helix is mapped into the 2017 Research Priority Roadmap that includes grand challenges, translating R&D into business values, talent development and global prominence.

2.2 Niche Area of UiTM versus Sustainable Development Goals

Simply put, niche area is the area of strength of a certain group of people. In 2015, the Ministry of Higher Education of Malaysia has requested each public university to declare their niche area, based on their specific expertise make up. Universiti Teknologi MARA (UiTM) has declared the following as its niche areas, as approved by the universities Boad of Executives:

- Sustainable Technology & Economy
- Social Engineering & Creative Media

Both niche areas are in line with the philosophy and direction of the United Nation who has introduced Sustainable Development Goals (SDG), mooted during the United Nations Rio+20 Summit in Brazil in 2012. SDGs for the people and planet are concerned about planetary stability that must be integrated with the United Nations targets to fight poverty and secure human well-being. There are seventeen SDGs as illustrated in **Figure 1**, excerpted from the United Nation Development Program (UNDP) booklet [3]. The United UNDP is the United Nation's global development network, advocating for change and connecting countries to knowledge, experience and resources to help people build a better life. They exist in 166 countries, working with each nation on their own solutions to global and national development challenges.

Figure 1. Seventeen Sustainable Development Goals (SDG) gazetted by the United Nation in the fight against poverty

2.3 Internet of Things (IoT)

In today's world where physical objects are seamlessly integrated into the information network, and where the physical objects can become active participants in business processes, services are available to enable interaction between these 'smart objects' over the Internet. This creates a situation where internet connectivity between consumers, industries and government generates a virtual network that opens up a whole new mode of communication and interaction. Parties may query and change the state of any information associated with them, taking into account security and privacy issues.

Teaching and learning, business transactions, administration, social interaction and all forms of communication do not only occur through the traditional physical means, but also increasingly becoming more virtual, happening through the cloud and via a whole range of internet connected communication tools such as desktop computers, laptops and handheld gadgets. Internet of things is the situation or phenomenon when a huge amount of connected data is being generated and transferred from one entity to another virtually.

2.4 Fourth Industrial Revolution (IR 4.0)

The manufacturing industry has undergone progressive changes or revolution throughout the years. The first revolution is the machination, using water power or steam power. The second industrial revolution occurred on the advent of electricity when assembly lines powered by electricity enabled mass productions. The third industrial revolution was when automation was made possible through the use of computers and programming.

With the ubiquitous internet connectivity, now machines are communicating with one another in the latest trend of automation and data exchange in manufacturing technologies. This is the fourth industrial revolution currently happening that includes cyber-physical systems, the IoT, cloud computing and cognitive computing. For all means and purposes, with each revolution, the dependence of manufacturing industries on manual labour and physical human interactions became less and less important. On the one hand, this trend increased productivity and efficiency, shortened production time and minimized human errors, but on the other hand, it also put many workers at risk of losing their jobs and livelihood.

2.5 Big Data

IoT has an intrinsic characteristic that inherently causes a lot of data to be created at a fast pace on a daily basis. Big data is the term coined to refer to a new data ecosystem with data set so large or complex that traditional data processing application software is rendered inadequate to deal with them. The data is generated and used by the industries, scientists, businessmen, advertising executives, practitioners of medicine, governments, consumers etc. It is increasingly used to spot business trends, prevent diseases, combat crime, social engineering, developing and refining war tactics and so on.

As with any new phenomenon, big data presents its own unique set of challenges that includes data capture, storage, analysis, search, sharing, transfer, visualization, querying, updating and information privacy. A new breed of ICT experts with knowledge and skills to address all the challenges and issues is urgently needed to cater for the rapidly growing needs and demands of the consumers and stakeholders of big data. Researchers must respond to these needs to remain relevant and referred.

2.6 National Transformation 2050 – Transformasi Nasional 2050 (TN50)

'Transformasi Nasional 2050' or TN50 is an initiative to plan for the future of Malaysia in the period 2020 to 2050. From the vision of becoming a developed nation, we should strive to be amongst the top countries in the world in economic development, citizen well-being and innovation. From the humble and uncertain beginning of Malaya, the nation has grown and evolved significantly into a modern economy and society.

Beginning with the era of the New Economic Policy (1971-1990) where the goal was to eradicate absolute poverty irrespective of race and eliminating identification of race by economic function, Malaysia stepped into the Vision 2020 era (1991-2020), to be a developed nation, in our own mould. Now, Malaysia has to be ready to face the next 30 years. In the *Transformasi Nasional* 2050 era (2021-2050) Malaysia aims to become a top 20 nation in economic development, social advancement and innovation.

There are seven thrusts of the TN50:

- Pure values/attributes – A community with open spirit and inclusiveness
- Governance – A public administration model that is effective and balanced
- Environment – A safe, sustainable and efficient environment that provides for a healthy life
- Education – A relevant and wholesome education ecosystem that has continuous learning environment
- Employment – A competitive and advanced economy driven by highly able workforce
- Lifestyle – An energetic, active and socially friendly lifestyle
- Malaysia on the global platform – A nation that has significant influence in the international arena

The secretariat for TN50 is the Ministry of Youth and Sport with the target groups of youth between the ages of 18 to 40 years old.

2.7 Bottom 40% (B40)

A report entitled Household Income and Basic Amenities Survey 2016 [4] by the Department of Statistics of Malaysia has categorized Malaysians into three different income groups namely (a) Top 20% (T20), (b) Middle 40% (M40), and (c) Bottom 40% (B40). The change in values of the bar for each group's income level is one of the indicators of economic growth of the country. Based on the 2016 monthly household earning, the latest definitions for T20, M40 and B40 is shown in Figure 2.

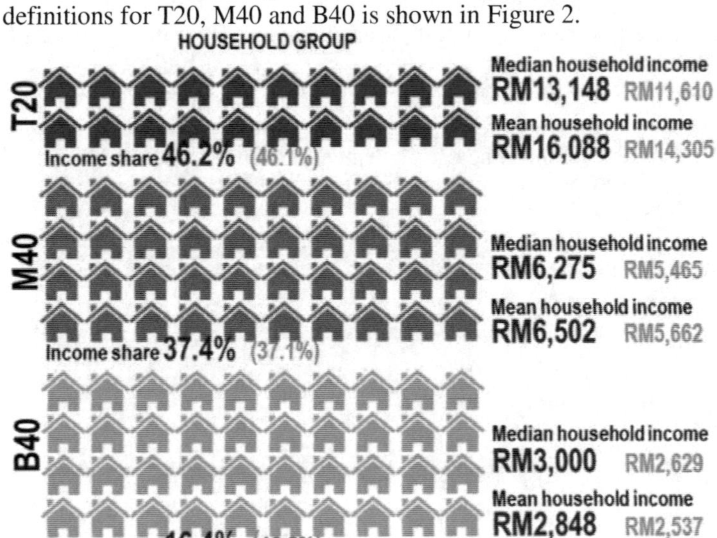

Figure 2. Categorization of Malaysian population into T20, M40 and B40 based on 2016 household monthly income

In the Tenth Malaysia Plan (2011-2015) several national major issues and challenges have been identified. These were socio-economic disparity, low education attainment and skills acquisition, low skilled and low value-added jobs, lack of entrepreneurship capabilities, increase in cost of living, high indebtedness, fragmented and silo implementation of social safety net programs, inability to own houses and limited access to quality healthcare services. Therefore, as a way forward, in the Eleventh Malaysia Plan (2016-2020), three issues have been give due attention and importance. These are raising the income and wealth of B40 households, addressing increasing cost of living and enhancing the delivery system in B40 programs.

The B40 programs are created to address pockets of poverty and socio-economic imbalances among the bottom 40% of the society with the aim of elevating the B40 households towards the middle class society.

3. Recommendation

Researchers need to contribute to the community at large to remain relevant. To tackle the issues highlighted above, the most effective approach is transboundary where resourceful researchers with various expertise and experience come together to try to solve the problems or elevate the current achievements in a concerted manner. Disjointed silo efforts would only result in misaligned solutions that by themselves present a set of problems and cause wastage of precious resources.

Groups of academics and researchers need to synergize and crosslink on formal or informal platforms to produce effective and impactful translational outcomes that benefit various stakeholders directly with quick wins, and in the long run would achieve far reaching objectives for the growth of the nation.

4. References

[1] Rey, J. (2011) Frontier Research: Bringing The Future Closer, Issue 05/June 2011, Journal of Lyncnos, Forum Section, Fundación General CSIC. http://www.fgcsic.es/lychnos/en_en/past_issues/lychnos_05/forum (accessed on 23 December 2017)
[2] 2017 New Year Address of the Minister of Higher Education of Malaysia.
[3] Sustainable Development Goals, United Nation Development Program (UNDP) Booklet.
[4] Report of Household Income and Basic Amenities Survey 2016, Department of Statistics of Malaysia, 9 October 2017.

IOP Conf. Series: Earth and Environmental Science **117** (2018) 012050 doi:10.1088/1755-1315/117/1/012050

Focus group discussion in built environment qualitative research practice

[1]D Omar

Faculty of Architecture, Planning and Surveying, Universiti Teknologi MARA, Puncak Alam, Selangor, Malaysia

*dasimah629@salam.uitm.edu.my; dasimaho@yahoo.com

Abstract. Focus groups discussion is a useful way in built environment for qualitative research practice. Drawing upon recent reviews of focus group discussion and examples of how focus group discussions have been used by researchers and educators, this paper provides what actually happens in focus group discussion as practiced. There is difference between group of people and topic of interest. This article examines the focus group discussions as practiced in built environment. Thus, there is broad form of focus group discussions as practiced in built environment and the applications are varied.

1. Introduction

The data are numerical information needed to make a good decision in certain situations [1]. In fact, important data is needed in the decision-making process. There are many methods used to obtain the required data. The data is obtained from various sources and collected through various techniques. Data are grouped into two types, primary data and secondary data. Primary data is data obtained or collected by researchers directly from the data source. Techniques that can be used to collect primary data includes observation, interview, focus group discussion (FGD) and through questionnaire. Secondary data are data collected from documents such as journal, books and any published articles.

This paper mainly focuses on the understanding towards FGD as a basis in determining the qualitative research practice in built environment. The qualitative research is an exploratory rather than conclusive type of research and helps in understanding not only what people think, but also how and why they think that way [2]. The process of data collection in qualitative research gives researchers the opportunity to get the following data:

- Unexpected data by researchers
- In-depth data can explain the issues of research topic [3].

FGD can be useful to gather certain types of data or when information is difficult to obtain in certain circumstances using other methods for data collection [4]. According to Overseas Development Institute [5] 'the FGD in built environment is a good way to gather together people from similar backgrounds or experiences to discuss a specific topic of interest' [1]. Focus groups are rarely used in isolation. FGD is an integral part of gauging public perceptions. In obtaining qualitative information, FGD may be held with a cross section of community groups from a small group of participants on particular themes [6].

2. Literature Review

There are five (5) characteristics/features of focus group interviews [7]:

- people, who
- possess certain characteristics
- provide qualitative data
- in a focused discussion
- to help understand the topic of interest [p.6].

Focus group discussion (FGD) is one of the most popular qualitative research methods [8]. According to [9] in her study 'FGD is a structured discussion used to obtain in-depth information (qualitative data-insight) from a group of people about a particular topic' [p.4].The aim of FGD is to study a topic in-depth and intensively. It is a discussion guided by the moderator according to the prepared interview guidelines [8]. FGD is not open to public meetings, because the selection of participants whose meetings present characteristics which it is important to ensure that participants in group have something in common to each other [10].

FGD is a group discussion of eight (8) to twelve (12) participants with guidance from a facilitator, which discuss a certain topic among participants [11]. A focus group is not a group interview; focus groups are successful when participants are able to talk to each other about the topic [12]. In addition, it provides opportunities to share and make comparisons about the participants' experiences, idea and views as well as provide opportunities to talk about something [13] which is effective in supplying information. However, the facilitator's expertise is very important to stimulate and support discussion but cannot act as an expert on the topic [4].

FGD is to gain a deeper understanding of a topic, such as motivation, behaviour, feeling, decision-making strategy, or opinion of a particular person on an issue or topic [9]. 'FGD techniques can be used to develop relevant research hypotheses by exploring in greater depth the problem to be investigated and its possible causes' as explained by [4 p.61].

According to [14], the process of FGD is as shown in Figure 1. It involves identification of goal/objectives, identify questions, identify people, select time, place/environment, conduct research, evaluate findings/data and preparation of report. Meanwhile, Figure 2 shows the FGD in practice that consisting moderator, observers and participants. Generally, FGD in practice starts with schedule groups, create plans, invite participants, moderate and report [14].

Figure 1. The process of FGD.
Source : Glynn, Shanahan and Duggan (2015)

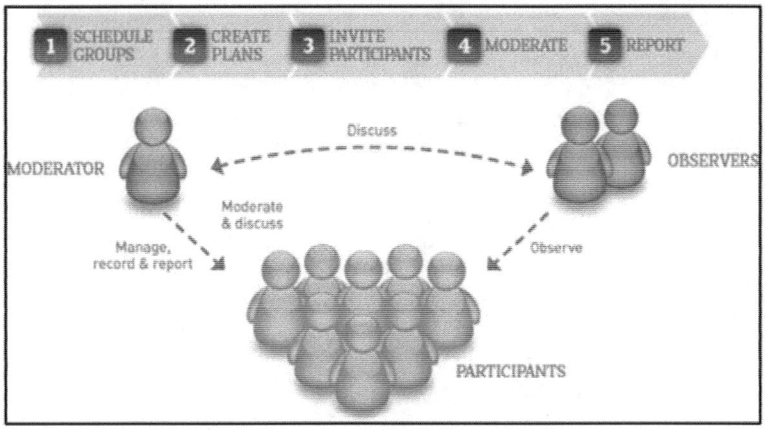

Figure 2. Focus groups in practice.
Source ; Glynn, Shanahan and Duggan (2015)

3. Research Methodology

This paper examined the methods applied during the data collection applying seven (7) Focus Group Discussions (FGDs) practiced in two different built environment qualitative researches. Based on the previous researches, FGDs have been applied to obtain in-depth information on the research. The first research project with four (4) FGDs was conducted for Longterm Research Grant Scheme (LRGS) -2013-1016 on the topic of Recreation Facilities for Youth in Malaysia. The respondents were youth, community leaders and other stakeholders. The second research project conducted three (3) FGDs for The Research on the Water Bodies in Selangor State : Ex-Mining Land (2016-2017). The group members representing the stakeholders were from government agencies, town planners from Local Authorities and Land Administration Officers from Land Offices.

3.1. FGD for Long-term Research Grant Scheme

In the LRGS research, four separate FGDs for youth and stakeholders were carried out to allow respondents to cooperate in giving their most precise answers. The youth participants were selected to share their opinion and to report on the imperfection of physical environment around them. The youth were students, employed and unemployed and some are married people. In this study, the youth need to respond so as to achieve the required research aim. They were selected randomly according to several criteria which representing both female and male gender, age between 15 to 25 years old and consisted of the Malay, Chinese and Indian races. The LRGS research started with the FGD for a case study of Lembah Pantai Kuala Lumpur which represented the urban area and followed by Kota Samarahan as the rural case study area [15]. These FGDs were held between September 2014 and November 2014 at different venues. The first session for Lembah Pantai was at community hall of PPR which is accessible to youth in the afternoon between 3-5pm. The time was appropriate for the youth as most of them were in schools in the morning. The second FGD with stakeholders group was held at 10am-12pm at Armada Hotel in Petaling Jaya which is accessible to all respondents.

The focus group discussions (FGD) were held in two sessions with two groups involving twenty-one (21) respondents. The first meeting was attended by 12 (twelve) youth who were the residents of the low-cost public housing in The Federal Territory of Kuala Lumpur. This site selected was the *PPR* Kerinchi (*Projek Perumahan Rakyat* Kerinchi), Lembah Pantai. The youth participants were chosen from each *PPR* block and they were to express their opinion and share their feeling regarding their neighbourhood living environment. The second FGD representing the stakeholders consisted of nine (9) participants. Among them were three (3) the *PPR* Residents' Committee members, a Town Planner from Kuala Lumpur City Hall (*DBKL*), a Town Planner in Federal Department of Town and Country Planning, Headquarters, Kuala Lumpur and four (4) members of the Malaysian Youth Council Committee. Meanwhile, the third and fourth FGD were held at the Universiti Teknologi MARA, Kota

Samarahan Sarawak. The morning session was with the stakeholders while the afternoon was with the youth [15]. The morning session consisted of twenty-two (22) related stakeholders including professional such as policemen, health officers, planners, parents, Malaysian Youth Council Committee and residents committee's members [15], while the afternoon session consisted of twenty-four (24) youth respondents.

3.2. FGD for the Water Bodies in Selangor State: Ex-Mining Land research

The research on The Water Bodies in Selangor State: Ex-Mining Land (2016-2017) was carried by conducting three FGDs. The first FGD was held on the morning on the 10th May 2016. It was held at the Mines Resort. It was attended by The Director and officers of the Selangor State Town and Country Planning Department, five (5) research consultants, an officer from *Institut Penyelidikan Hidraulik Kebangsaan Malaysia* (NAHRIM) and fifty-one (51) officers representatives from related departments and agencies. The second FGD was held on 24th August 2016 at The INTEKMA Resort and Convention Centre. The respondents were twelve (12) Town Planning Officers from nine (9) Local Authorities in the State of Selangor which involved in the research. Town Planning Officers from the Shah Alam City Council were also invited to share their experiences. The third FGD was conducted on 9th September 2016 also at The INTEKMA Resort and Convention Centre. This FGD group consisted of five (5) Land Administration Officers from Land Offices in the State of Selangor.

Figure 3 below shows the research process where all the three (3) FGDs for The Research on Water Bodies in Selangor: Ex -Mining Land (2016-2017) were held during the second stage of the study.

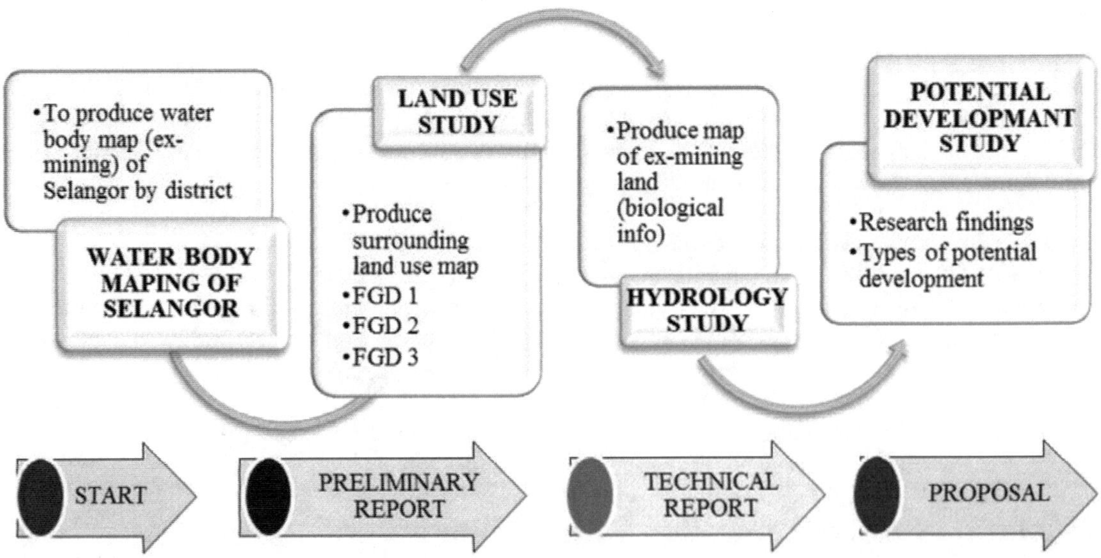

Figure 3. FGD 1, 2 and 3 for The Research on Water Bodies in the State of Selangor: Ex -Mining Land (2016-2017)

4. Result and Findings

4.1.1. LRGS (2013-2016)

The analysis found that from the overall understanding and review by all FGD sessions the outdoor features seem to be limited in number and are not properly maintained [p.4]. The research on public users was to highlight 'space' as a keyword to interpret respondents' feeling and understanding regarding their surrounding environment [16]. Figure 4 shows the space thematic analysis of the youth neighbourhood living environment [16].

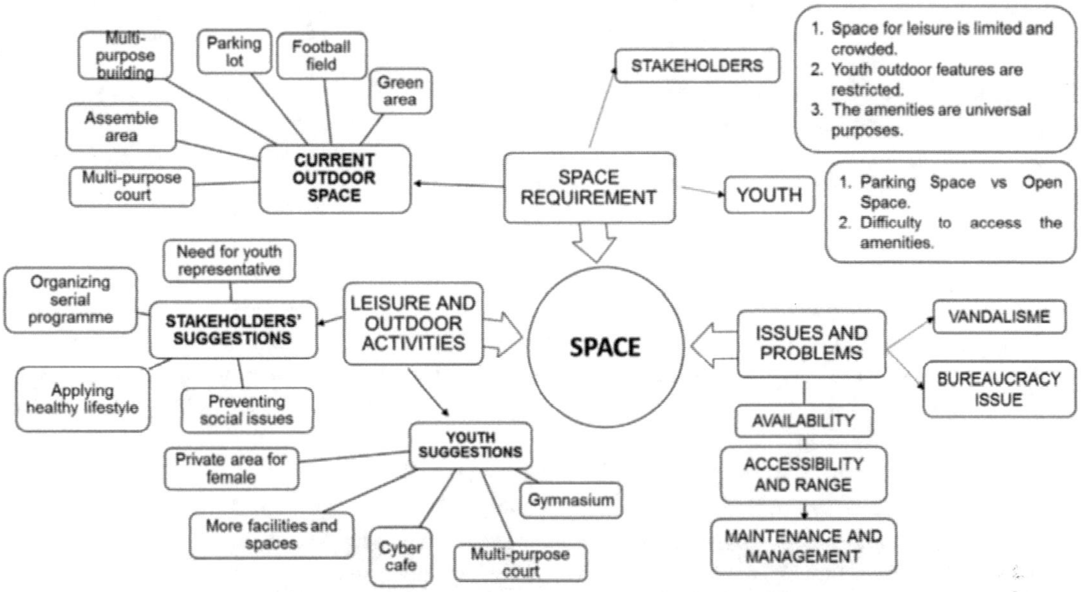

Figure 4. Space thematic analysis [16, p.4].

The space includes soccer field, playground, multipurpose courts and any green area appeared proximate to the neighbourhood surrounding [p.4]. The assessment was carried out on other related issues such as accessibility and range; maintenance; space provision; outdoor features requirements; and suggestion from all respondents of the FGDs.

According to the respondents from the stakeholders group, they stated that space for leisure was limited and crowded, youth outdoor features were restricted and the amenities were for common purposes. However, the youth focus group discussions concluded that parking space were treated as open space and they had difficulty to access some of the amenities. Table 1 show all related issues and problems raised during the FGD sessions.

The FGDs agreed that youth were directly related to the problems identified [18]. The FGDs showed that the youth should be given the chance to voice out their opinions and ideas. Their needs should be highlighted during the neighbourhood development planning for better youth participation in the neighbourhood community. In addition, the respondents from the Youth Association suggested that the involvement of youth should be encouraged. They stated that more attention should be given to the youth. This can solve the problem raised by the residents association committee that conflict with the request made by youth. Based on the discussions [18] the sense of neighbour and family are needed as a preventive measure on the issues and problems that occur in the neighbourhood community [p.6].

Table 1. Issue and Problems

Issues	**Problems**	
	Youth	Stakeholders
Availability	The existing outdoor spaces were less compared to the number of users.	Provision of facilities and amenities were limited due to space availability.
Accessibility	The location of spaces is strategic but the pathways were blocked by physical barriers.	Access to some outdoor features being blocked by residents' cars, motorcycles, and stalls.
Comfortable	Female youth respondents were requesting for safety and separate space for leisure.	The existing features and space for leisure and physical activities were shared by all residents.

Maintenance	Some outdoor features were broken and damaged and were not fixed or not maintained properly.	Maintaining the outdoor features was not continuously practised.
Bureaucracy	Residents' committee members were focusing more on parking and cleanliness issues.	Youth should find their own way to have leisure and should not be given special spaces to avoid social issues.

Source: Omar et al. [16, p.5]

4.1.2 The Research on the Water Bodies in Selangor State: Ex-Mining Land (2016-2017)

Three (3) FGD sessions were held during the study period. The first FGD session was for stakeholders from government agencies and as the introduction for The Research on the Water Bodies in Selangor: Ex -Mining Land (2016-2017). The topic of the session was on the information and experience on the ex-mining land re-development programme by each government agency. Based on the overall conversation recorded, listed below are the issues that had been highlighted by respondents from the first FGD session:

- There were no guidelines to bathymetric measurements
- All location of ex-mining lands need to be analysed
- There are parameters and standard for the determination of water quality
- NAHRIM and LUAS have guidelines for water sampling and sediment
- The information in the Local Plan was inconsistent
- Illegal aquaculture activities at the ex-mining land
- GIS attribute for mining and hydrography
- There are acts and guidelines relating to ex-mining land

The second FGD session was held for more detail information for the preparation of the Technical Report. This FGD involved with twelve (12) Town Planners from Local Authorities. The results expected from the FGD was on the issues and technical input on town planning aspects, development potential and legislative proposals, policies and guidelines for ex-mining land. Each site of the water body was checked meticulously by the town planner from the related authority where the water body is located. Below is the list of issues raised during the FGD session.

- Land use should be based on zoning map rather than current land use map
- The use of zoning map to facilitate the future development to be implemented by Local Authorities
- Land gazetted by agency such as *Lembaga Urus Air Selangor (LUAS*) should be identified
- The location of each water body of mining land need to be reviewed for accuracy
- The data to exclude type of soil and water quality
- The land title must be detail and to list all the activities
- North point should be simple and on the right side of the map
- Should get the location of mining land from land office

Based on the third FGD session, the topic highlighted the issue on information about land title for each water body (ex-mining land) which involved five (5) the Land Administration officers from Land Offices. The aim of this session was to determine the status of land title of each water body (ex-mining land) by district. The ownership status was needed in this research to confirm whether the land belongs to government or private ownership. In addition, from the discussion the list of lot numbers for each water body (ex-mining land) was completed with the cooperation of those involved. During the FGD session the potential development for the ex-mining land such as recreation, water reservoir, agriculture and township was also discussed.

5. Conclusion

The paper highlighted the Focus Group Discussions as practiced in the Built Environment Qualitative Research. Generally, the participants of FGDs were selected from related target groups. The FGDs' results should be used to support the aim of the particular research. The different FGDs were for the different purposes. The different participants were for different FGDs in order to get clear, detail information and views. Thus, there is broad form of FGDs as practiced in built environment and its application is varied.

References

[1] Mohamad, O. *Kaedah pengumpulan data*. Available from:https://www.scribd.com/document/356282946/Kaedah-Pengumpulan-Data [Accessed on 10th November 2017].

[2] Suvey Monkey. *Running a survey vs. a focus group: what's the best way to collect customer reviews?*. Available from:https://www.surveymonkey.com/mp/running-survey-vs-focus-group-best-way-collect-customer-reviews/. [Accessed 18th October 2017].

[3] Mack N, Woodsong E, MackQueen K M, Guest G and Namey E. 2005 *Qualitative research methods: a data collectors field guide, family health international, research triangle park.* NC: FHI.

[4] Degu Gand Yigzaw T. *Research methodology*. Available from: https://www.cartercenter.org/resources/pdfs/health/ephti/library/lecture_notes/health_science_students [Accessed on 12th November 2017].

[5] Overseas Development Institute. *Research tools: Focus group discussion.* Available from: https://www.odi.org/publications/5695-focus-group-discussion [Accessed on 10th November 2017].

[6] NSW Government. *Recreation and open space planning guidelines for local government.*Available from: http://www.planning.nsw.gov.au/Portals/0/StrategicPlanning/UrbanRenewal/Attachment 2 Open Space Guidelines for Local Government.pdf [Accessed on 14th November 2017].

[7] Krueger R A and Casey M. A. 2009 *Focus groups: a practical guide for applied research.* United States of America: Sage Pulication INC. .

[8] Temkin J. *Advantages and limitations of focus groups.* Available from: https://money.howstuffworks.com/business-communications/how-focus-groups-work1.htm [Accessed on 14th November 2017].

[9] Gerritsen A. *Focus group discussions –a step-by-step guide.* Available from: https://www.slideshare.net/AnnetteGerritsen/fgd-manual-14-april-2011 [Accessed on 15th November 2017].

[10] Freitas H, OliveiraM, Jenkins M and Popjoy O. *The focus group, a qualitative research method.* Available from: http://gianti.ea.ufrgs.br/files/artigos/1998/1998_079_ISRC.pdf [Accessed on 15th November 2017].

[11] Funded Projects in Asia and M & E Toolkit. *Using focus group discussions to complement survey findings.* Available from: https://asia.ifad.org/c/document_library/get_file?uuid=22249225-0c00-4fe6-96e4-2936efebc5af&groupId=627927 [Accessed on 13th November 2017].

[12] Minardi J A. *Focus groups: how to develop a 360* view of your target audience.* Available from: https://www.slideshare.net/susanbergson/focus-group-presentation-2851883 [Accessed on 13th November 2017].

[13] Duggan. Steps to *take in writing a recommendation report.* Available from: http://work.chron.com/steps-writing-recommendation-report-28828.html [Accessed on 11th November 2017].

[14] Glynn P, Shanahan M and Duggan D. 2015 *Focus groups.* Available from: https://www.slideshare.net/johnglynn940/focus-groups-presentation [Accessed on 12th November 2017].

[15] Omar D, Omar K A, Kamaludin Z, Othman Sand Yusoff Z M 2016 *American Scientific Publishers Advanced Science Letters*

[16] Omar D, Omar K A, Kamaludin Z, Othman S and Yusoff Z M 2016 *ASEAN-Turkey ASLI (Annual Serial Landmark International) Conferences on Quality of Life 2016*
[17] Omar D, Omar K A, Othman S and Yusoff Z M 2017*AicQoL2017Bangkok Journal*
[18] Omar D, Omar K A, Kamaludin Z, Othman Sand Yusoff Z M 2016 *ASEAN-Turkey ASLI (Annual Serial Landmark International) Conferences on Quality of Life 2016*